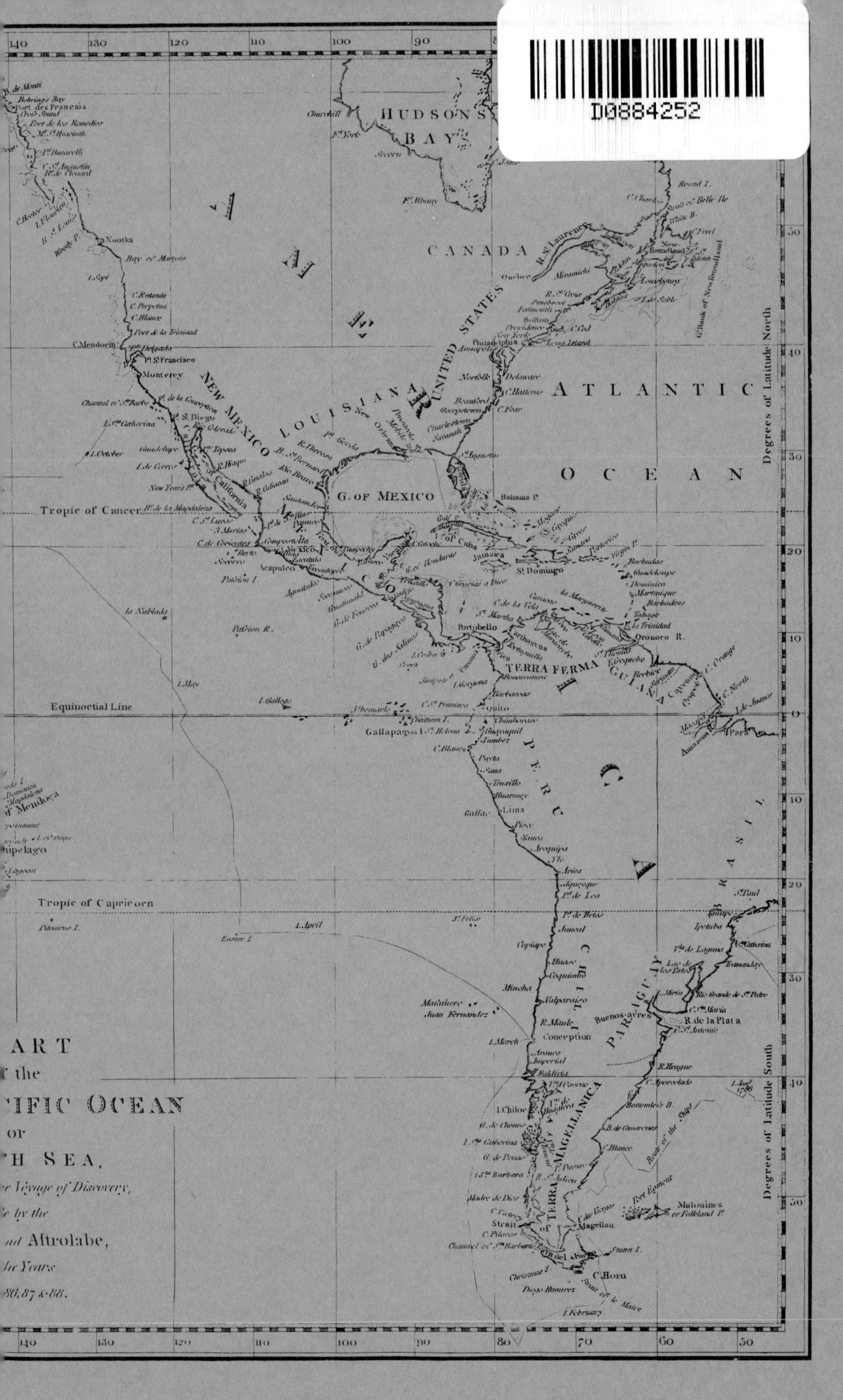

HUDSONS BAY
CANADA
UNITED STATES
LOUISIANA
NEW MEXICO
G. OF MEXICO
ATLANTIC
OCEAN
TERRA FERMA
GUIANA
PERU
BRASIL
CHILI
PARAGUAY
TERRA MAGELLANICA
Tropic of Cancer
Equinoctial Line
Tropic of Capricorn
Nootka
Monterey
Acapulco
Mexico
Portobello
Quito
Lima
Valparaiso
Conception
Buenos-ayres
R. de la Plata
Strait of Magellan
C. Horn
Staten I.
Malouines or Falkland I.
Gallapagos
Juan Fernandez
Easter I.
Philadelphia
Quebec
Jamaica
St. Domingo
Bahama I.
Oronoco R.
Degrees of Latitude North
Degrees of Latitude South
ART
the
CIFIC OCEAN
or
H SEA,
e Voyage of Discovery,
e by the
nd Astrolabe,
he Years
86, 87 & 88.
son Paternoster Row, London.

Whales & Destiny

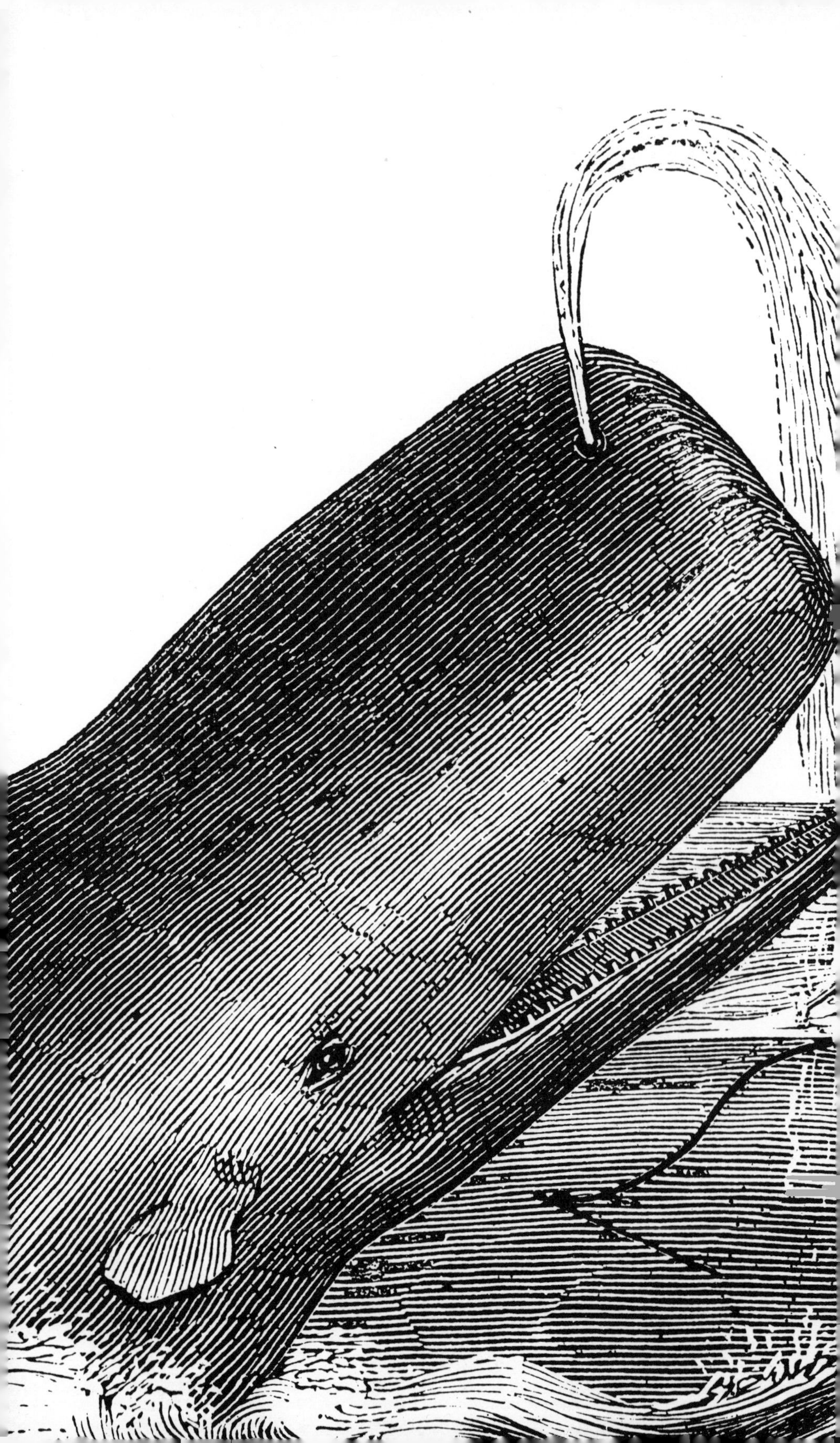

Whales & Destiny

The Rivalry between America, France, and Britain for Control of the Southern Whale Fishery, 1785-1825

Edouard A. Stackpole

The University of Massachusetts Press

LIBRARY OF CONGRESS CATALOG CARD NUMBER 72–77567

PRINTED IN THE UNITED STATES OF AMERICA

DESIGNED BY RICHARD HENDEL

PUBLISHED WITH THE COOPERATION AND

SUPPORT OF THE NANTUCKET HISTORICAL TRUST

In grateful acknowledgement of
grants from The New England Society in
The City of New York which made possible
research trips to England and Wales
in preparation of this book,
The Nantucket Historical Trust, and to
Charles F. Batchelder, Whaling Historian,
for his many helpful suggestions
on the manuscript.
And to my wife, Florence T. Stackpole,
for her help in copying records and
for her unfailing encouragement

Contents

LIST OF ILLUSTRATIONS ix
FOREWORD xi
1. A WHALING KINGDOM WITHOUT A COUNTRY 1
2. LONDON GAINS A NEW INDUSTRY 15
3. WHALEMEN'S EXODUS—THE TIDE SETS 27
4. THE PILGRIMAGE OF WILLIAM ROTCH 37
5. LONDON TAKES THE LEAD 51
6. QUAKER MERCHANT AND TITLED LORDS 59
7. POLITICS AND WHALE OIL 71
8. BRITAIN CONSOLIDATES AN INDUSTRY 87
9. THE SEAPORT WITH THE "BACK DOOR" 97
10. 'ROUND THE CAPES—AND NEW WORLDS 113
11. WHALE OIL DIPLOMACY AND ADVENTURE 133
12. STORM OF REVOLUTION—DUNKIRK 1791–1792 159
13. NEW WORLDS FOR WHALESHIPS AND WHALEMEN 175
14. QUAKER WHALERS' HAVEN AT MILFORD 197
15. ONCE MORE THE TIDES OF WAR 217
16. THE HOME-COMERS AND A NEW HOME PORT 249
17. THE SOUTH SEAS—OCEAN HOME OF WHALESHIPS 263
18. MEETING NEW HAZARDS AND CHALLENGES 285
19. THE PIONEER PORT RESURGES 315
20. THE WHALESHIPS GO TO WAR 335
21. CAREERS AND ADVENTURES OF THE MIGRANT WHALEMEN 353
22. NEW ENGLAND REGAINS THE ASCENDANCY 371
APPENDIX 385
INDEX 409

List of Illustrations

1. The Town of Sherborn 2-3
2. Ship *Maria* 11
3. Bill of Lading for the Ship *Maria* 20
4. Bill of Exchange on Buxton and Enderby 28
5. Captain Hezikiah Coffin 38
6. Map of Nantucket 46-47
7. The Rotch Counting House 56
8. William Rotch 65
9. Charles Jenkinson, Lord Hawkesbury 72
10. Captain William Mooers 80
11. "Mill Cove," Dartmouth, Nova Scotia 91
12. Hive of Nantucket Whaling Masters 99
13. Whaleship *Edward* of Dunkirk 107
14. Captain John Hawes 117
15. Captain Matthew Jones 124
16. Port Jackson, New South Wales 134
17. Whaleship *Sukey* of Nantucket 142
18. Ship *George* of Nantucket 151
19. Letter by Peter Macy 160
20. New Bedford 169
21. Home of Captain Silas Jones 177
22. Ship *Hero* of Nantucket 184
23. Whaleship *Syren* of London 192
24. Captain Frederick Coffin 202
25. Table Bay and Cape Town 212
26. Letter from Captain Frederick Coffin 222
27. Friends Meeting House, Milford Haven 232

28. A. Milford Haven, Wales (1794) 239
28. B. Milford Haven, Wales (1812) 239
28. C. Charles Francis Greville 240
28. D. Lady Hamilton 240
28. E. "Castle Hall," Milford Haven 241
29. "The Priory," Milford Haven 250
30. Order Approving a Report 258
31. Head of a Harpoon from the *Ospray* 268
32. Chart of Cloudy Bay, New Zealand 278
33. Captain Reuben Joy 289
34. Ship *Atlas* of Nantucket 300
35. The Henry Coffin (Carlisle) House 309
36. Captain Eber Bunker 321
37. The "London Club" 332
38. The Thomas Macy House 342
39. Obed Macy, Nantucket Historian 343
40. Captain Zophar Haden 365

Foreword

IN THE history of international maritime affairs one of the least-known periods marked the rivalry between the young United States, England, and France for control of that lucrative industry known as the Southern Whale Fishery. Beginning in the decade immediately following the Revolutionary War it continued well into the nineteenth century. It was a struggle which spread across the seven seas; an economic rivalry, concerning merchants, whaleships, and whalemen, waged against the better-known political and social background of those times.

The Southern Whale Fishery was originally an American Colonial maritime industry, founded by the Quaker whalemen of the Island of Nantucket in the Massachusetts Bay Colony. Now, in these post-war years, it was to undergo its great crisis. Events governing these years of trial were also to have a profound effect on political as well as maritime developments in both Europe and America. Despite these adventurous decades, when international affairs were sharply affected, it is one of the most neglected of chapters in maritime history.

The economic contention for the control of the Southern Whale Fishery—so-called to distinguish it from the Northern or Greenland Whaling Industry—found the newest of countries, the United States, facing the challenge of the mature European nations, England and France. Serving as pawns in the economic struggle were the leading whalefishermen of their time, the mariners and merchants of that tiny Atlantic Island of Nantucket. Caught up in the crosscurrents of the rivalry these Quaker whaling folk found themselves entering a series of adventures which developed into as much a search for their own destiny as their continuing pursuit of the great sperm whale.

Within the historical record there is not only the narrative of the Nantucketers' dual pursuit but a documentation of the range of international factors which shaped the remarkable story. As we follow their adventures one of the outstanding accounts in maritime history is revealed—whalemen who wandered like a hundred Ulysses through a watery world until many found their way back to their Island home, sustained by the pride of a tradition that had been founded by their fathers.

This whaling saga also includes the stories of some unusual individuals, particularly a Quaker merchant prince named William Rotch, Sr., a man far ahead of his time, who advanced a concept of internationalism that was both original and daring.

At the outset it was an economic circumstance which began the chain of adventure—a demand for whale oil (principally sperm oil, the peer of such oil) needed for lighting homes and cities, as well as for lubricating the new machinery of the Industrial Revolution. It was a period rightly termed the first "Age of Oil," resulting from a joint social and industrial progress, when the world, entering a new age of enlightenment, needed the products of the whaling industry as never before. *Whales and Destiny* is a recounting of the events of this period.

EDOUARD A. STACKPOLE
Nantucket Island
January 1972

But, though the world scouts us
as whale hunters, yet does it unwittingly
pay us the profoundest homage;
yea, an all-abounding adoration!
for almost all the tapers, lamps, and
candles that burn around the globe,
burn, as before so many shrines,
to our glory!

HERMAN MELVILLE
"The Advocate," *Moby-Dick*

1. A Whaling Kingdom Without a Country

THE American Revolution had created a new nation—the United States of America. But there was one small outpost of New England that still considered itself a part of the mother country—old England—until the Treaty of Paris decreed otherwise. This was the Island of Nantucket, well off shore from the southeastern coast of Massachusetts, which, through its development of deep-sea whaling, had become more closely allied economically with England, principally London, by sale of oil and candles, than with the Colonial ports of America.

During the eighteenth century Nantucket was a maritime phenomenon, with its shire town of Sherborn not only a leading Colonial port but the outstanding whaling center of the Atlantic world. With skill and courage acquired by years of experience, the seafarers and merchants of this isolated maritime community became the Colonial pioneers of whaling in "ye deep," and eventually developed a completely new phase of an old industry, that came to be known as the Southern Whale Fishery.[1]

The Revolution found the Island caught between two fires, Continental economic blockade and Royal Navy ship blockade stopped its trade and captured its ships. The war's bitter aftermath had proven equally devastating. Nestled in the inner curve of Nantucket harbor, the town of Sherborn presented an aspect of gloom: its once bustling waterfront unnaturally quiet, its famous whaling fleet nearly destroyed; its fleet of sloops, schooners, and brigs decimated. What did the Treaty of Paris promise? Could this new, untried American nation provide a market for whale oil to match London and Amsterdam? The political gulf separating the Island from the main was of far greater measurement than Nantucket Sound.

1. Obed Macy, *The History of Nantucket* (Boston, 1835), pp. 35–40.

1. The Town of Sherborn in the Island of Nantucket
From *The Portfolio* (1810), sketch by Joseph Sansom
[Courtesy the Nantucket Historical Association]

Kingdom Without a Country 3

4 Whales and Destiny

Playing a dominant role in the life of Nantucket was the Society of Friends, or Quakers as they were commonly called. The Island Society was a new breed of Quakers, influencing not only the religious factors but guiding the mercantile affairs and exerting a powerful control over social activities. Equally dominant in American whaling were the Quaker whalemen at sea.

During the ebb and flow of the Revolution, only the unrelenting perseverance of the Quaker leaders maintained the Island as a neutral in the war. By extraordinary measures these leaders were able to arrange pacts with the Royal Navy and the Continental Congress whereby some of the Island whaleships were permitted to complete voyages. This tenuous hold on the industry had kept their hopes alive.

As an aftermath to war there was now disillusionment. The split in the ranks of the Friends was not easily healed. Certain militant young Islanders were not forgetting the Loyalist domination of Island affairs during the conflict. While the whaleships' voyages were quickly resumed, the oil promptly refined, and the spermaceti candles once more produced, Nantucket's famous products were no longer finding a lucrative market in Britain. A tariff wall had been quickly erected—an import duty of £18, three shillings per ton on American oil and candles—making a profit impossible and a loss certain.[2] Equally disheartening was the lack of a strong American market for the Island's export.

The latter condition was a result of two factors. To counteract the British duty on American oil several states, notably Massachusetts, now offered bounties for oil cargoes brought in by American vessels, and several ports had sent out ships. Due to the war years the number of whales along the coast had increased, not having been hunted, and became easy prey for the newcomers. This brought a glut on the American market, especially in Boston, Philadelphia, Baltimore, New York, and Charleston—places where there had been before the war good markets for Nantucket oil.[3]

The shortages of the war had also brought to the fore a number of

2. *Records of the Privy Council*, Public Record Office, Chancery Lane, London, England (hereafter cited as P.R.O.), P.C. 2/131, p. 258.
3. Alexander Starbuck, *The History of Nantucket* (Boston, 1924), p. 384.

substitutes for whale oil, in lighting and lubrication and processing hides and wool. Tallow dips had become crude substitutes for spermaceti candles, lard-oil for lamps, vegetable oils for tanning and treatment of wool.

The situation during the war had been one of exposure to known dangers; now the economic conditions brought apprehension and a new despair. Whaling was the Island's sole industry; every phase of the shore-side activities was directly related to this industry—sail lofts, cooper shops, rigging lofts, chandleries, refineries, candle-houses. The ships at sea and the workers ashore were allied as closely as the native families —sharing a common heritage.

The strong Loyalist sentiment in Nantucket kept alive the hope that England, the Mother country, politically as well as economically, might be a possible aid. A number of incidents during the critical period before the war had shown how close an understanding existed in London for the Island's problems. Among the coercive measures by which Parliament had hoped to break the spirit of Colonials (especially those militants in Massachusetts Bay) was the Restraining Act of 1775, a feature of which dealt with curtailment of trade and fishing—the latter including whaling. The House of Commons on March 2, 1775, conducted a hearing for certain American remonstrants. Among those appearing was Captain Seth Jenkins of Nantucket, who gave not only comprehensive answers to questions but presented a picture of whaling from the Colonies, which commanded the attention of not only the members of Commons but the visitors and interested folk in the galleries.

Captain Jenkins' remarks were both informative and instructive. But they held a further touch of prophecy. His listeners learned that, at this time (1775), the Nantucket whaling fleet exceeded one hundred vessels, practically all being owned by members of the Society of Friends, the majority faction in the Island's inhabitants; that these craft had "lately extended their sperm whale fishery as far as the Falkland Islands, and were sometimes twelve months on a voyage," and were the only Colonial mariners fully engaged in this new branch of whaling known as the Southern Fishery.

Jenkins pointed out that Nantucket sent most of its oil and candles to England, and the prohibition of the business would not only cause hardship and suffering to the Island, but also have a marked effect on trade with certain merchants in London and Bristol, as the Islanders purchased quantities of manufactured goods from them.

"As there is seldom more than three months' supply of provisions in Nantucket," Captain Jenkins stated, any curtailment of whaling would find the Nantucketers "obliged to emigrate to the southward," as, he pointed out with a startling frankness, "they would not . . . live under the military government of Halifax."

The element of prophecy was proven some years later. Captain Jenkins became a leader in a migration of Nantucketers who settled at a sleepy Dutch hamlet on the Hudson River and founded the city of Hudson, New York.

Among those impressed by Captain Jenkins' forthright remarks were Robert Barclay (a member of the Society of Friends), and Edmund Burke, a great orator and active supporter of the Colonial cause, whose eloquent speech "On Conciliation With America," contained that ringing and oft-repeated tribute to the American whalemen. When Burke referred to the fact that neither the perseverance of Holland, nor the activity of France, nor the sagacity of Britain had "carried the most perilous mode of hardy industry to the extent to which it is pushed by this recent people," he was directly referring to the Quaker whalemen of Nantucket.

The Greenland or Northern Whale Fishery brought back only the "black oil" of the Arctic right whale. The Southern Whale Fishery was a new branch of whaling, principally devoted to the capture of the great sperm whale, whose oil was worth twice the other, and its chief exponents were the Nantucketers, and it was an American enterprise. There were no British-based vessels then engaged in this type of whaling.

In marketing their sperm oil and candles, the Nantucket merchants dealt with such British firms as Enderby and Sons, George Hayley, Champion and Dickason in London, and Richard Champion in Bristol, who also had connections with Boston, Newport, and Providence business houses. At this time none of the British merchants had shown any evidence that they planned to enter their own ships into the Southern Whale Fishery. There was little reason for them to do so, as they were receiving the marketable product from the New Englanders, more particularly from the Nantucketers.

To Francis Rotch, brother of William Rotch, Sr., should go the credit for advancing the idea of establishing the Southern Whale Fishery in Great Britain from London. Early in September 1775, nearly six months after the opening of hostilities of the Revolution, Francis Rotch, together with Leonard Jarvis of Dartmouth, Richard Smith of Boston, and Aaron Lopez of Newport, organized a fleet of whaleships to sail from Nantucket and Bedford (the little town only recently launched by his father, Joseph Rotch, and Joseph Russell) and proceed directly to the South Atlantic, as far as the latitude of the Falkland Islands. The vessels—numbering sixteen—were to avoid "all touch" with the South American coast, and if provisions were needed they were ordered to go into Port Egmont, in the Falklands. Francis Rotch, himself, planned to go there also during this particular period, to supervise the operation of the fleet.[4]

Of crucial importance, however, was the nub of the entire plan. Instead of returning from the Falklands to their New England ports, the whaleships in the fleet were to proceed directly to London and deliver their oil. The determining factor in the scheme was that both Rotch and Lopez planned to go to London, "with the intention of residing there and fitting out such vessels for the whale fishery as proper men could be obtained for . . . all the oil and other products of the Fishery will be shipped . . . Immediately for London . . . no other market calling for the quantity or affording so good a price for the proceeds of the Sperm Fishery." In the instructions to the captains, it was emphasized that if they fell in with any vessels bound for England they should write "in a particular manner" to George Hayley, Esq., Merchant, of London, and "let him know your particulars & success."

The bold plan came close to succeeding, and but for the fortunes of war the British would have had their Southern Whale Fishery established by the very people they would call upon a decade later. The Royal Navy captured several of the Falklands bound whaleships, five ships being taken by H.M.S. *Renown* and H.M.S. *Experiment* halfway to their destination and rendezvous. Only by the efforts of Francis Rotch, who went to London, were they finally released from Portsmouth, where they had been interned. In his petition concerning the fleet Rotch carefully pointed out: "Your Petitioners (are) the first to dare attempt to

4. Edouard A. Stackpole, *The Sea Hunters* (Philadelphia, 1953), pp. 130–131.

establish so valuable a branch of the Fishery in Great Britain without any previous requisition from Government. . . ."[5]

Another of the ships—the *Britannia*—which did obtain a cargo of oil was captured by none other than Lieutenant John Paul Jones in the sloop *Providence*, and a prize-master put on board. In his report of the capture, Jones wrote: "She [the *Britannia*] is a Nantucket Whaler, and appeared by the voluntary testimony of the Master, Mate, etc., to be the property of rank Tories, who had ordered their oil to be carried to the London market, and the amount of it to be shipped out in British goods to Nantucket."[6]

While in London Francis Rotch had become even closer to the firm of Champion and Hayley, and, with the death of George Hayley, his association continued as that of the deceased partner's chief clerk. That he was not as strict a member of the Society of Friends as was his elder brother may have been gleaned from his open relationship as the consort to the Widow Hayley, whom he accompanied on many journeys in the countryside outside London. Despite his social activities he maintained his "finger on the pulse" of whaling, and was an important source of information on the London and French whaling markets—information which Rotch passed along to his brother William Rotch, Sr., who was in Nantucket.

The fates were never kind to Francis Rotch. Despite all the care in planning, the remnant of the Falklands whaling fleet could not import enough oil to sustain the operation. The intensification of the war blocked all further efforts by the promoters.

But Francis Rotch did manage one significant accomplishment—he outfitted some of the ships in London, and, from that port, sent out two former Nantucket whaleships. This unprecedented action was prophetic, but not in the way Rotch had anticipated. Several London firms had long been associated with Nantucket in the importation of oil and candles—firms such as Champion and Dickason, George Hayley, and Samuel Enderby and Sons. They were quite familiar with the quality of

5. *Records of of the Colonial Office*, Francis Rotch to the Hon. Lord North, First Lord Commissioner of His Majesty's Treasury, P.R.O., C.O. 5/146, p. 63b.
6. John Paul Jones to Marine Committee, Continental Congress, *American Archives*, Washington, 5th Series, Vol. II, p. 171.

the Nantucket product, and had a marked respect for the reputation of the Nantucket whalemen. It was a natural step for them to consider organizing the Southern Whale Fishery fleet from London.

It is generally credited that Samuel Enderby, in London, first thought of sending a British fleet of whaleships into the Southern Whale Fishery. He later managed this by following Francis Rotch's initial venture—but using British registered vessels manned by Nantucket crews, the ships to be protected by the British Navy.

The importance of these British Southern Whalers was not lost on one of the American observers then in Europe—John Adams. During his residence in Paris in 1779, Adams learned from correspondence with American agents in London that Enderby's South Sea whalefishery was being carried on successfully in voyages to the Brazil Banks. Adams noted that insofar as these whaleships were involved—"South Seasmen"—"all the officers and men were Americans."[7]

The facts were significant. "South Seasmen" then meant whaleships sailing into the South Atlantic in contrast to the regular British northern or Greenland fleet, which traditionally sailed to the Arctic for right whale oil and bone. The South Seasmen were sperm whalers, and chief of these were the Nantucketers. Unable to follow their profession at home, the Islanders who had gone to London during the war were eagerly received there and given vessels as commanders and officers.

It might be assumed that these men were traitors. But history presents them as they actually were—inhabitants of a world of their own, previously a self-sustaining independent world, only comparable to one of the ancient city-states of the Mediterranean. Their industry took them away from the life of the colonial world "on the main"; their religion, as members of the Society of Friends, made war contrary to their convictions; as Thomas Jefferson stated, they were a race apart from the mainstream of American or European life.[8]

On his return to New England, John Adams wrote the Massachusetts General Court under date of September 12, 1779, that the British Southern Whale Fishery numbered seventeen vessels, sailing from London. On the advice of his French associates in Paris, Adams suggested that a Conti-

7. John Adams, *Massachusetts Archives*, Vol. 210 (Boston, Mass.), p. 16.
8. Thomas Jefferson, *Report of The Secretary of State On The Subject of The Cod and Whale Fisheries* (Philadelphia, 1791), p. 10.

nental frigate be dispatched to the South Atlantic where the British new whaling fleet would be easy prey. Adams also noted that the capture of these whaleships would not only yield valuable oil and bone but "at least 450 of the best kind of seamen would be taken from the hands of the English, and might be gained into the American service to act against the enemy. . . . This State has a peculiar Right and Interest to undertake this Enterprise as almost the entire fleet belongs to it." Adams' recommendations were referred to the Continental Congress—but never acted upon.[9]

When the Revolution was finally concluded by the Treaty of Paris in 1783, Obed Macy, the contemporary Nantucket historian, noted the effect of the Revolution: "The town exhibited the appearance of a deserted village rather than a flourishing seaport. . . . About this time many young men came home from different parts where they had been confined as prisoners. Some of them had been absent so long, without being heard from, that their connections had relinquished all hope of ever seeing them again. . . ."[10]

Now, a new feeling sprang up in Sherborn. In November 1782 word came from London, through business connections, that England was ready to sign a provisional treaty of peace with the Colonies. William Rotch immediately decided to gamble on the truth of the report. Two of his ships, the *Bedford* and *Industry*, had just returned from successful voyages to the Brazil Banks. The oil was landed, quickly refined, and the ships reloaded and dispatched to London. It was December 1782 when the *Bedford* sailed, and the winter passage was protracted, but the ship crossed the Atlantic safely under command of Captain William Mooers.[11]

The *Bedford* reached the English coast late in January 1783, arriving in the Downs on February 3, and proceeding up the Thames from Gravesend the next day. After a night at anchor off Greenwich, the *Bedford* continued her passage and on the morning of February 6 she came up to London, anchoring just below the Tower. From her main-masthead she displayed the bright new flag of the newest nation—the Stars and Stripes of the United States. It created an instant sensation.

9. John Adams, *Massachusetts Archives*, Vol. 210 (Boston, Mass.) p. 16.
10. Macy, *The History of Nantucket*, pp. 120–121.
11. Starbuck, *The History of Nantucket*, p. 383.

2. Ship *Maria*
Built in 1782 at Pembroke, Massachusetts, for William Rotch, Sr.
[Courtesy the Nantucket Historical Association]

The *Bedford*'s arrival posed a problem. Just as William Rotch realized, although a state of war still existed (the actual treaty of peace would have to await the termination of hostilities with France), the *Bedford* was carrying a commodity which London desperately needed—Nantucket sperm oil. Customs officials were perplexed by this unprecedented situation, but finally permitted the entry of the cargo.

The appearance of the *Bedford* and the display of her flag were duly recorded in several British journals. One of them described the incident in this way:

The ship Bedford, *Capt. Mooers, belonging to Massachusetts, was reported at the custom house on the 6th inst. She was not allowed regular entry until some consultation had taken place between the Commissioners of Customs and the Lords of Council, on account of the many acts of Parliament yet in force against the Rebels in America. She is loaded with 487 butts of whale oil, is American built, manned wholly with American seamen, wears the Rebel colors, and belongs to the Island of Nantucket in Massachusetts. This is the first vessel which displayed the thirteen rebellious stripes of America in any British port. The vessel lies at Huntly Down, a little below the Tower, and is intended immediately to return to New England.*[12]

Many Londoners came down to the riverside to see the *Bedford*, and several went on board. Among the interested visitors was Madame Hayley, widow of George Hayley, whose firm had carried on extensive business in Nantucket whale oil before the war. Madame Hayley was the sister of that controversial Londoner John Wilkes. As a business associate of the Hayley firm in London, Francis Rotch of Nantucket, in close touch with his brother William on Nantucket, probably had much to do with the *Bedford*'s voyage into history. He was with Madame Hayley from 1777 to 1785.[13]

The brig *Industry*, under Captain John Chadwick, did not leave Nantucket until January 20, 1783, arriving in the English Channel "after a long and tedious passage." Captain Chadwick put in at Portland on March 10, having previously spoken with a "Dutch galeot that told us

12. *The Gentlemen's Magazine*, Vol. 53 (London, England, 1783), p. 173.
13. John M. Bullard, *The Rotches* (New Bedford, 1947), pp. 54–55.

the Happy News of Pease."[14] Off Dover, the *Industry* obtained a pilot, then proceeded up the Thames to London, becoming the second vessel from America to reach that port with whale oil before the signing of the Treaty of Paris.

The moment of joy which greeted the news in Nantucket combined relief at the war's end and hope of renewed trade with London. But this was followed by dismal tidings. Almost immediately after the Treaty an import duty of £18/3 was placed on American oil. The news was a vital blow to the Nantucketers.

With the establishment of the import duty on American oil the British government gave the London merchants both a protection and an incentive. John Adams, on the eve of coming to Britain as the first Ambassador from the United States, was deeply shocked by the news, while his colleague in Paris, Thomas Jefferson, appraising the situation a few years later, remarked that England intended to "found their navigation on the ruins of ours."

Both the London merchants and the Nantucket whalemen were all too aware of the significance of the high duty barrier. The London contingent held the trump hand; they knew Nantucket's loss of her chief market meant her whalemen must seek a means of carrying on the only industry they possessed. London now provided the openings.

14. Starbuck, *The History of Nantucket*, p. 384.

2. London Gains a New Industry: The Southern Whale Fishery

WHEN that enterprising London firm of Samuel Enderby and Sons seized the opportunity to send out the first British-owned ships to enter the Southern Whale Fishery, it was a venture based on a well-calculated risk. Samuel Enderby, the senior member of the firm, having the advantage of a long business acquaintanceship with the Nantucket whale oil merchants, especially the House of William Rotch and Sons, and having personally observed the situation concerning the Falklands Fleet of Francis Rotch, planned carefully. The timing of the Enderby venture was excellent.[1]

The fact that the Enderby firm was committed to the enterprise by late in 1775 is shown by a letter Samuel, Sr., wrote to another of his American business associates, Nathaniel Wheatley of Boston. Due to the British occupation of that city, Wheatley at this time was in Providence, and Enderby's letter to him there clearly reveals that the project, or "grand scheme," was well advanced:

Should you get the [*ship*] Pitt *to Nantucket . . . you had better sell her. . . . If you come over in the spring, I would wish you to bring over 30 good whalemen, as I have a grand scheme in view; am afraid you will not be able to hear from me again as all communication is now cast off. . . . I have bought Bartlett a fine new ship of 370 tons, . . . Richard and Alexander Coffin are still here . . . undetermined in regard to staying or removing.*[2]

Engaging the services of the experienced Nantucket whalemen was to become the key factor in implementing Enderby's bold plans. Some of

1. Charles Enderby, *Proposal For Re-Establishing The British Southern Whale Fishery* (London, 1947), p. 6.
2. Samuel Enderby, Sr., to Nathaniel Wheatley, Colonial Office Records, P.R.O., C.O. 5, Vol. 40, p. 406. Dated December 6, 1775.

these men were already in London; others were invited, as his letter indicates. That the Rotches had not abandoned the Falkland Islands as a base during the Revolution is revealed by an entry in the London customs, showing their ship *Nancy*, Captain Scott, had been granted a permit to sail thence.

But, to the initial advantage of the Londoners, it was the Enderby ships which succeeded, having protection from the Royal Navy. Obtaining the services of the Nantucket whalemen meant that the London whaleships were to follow the voyages to the South Atlantic where Nantucket ships had sailed for decades.

As an example of the knowledge of these whaling grounds, which the Americans had gained, a letter of instruction from the Colonial merchant to his captain reads:

> *We flatter ourselves you will be able to get upon the whaling grounds at Brazil sometime in November* [*1775*], *and we would recommend your cruizing Southward after Spermaceti Whales off and about soundings, until you get into the latitude of the Falkland Islands. If, in the forepart of your cruize you do not find these aplenty, we have no doubt of your meeting them near the Islands. . . . We have certain and very late information that between the Latitude of 45° and 48° South, near Soundings, there is a great plenty of Spermaceti Whales. . . .*[3]

The Enderby ships returned with valuable cargoes and thus the firm prospered. At the Enderby counting house on Lower Thames Street, at Paul's Wharf, Samuel Enderby, Jr., with his two brothers Charles and George, joined their father as partners. In 1783, shortly after the Treaty of Paris, Samuel Enderby, Jr., sailed for Boston "expressly to get information about the Fishery and to engage Nantucket men to come to England."[4]

Other London merchants now followed Enderby's example, notably Alexander and Benjamin Champion, Thomas Dickason, and John St.

3. Francis Rotch and Aaron Lopez to Captain John Lock, September 4, 1775. Colonial Records, P.R.O., C.O. 5, Vol. 122, p. 38.
4. *Records of the Lords of The Committee of Council Appointed For The Consideration of All Matters Relating To Trade and Foreign Plantations* (hereafter cited as *Board of Trade Records*), P.R.O., B.T. 6/93.

Barbe. The founder of the Champion firm, Alexander, Sr., had been connected with the George Hayley Company, and was closely associated with Francis Rotch. George Hayley was a "merchant of eminence," who became President of the famous insurance firm, Lloyds of London. Dickason had purchased Nantucket whale oil over a long period. John St. Barbe, while a newcomer to the scene, proved an able entrepreneur. A contemporary account stated he had been a Lieutenant in the Royal Navy, "also, an adventurer in this Whale Fishery . . . of a very active and enterprizing, adventurous disposition, and seems very sanguine in pursuit of it."[5]

The success of the London whaling fleet during the Revolution had alarmed John Adams. As noted, Adams, then in Holland, wrote to Congress describing the participation by the Nantucketers, and urging naval action for the possible capture of the British whaleships.[6] Now, with the war over, the pattern of these voyages had been set, and the British advantage was pursued quickly.

Recognition of the role the Enderby firm had played in inaugurating England's direct participation in the Southern Whale Fishery was established early, as a document in the records of the Board of Trade, at Whitehall, states:

He was the first adventurer from Britain in this Southern Whale Fishery and had persevered in it during the war, has had the greatest experience and now has the largest concern in it, having 5 ships employed and is fitting out 2 more. . . . Mr. Enderby has now [1785] about 30 Nantucket men in his employ; all the other adventurers have followed his example, and they say they will continue to employ their men upon shares.[7]

The latter reference shows that the British had now adopted the American system of settling a whaling voyage, with the crew receiving their pay by shares in the cargo, or by "lays," as the Americans termed it, rather than by wages, as in the merchant service.

5. Ibid.
6. Massachusetts Colonial Manuscript Resolves 6,6,2/6.
7. *Memorandum, Board of Trade Records*, P.R.O., B.T. 6/93.

Britain's policy of placing a high protective tariff on the whale oil imported from America provided the London whaling merchants with their great opportunity. To further increase this advantage Messrs. Enderby, Champion, and St. Barbe, acting for the other whaling firms in London, were able to persuade the powerful Board of Trade to reestablish a bounty of forty shillings per ton for all British whaleships. As an added inducement in recruiting crews, the young apprentices were to be exempted from impressment into the Royal Navy. This was a recognition that whaling was as much a nursery for seamen as it was an industry.

By 1785, the uses of whale oil had brought about a demand far exceeding the supply. Britain, despite the loss of her American Colonies, was on the verge of a new era of prosperity, made possible by the development of her manufacturies. The Industrial Revolution had sparked the remarkable growth of her textile industry, with the invention of the flying shuttle, the spinning jenny, power loom, etc.

Whale oil was needed for lubrication as well as lighting. However, Britain's as well as Europe's chief use of this oil was for lighting. One of the most significant uses came with the increased practice of lighting the streets of cities. In London, the oil for this purpose was obtained by contract with companies, and it was noted that one contractor "by singular arrangement, agreed to pay the city £600 per year for a monopoly. In return for this, they were empowered to levy a rate of 6 shillings a year for all housekeepers . . . and from all who owned houses a levy of over £10 per annum, unless they hung out a lantern or candle before their door, in which they were exempt from public lamps."[8] The increase in systemized street lighting in London during the 1765–1775 period had brought about a sharp decrease in crimes after dark, and the use of whale oil became an important factor. Sperm oil from Nantucket found a ready market for street lighting, being far superior to the black oil of the Greenland or Northern Fishery. By 1780 there were more lights on Oxford Row than in all Paris.

Improvement in house lamps after the Revolution also gave sperm oil an extra premium, as the superior quality of the oil was well known. The

8. William F. Leckey, *A History of England In The 18th Century* (New York and London, 1878), Vol. 1, p. 527.

London whaling merchants knew they had a guaranteed market for their ships' cargoes. Reports to the Board of Trade showed that, before the Revolution, some five thousand tons of whale oil were imported from America, "two thirds from Nantucket." London, alone, spent £300,000 annually on the purchase of oil for her street lamps.

On his arrival in London in 1785, John Adams, the first Ambassador to the Court of St. James from the United States, tried to negotiate a commercial agreement with Britain whereby the import duty on American whale oil would be reduced, and a former strong market restored. Rebuffed at every turn, Adams became quite depressed. In letters to his New England friends he not only deplored the British restrictions on trade but warned there was little likelihood of any change.

Adams strongly urged that everything possible be done to revive the American Whale Fishery, "which is our glory." In a rare, almost lyrical paragraph, he described the superior properties of spermaceti oil, "which preserves its fluidity in the cold, gives a clear and glorious flame, burns till 9 o'clock in the morning, and is cheaper." He felt that every home should have some of the "beautiful spermaceti candles."[9]

Following his unsuccessful efforts to win concessions from Britain's ministers, John Adams decided to appeal directly to William Pitt the Younger, the Prime Minister. The two men made a sharp contrast. Pitt, the aristocrat, tall, thin-faced, aloof; Adams, the typical Yankee, short, stout, round-featured. At first disheartened by Pitt's detached manner, Adams, with characteristic straightforwardness, plunged into his appeal. When the question of shipping and the high duty on American oil came up there was a sharp exchange. Pitt remarked Americans could not think hard of the English for encouraging their own shipwrights and their own whale fishery. Adams, in reply, stated: "By no means, but it appeared unaccountable to the people of America that England should sacrifice the general interests of the nation to the private interests of a few individuals interested in the manufacture of ships and in the whale-fishery, so as to refuse these remittances from America in payment of debts and for manufactures which would employ so many people."[10]

9. John Adams to Cotton Tufts, *Works of John Adams*, ed. Charles Francis Adams (Boston, 1853), Vol. 8, p. 295.
10. *Works of John Adams*, Vol. 8, pp. 308–309.

SHipped by the Grace of God in good Order and Condition, by Micajah Coffin
in and upon the good Ship called the Maria
whereof is Maſter, under God for this preſent Voyage, William Moores
and now riding at Anchor at Nantucket
and by God's Grace bound for London to ſay,
Twelve barrels of Whalebone contents thirty seven hundred two quarters & ten pounds Nett on the proper acc.t & Risque of Ebenezer & Simeon Thomas of Norwich in the State of Connecticut Merchants

being marked and numbered as in the Margin, and are to be delivered in the like good Order, and Condition, at the aforeſaid Port of London (the Danger of the Seas only excepted) unto Rogers & Bromfield Merchants or to their Aſſigns, he or they paying Freight for the ſaid Goods forty five shillings sterling pr Twenty hundred weight
with Primage and Average accuſtomed. In Witneſs whereof the Maſter or Purſer of the ſaid Ship hath affirmed to three Bills of Lading, all of this Tenor and Date; the one of which three Bills being accompliſhed, the other two to ſtand void. And ſo God ſend the good Ship to her deſired Port in Safety. *Amen.* Dated in Nantucket June 22. 1785

Contents unknown to Wm. Moores

3. Bill of Lading for the ship *Maria*, Nantucket, 1785
[Courtesy the Nantucket Historical Association]

In a general discussion of the whaling industry, Pitt suddenly asked if it were not true that measures were being taken to sell American whale oil in France. Adams knew this to be true but that no agreement with the French had as yet been reached. He returned quickly to the original theme of the meeting, declaring:

We are all surprised, Mr. Pitt, that you prefer darkness and consequent robberies, burglaries and murders in your streets to the receiving, as a remittance, our sperm oil. The lamps around Grosvenor Square, I know, and in Downing Street, too, I suppose, are dim by mid-night, and extinguished by two o'clock; whereas, our oil would burn brightly until 9 o'clock in the morning and chase away, before the watchman, all the villains and save you the trouble and danger of introducing a new police into the city.[11]

11. Ibid.

Pitt listened attentively, but his characteristic coolness chilled Adams' expectations. He told his American visitor that he had reliable information that Britain had so advanced her own Southern Whaling fleet as to be confident in soon being able to bring in sufficient sperm oil in her own ships. Adams suspected this but was not willing to believe there was a large enough fleet out of London to make possible an importation of such a quantity of oil. As he had written to the Massachusetts General Court, the Nantucket whalemen, forced to sail in British ships, were providing Britain with the nucleus of a Southern Whale Fishery, and he realized the potential threat to American whaling.

In the Board of Trade, which Pitt had just created (August 1786) to replace the Committee of the Privy Council, the leading figure was Charles Jenkinson, a man experienced in both the Treasury and Foreign Plantations affairs of the King's several ministries before the younger Pitt took office. Adams described Jenkinson as "a great foe of America," a fact that was generally known. On one occasion during the Revolution (May 1777) Jenkinson stated he would "never depart from his original opinion that if America was to remain a part of the British Empire she ought most certainly to bear a proportionate share of the expense of general protection." Now, he was equally sure that the Colonists, having been declared independent, and no longer entitled to be considered as British subjects, or as Englishmen, were in the same category as a foreign country.

Pitt's reliance upon Jenkinson's judgment was not surprising, as the latter had become one of the best informed and strongest figures in English commercial affairs. His advocacy and support of the high tariff on American whale oil was founded on a thorough knowledge of the situation. He was fully aware that the Southern Whale Fishery, which Nantucket had founded, nurtured, and brought to a state of great success in a half century, was made so by the fact that London was her chief market. Now that the market was closed the Nantucketers had lost their ascendancy and, he believed, would soon lose their fleet by the simple fact that they would accept invitations to come to England so that they could follow their familiar pattern of industry.

Jenkinson had great confidence in the London whaling merchants—and they had equal confidence in the man who was to be their steadfast supporter.

The "closed door" policy as regards commercial relations which Britain adopted so swiftly after the Treaty of Paris, had more reason than mere vindictiveness or distaste for the new United States. It was rather a realization that America was Britain's great commercial rival. The flame of success of the Colonies' cause by war was now being reduced in England to a faint glow by the cause of peace. America no longer had her strong advocates in Parliament, as those men who had called for the rights of Englishmen in the Colonies could no longer consider this as a legitimate reason to support American commerce.

The advent of Pitt the younger and his cabinet had brought into power a group of strong men whose chief concern was the strengthening of the British Empire. One of the maritime industries which was to contribute to this strength was the Southern Whale Fishery, and Charles Jenkinson was to make this an almost personal project.

Adams had good reason to feel discouraged over the prospects for negotiating a commercial treaty. News from home further embittered him, as he recognized all too well the commerical factions which endangered the young nation. Rival navigation laws between the States, smuggling, disorganized efforts to establish trade with the West Indies and Europe, and other matters made thirteen disunited states highly suspicious of one another. The need for a strong central government was critical.

An Englishman in America at this time wrote in his diary:

The British merchants had great claims on the Americans whose Government, during the War, confiscated the Loyalist property and invited the American debtors to pay the balance they owed into the Public Treasury—when a receipt was given which exonerated them from future obligations. The unprincipled took advantage of this, which, I am sorry to say, was universal, except with the Quakers and truly conscientious men. In this state of commercial derangement; with the disorganization of the people and the laws, and all the consequent changes from the Revolution, nothing agreeable could be expected.[12]

12. Joseph Hadfield, *An Englishman in America* (Toronto, 1933), p. 1.

Of this John Adams was all too aware. While he pressed for the British to follow up their agreement in the Treaty of Paris, which called for the abandonment of her forts along the Canadian border, he was fully cognizant of the fact that the fourth article in the Treaty, respecting the recovery of debts by the Loyalist and British merchants, had not been faithfully observed by the United States.

It was a grave point and Adams felt it keenly. Only a few months before (May 1786) he had advised Congress "that every law of every State which concerns either debts or loyalists, which can be impartially construed as contrary to the spirit of the Treaty of Peace, be immediately repealed, and the debtors left to settle with their creditors or dispute the point of interest at law."[13]

This was to become one of the greatest stumbling blocks and was to interrupt many efforts to establish mutual trust—that necessary ingredient in negotiating any commercial treaty between the mother country and the vigorous offspring. Adams knew many of the Loyalists as his personal friends and deplored repudiation of just debts on the part of their creditors.

The honest New Englander was not to have his pleas heeded, and this, with another of the controversial points—the West India trade—was left to rankle in the mercantile relationships of the new and old nations. Much of the attitude of Nantucket whaling merchants was affected by the repudiation of debts owed to their British merchant friends by colleagues on the American mainland. This loss of confidence in the Congress and its ability to govern had much to do with deciding many of the Nantucket leaders, as well as the whalemen, to seek places outside the country where they could again make whaling a paying trade.

Many of these wandering whalemen spent years of self-imposed exile before returning to their Island home; others never did return; still others made their homes in other parts of the world for the rest of their lives. But the story of their voyaging, their searching, and their successes and failures is as much a part of America's maritime history as it is of their own personal adventuring.

Having tried in vain to bring about an understanding leading to the introduction of a commercial treaty with Britain, John Adams found his

13. *Works of John Adams*, May 13, 1785, Vol. 8, p. 249.

frustration finally giving way to calmer appraisal. He wrote his friend John Jay:

Britain has ventured to begin commercial hostilities; I call them hostilities because their direct object is not so much the increase of their own wealth, ships or sailors, as the dimunition of ours. A jealousy of our naval power is the true motive, the real passion which actuates them. They consider the United States as their rival and the most dangerous rival they have in the world. . . . They think they forsee that if the United States had the same fisheries, the same carrying trade, and the same market for ready built ships, which they had ten years ago, they would be in so respectable a position, and so happy in their circumstances that their own seamen, manufacturers and merchants, too, would hurry over to them.[14]

And then he made one of his observations which contained, however regretfully expressed, a typical Adams prophecy:

If Congress should enter in earnest with this commercial war, which may break out into a military war, . . . it is to be hoped that the people and their councils will proceed with all the temperance and circumspection which such a state of things requires. I would not advise to this commercial struggle if I could see a prospect of justice without it, but I do not; every appearance is to the contrary.[15]

In London, at this time, Francis Rotch had resumed his association with the Hayley firm, becoming chief adviser to Madame Hayley, who had continued the business upon her eminent husband's death. The Nantucketer closely watched the success of the Enderbys, keeping his brother William, on Nantucket, informed of the efforts of the London whaling merchants who were to petition for renewed bounties on the quantity of oil brought in by the new branch of the industry—the Southern Whale Fishery.

Francis Rotch had not been idle in his own plans to circumvent the high duty on Nantucket oil and candles, and, through his London residency, sought to disguise the imported American oil, and apparently was

14. John Adams to John Jay, *Works of John Adams*, Vol. 8, p. 290.
15. Ibid.

successful in a number of instances. The London merchants, alert to these activities, warned the Board of Trade, citing one particular case:

A large ship named, the Rising States [*actually the ship* United States], *Captain Scott, cleared out as a British ship from London to Boston. There she delivered her cargo and fitted out for the Southern Whale Fishery, and sailed from thence with another Capt. named Coffin* [*Captain Uriah Coffin, of Nantucket*], *and a supercargo named Hussey* [*Captain Benjamin Hussey, of Nantucket*], *both Americans and the best part of her crew American fishermen, and had two small tenders with her. At the Falkland Islands 100 tons of oil taken by these vessels was shipped on board a British ship, commanded by one Massey* [*Macy*] *and brought to London, to be entered as oil taken by a British ship, but it seems to be under seizure.*[16]

These efforts to "crack" the British tariff wall were thwarted, for the most part, by the vigilance of the London whaling merchants and the Customs. The effect on Nantucket's fortunes was disastrous. A letter from James Bowdoin of Boston to John Adams, written in January 1786 summed up the situation. Before the Revolution, it was pointed out, the Nantucket whalemen produced three thousand tons of whale oil annually, most of which went to Britain, at an average sale price of £35 per ton, totalling some £105,000. A considerable quantity of British manufactured goods was then purchased by the Nantucketers for the American market. Bowdoin observed that both nations gained by the enterprise of the Nantucketers without risking financial outlays for insurance or carrying charges.[17]

Thus, with their most lucrative market closed, the Nantucket whalemen were faced with the alternative of seeking another port where they could be gainfully employed. The London merchants provided the opportunities. In the balance was the control of the Southern Whale Fishery —a prize of considerable economic magnitude.

In one of his letter to Bowdoin at this time, Adams noted that "These Americans [Nantucketers] are to take an oath that they are to settle in England before they are entitled to the bounty."

16. *Memorandum, Board of Trade Records*, P.R.O., 6/96.
17. James Bowdoin to John Adams, January 12, 1786, *Works of John Adams*, Vol. 8, pp. 363–365.

He also wrote he had informed Congress that nothing could be expected from Britain in the way of trade agreements with the United States, warning: "No man can do a greater injury than by holding up to their hope that we will receive any relief by taking off the duty on whale oil, or by admission to the West India Islands. They will infallibly be deceived if they entertain such expectations."[18] The British West India Islands had been a strong market for Nantucket sperm candles, as well as oil, and the loss of this potential was another economic blow.

The Nantucket whalemen migrating to London were, for the most part, in those crucial years, captains, mates, harpooners (boat-steerers), and coopers. They were the specialists in the sperm whale industry. Many of them were Quakers, and they were welcomed by such Quaker Meetings as that of Ratcliff.

These Island whalemen were as much at home on the sea as on the land, and they had to pursue the only means of livelihood they knew. Nantucket was like a lodestone, to draw their thoughts ever homeward, and their ultimate fate would never actually separate them. However, in the exigencies of this crisis, it is unlikely that they fully realized that in their pursuit of the whale they were pursuing as well their own destinies.

18. John Adams to James Bowdoin, May 9, 1786, *Works of John Adams*, Vol. 8, p. 389.

3. Whalemen's Exodus: The Tide Sets

FACED with the economic crisis through the loss of her London market on the one hand, and with the exodus of her whalemen on the other, Nantucket found herself confronting the year 1785 with understandable apprehension. That sharp cleavage in the community caused by the Revolution was still in evidence, but the exigencies of the times brought together the factions. Again, the leadership of the Quaker merchants prevailed. A decision was made which, in light of the circumstances, was a natural one, but against the larger backdrop of that day was startling.

It was decided to petition the state for the establishment of Nantucket as a "neutral" Island within the complex of the United States. If this were granted it might enable Nantucket to appeal to Britain for the entrance of whale oil and candles duty-free.

Action by the Island was prompt. Early in 1785, at a special Town Meeting, it was voted to prepare a "Memorial" containing this request for submission to the General Court of Massachusetts at its May session, the main point reading:

That a neutral state is the most convenient situation that the Town can be placed in for the benefit of the Inhabitants thereof under their present circumstances.[1]

As a committee to present the "Memorial" formally in Boston four of the leading Nantucket merchants were selected, all members of the Society of Friends—William Rotch, Sr., Timothy Folger, Richard Mitchell, Jr., and Josiah Barker.

In the "Memorial" it was emphasized that "the whale fishery cannot be preserved in this place, nor any part of the business carried on . . .

1. Starbuck, *The History of Nantucket*, p. 386.

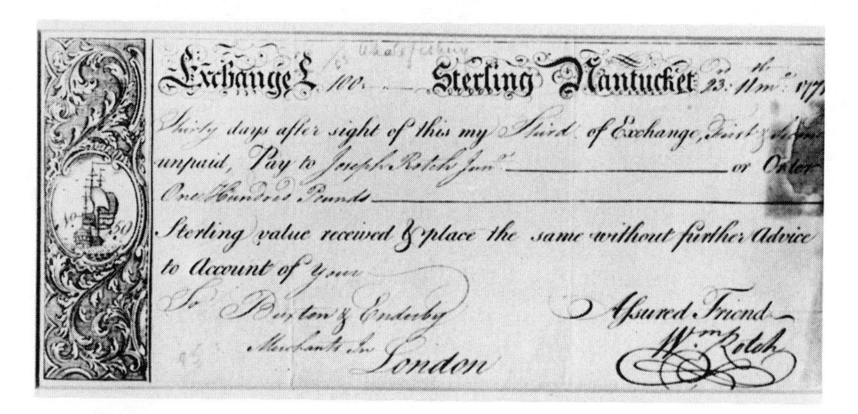
Exchange £ 100. Sterling Nantucket 23. 11mo. 177

Thirty days after sight of this my Third of Exchange, First &
unpaid, Pay to Joseph Rotch Junr. or Order
One Hundred Pounds
Sterling value received & place the same without further Advice
to Account of Your
To Buxton & Enderby Assured Friend
Merchants in Wm. Rotch
London

4. William Rotch, Sr.'s Bill of Exchange on Buxton and Enderby, London [Courtesy the Nantucket Historical Association]

without great loss, . . . and the only remedy is in placing the Island and its inhabitants in a state of neutrality."[2]

As was to be expected the Legislature refused to grant such an unprecedented request. A contemporary newspaper reported that Nantucket Island had made an astonishing petition requesting they be granted a "separation from their government to enable them to make a contract with England."

While public reaction to the "Memorial" on the mainland was mixed with astonishment and resentment, Nantucket felt justified in the attempt. To carry on the industry swift action was needed. The exodus of the whalemen was already setting strongly as a tide, and there was now a prospect that the year would feature such migrations to England and other places. A company had already left the Island and settled on the Hudson; there were already proposals for a group to go to Nova Scotia. The times demanded action. The move to Canada had a telling practicality; oil brought into the Provinces could be sold in England duty-free.

There was another factor involved in any transfer to Nova Scotia. In 1761–1762 a number of Nantucket families joined a company of New Englanders settling at Barrington. Toward the end of the Revolution

2. Macy, *The History of Nantucket*, p. 126.

there was a prospect of establishing a whaling port at Port Roseway, also on Nova Scotia's southern shore, and a number of Loyalists from Nantucket were interested in the plan.[3]

Despite these precedents the first major migration of the Nantucketers was not to Canada. Instead, a company led by Island merchants sought a new home well up the Hudson River in New York State. Prominent in this migration were Seth and Thomas Jenkins, who had suffered considerable loss in the Revolution by Loyalist "Refugee" raids; Captain Alexander Coffin, an outspoken supporter of the Continental cause, who had served as a dispatch bearer for John Adams in Amsterdam to the Continental Congress at Philadelphia; Marshall Jenkins of Edgartown, Martha's Vineyard, an active whaling master; Titus Morgan and John T. Thurston of Dartmouth, Massachusetts; John Alsop of Providence (where Thomas Jenkins had recently moved from Sherborn), and Hezidiah Dayton, also of Providence.[4]

These men, with their families and their friends, were convinced that Great Britain would not rest content with the loss of her American Colonies, and that sooner or later there would be a strong effort to resume the war and recover them. The Nantucketers wanted to reinvest in whaling but did not wish to be trapped again between the two warring factions. After investigating possible sites along Long Island Sound and around New York City, the group decided on establishing a settlement at the sleepy little hamlet at Claverack, on the Hudson, one hundred miles up the river from the open sea, but on navigable waters. The setting was quite in contrast to Nantucket's sandy shores. Girded by magnificent mountains, the river ran through a rich and fertile land.

In dealing with the Dutch owners of the land at Claverack the Nantucketers found a hospitable and reasonable people and the land was soon purchased. Word was quickly sent to Sherborn, and in the autumn of 1783 the family of Seth Jenkins, consisting of his wife Dinah Coffin and four children, and that of John Alsop and his wife and children, left Nantucket for their new homes, which they decided to name "Hudson."

3. Edouard A. Stackpole, The Migrations of The Nantucketers, in "Historic Nantucket," *Nantucket Historical Association*, Vol. 6 (October, 1958).
4. Walter V. Miller, "Hudson's Founding Fathers," in *The Chatham Courier*, June 11, 1964.

In the spring of 1784 the settlers formed a company similar to that of Nantucket Proprietary. In May of that year the other proprietors arrived in several vessels, with their families.

Captain Alexander Coffin took his family to Hudson in October 1785, bringing with him a number of other Island families, including the Joseph Barnards, the Jared Coffins, and the Josiah Coffins.[5]

This company of "Nantucket Navigators," as they were sometimes known, was one of the best prepared groups ever attempting a settlement. Their almost instant success may be attributed to their careful planning. First, their actual movement was by sea, so that their supplies, household goods, and furnishings could be transported as easily as their families. Second, they were bringing a fleet of brigs and sloops which were to serve as whaling as well as merchant vessels. Third, they had a company of artisans as well as whalemen—Titus Morgan was a shipbuilder; David Bunker, a miller; Stephen Paddock, a sailmaker; Thomas Jenkins, proprietor of a ropewalk; Peter Barnard, a carpenter and shipwright; Cotton Gelston, a surveyor; Shubael Worth, a schoolteacher; Seth Jenkins, proprietor of a hemp and duck (canvas) factory. Fourth, and far from least, the Proprietors had as cash assets the sum of $100,000.

As a Columbia County, New York, historian wrote, many families lived aboard the vessels until their homes were built. He went on to state:

Further evidence of the thoroughness, of the foresight and preparation that had been devoted to the undertaking is found in the prefabricated houses, framed and ready for erection, brought by many of the newcomers. Such a house was brought by Stephen Paddock, and is said to have been the first home for his family at the landing. This Paddock house was erected somewhere on what is now South Front Street.[6]

By mid-1785 the General Court of Massachusetts realized it could no longer close its eyes to the obvious. They had been informed that the Nantucket whalemen migrations had reached a critical state, and that the island's industry—which represented the heart of American whaling

5. Ibid.
6. Ibid.

—was in danger. The proposed move of the Nantucketers to Nova Scotia was particularly alarming. Fully aware of the gravity of the situation, the State legislature passed a resolution as follows:

Whereas this court, having a due sense of the high worth and importance of the whale fishery, and desirous of its preservations, not only to this State, but to the United States in general; therefore, Resolved that there be paid out of the Treasury of this Commonwealth, the following bounties upon whale oil:

For every ton of white spermaceti oil, five pounds;
For every ton of brown or yellow spermaceti oil, sixty shillings—
For every ton of whale-oil (so called) forty shillings.[7]

But the depression in the American oil market showed no signs of lessening, and the Nantucketers who had decided to remove were not convinced that the Massachusetts legislation would effectively change the picture. The Island merchants believed that recovery was to come only by regaining the British market through London, and the group planning on migrating to Nova Scotia knew this was the decisive inducement—sending oil duty-free to their former associates in England.

No strong central government then existed in the new government of the United States, and the individual States were already involved in tariff wrangles. Across the sea Britain continued to refuse American offers for a commercial treaty, and the import duty on whale oil continued—including the British islands in the West Indies.

The situation became even more complex at this juncture. Britain had gained more Nantucket shipmasters as the London merchants continued to hold out their flattering inducements, but now another European nation entered into the competition for American whale oil. France, the perennial rival of England, held out an opportunity, the overture appearing in the form of a letter from that well-loved friend of America—the Marquis de Lafayette—transmitted by a gentleman in Boston. A newspaper of the time reported that the French offer included

proposals from a company for taking a large quantity of Whale Oil of different qualities from America on very advantageous terms to the shipper, having obtained a remission of all duties in France in

7. Macy, *The History of Nantucket*, p. 128.

their favour, payment to be made in the produce and manufacture of that country. It is likewise proposed that an American company be formed to avail themselves of the benefit of the plan which promises much advantage to those who may engage in it, as well as the community from the encouragement it will afford to so important a branch of our commerce as the Whale Fishery. The original articles on the part of the French Company are duly executed and now in the possession of Mr. Samuel Breck, of Boston, who has also a sample of the kind of Whale Oil required. Any gentlemen, therefore, who desire to be interested in the business may be informed in the subject by said Breck, at his House nearly opposite the Mall.[8]

There was an excellent reason why the Nantucketers were not interested in the French overtures at this time. Thomas Jefferson, in a letter written to John Jay three years later, noted: "The French had not been insensible to the crisis as their counterpart was . . . to tempt the Nantucketers . . . to come to France. The British, however, had in their favor a sameness of language, religion, laws, habits and kindred. . . ."[9]

The second mass migration of the Nantucket whalemen was now being planned—a transfer to Dartmouth, across the harbor from Halifax, Nova Scotia. Two of the Nantucket leaders, Samuel Starbuck, Sr., and Timothy Folger, entered into communication with Governor John Parr of that Province. Starbuck and Folger were outstanding Loyalists, and were anxious to reclaim their status as British subjects. How many preliminary negotiations transpired is not known but one Canadian historian, writing of her research, believes that such evidence as there is points to the possibility of suggestions for removal having first come from Nova Scotia.[10]

Starbuck and Folger sailed for Halifax for a personal conference, arriving in July 1785. They were well received by the Royal Governor. The response to their petition was immediately favorable. Governor Parr

8. *The New Hampshire Gazette and General Advertiser*, October 28, 1785.
9. Thomas Jefferson Randolph, Editor, *Memoir, Correspondence, and Miscellanies*, Thomas Jefferson to John Jay, Vol. II, p. 519.
10. Margaret Ells, "The Dartmouth Whalers," *The Dalhousie Review*, Vol. 15, No. 1 (Halifax, Nova Scotia, April 1935), p. 86.

encouraged them to "spell out" their application in the form of "Enquiries." This was readily done. Most important to the Nantucketers were whether their ships would be registered as British "bottoms"; what they could expect in the allocation of land to be assigned them for settlement; and whether they would be hindered in the conduct of their religious lives as members of the Society of Friends.

Governor Parr's eagerness to clinch the prospect of obtaining a complete colony of Nantucket whalemen was shown by his prompt action in calling his Council together and receiving its approval. The Governor's action was praised by Richard Uniacke, Speaker of the Assembly, who wrote: "I apprehend there was not on the face of the earth a set of People who pursued a single Branch of business with the same spirit of enterprise . . . than the Inhabitants of that Island pursued the Southern Whale Fishery."[11]

These sentiments were equally endorsed by Richard Cumberland, the Provincial Agent. Agreements were quickly drawn up and signed and the terms approved by the Nantucketers.

The location for this transplanted colony was to be at Dartmouth, directly across the harbor from Halifax. Immediately returning home, Starbuck and Folger completed arrangements for the exodus. On September 20, 1785 the first contingent of the colony reached Halifax in three brigantines and a schooner, with their crews and everything necessary for the whale-fishery, having been preceded by Samuel Starbuck, Sr., with his son, Samuel, Jr., aboard the whaleship *Lucretia*, with their household goods on board. On "the 14th of 11th month (November 14), 1785," they "went over to live in the house of Thomas Cochrane built at Dartmouth; with them came Lydia Heath, William Bestick and Nell."

Thomas Cochran was a prominent merchant in Halifax, and was one of the chief supporters of the plan to encourage the Nantucket migration. The Cochran firm had sent out one whaleship, apparently with a Nantucket crew, and an editorial some years later in the *Nova Scotian*, a Halifax newspaper, commented on Cochran's support of Folger and Starbuck. The firm engaged in whaling for a number of years.[12]

11. Richard Uniacke, *Public Archives of Nova Scotia* (hereafter cited as *P.A.N.S.*), Vol. 107, M. 503, p. 161, July 27, 1785, enclosed in Parr to Sydney.
12. *P.A.N.S.*, Files of *The Nova Scotian*, June 15, 1825. Article on the Whale Fishery.

The Provincial Assembly voted £1,500 for the new settlers, with which to build their homes, warehouses, workshops, etc., and some two thousand acres of land were eventually allotted them. In a speech made before the Province's Assembly, presented on December 5, 1785, Governor Parr commented:

In respect to the latter [commerce] the Prospect of an establishment of the Whale-Fishery here, particularly affords a well founded expectation of great benefit and advantage—and I am confident you will not fail to give every suitable encouragement to a matter of so much concern. . . .

At this time Dartmouth was a little fishing village, virtually obscured by Halifax, across the harbor, that stronghold of Britain's navy. The arrival of the Nantucket colony was to provide strong commercial and religious factors previously unknown to the inhabitants.

The co-leader, Timothy Folger, who had returned to Nantucket to superintend the second embarkation, sailed for Halifax on June 7, 1786, with his family, including Samuel Starbuck, Jr.'s wife and children. Forty families, consisting of 164 people, exclusive of 150 whalemen, had removed to Halifax by mid-summer, 1786. The arrival of more Nantucketers gave the new settlement at Dartmouth a thriving appearance. At a point of land, the whalemen built their first warehouse using the cove, known as Mill Cove, as a basin for their ships. Some 347 tons of sperm oil belonging to Starbuck and Folger were brought from Nantucket, together with two-and-a-half tons belonging to Gideon Gardner. The whole was shipped immediately to London as Nova Scotia Oil—only the fifteen shillings Colonial import duty being required.

The little colony of Nantucketers, removed to a new land, amid strange faces and customs was to make its own especial history. In the stark simplicity of their faith as members of the Society of Friends they had resolved to exile themselves from their island home. To leave hearthstone and relatives, with little prospect of returning, must have brought forth heart-rending scenes. They sailed away, bearing household goods and furniture (and in many cases their new frame houses), on board their own ships. The extraordinary force of their motives for leaving Nantucket joined with their Quaker spirit to give them a determination of a strength hard to conceive today.

Across the Atlantic, in London, the close coterie of whaling merchants learned of the Nantucket migration to Nova Scotia soon after Parr's negotiations had been completed. Their resentment was to be expected. However, Richard Cumberland, the Provincial Agent for Nova Scotia in England, thoroughly approved the movement. In February 1786 he wrote to the Nova Scotian Assembly in Halifax, through the Committee of Correspondence, and expressed his reaction: "I observe with great satisfaction the public Attention which is paid to the important object of Whale Fishery . . . I have no reason to fear but Government here will take Every prudent Measure for the Prosperity of your trade." [13]

Less than two years before (December 2, 1784), Cumberland had written from London: "As there is little market for whale oil but ours, the Americans must abandon the trade if the duty is enforced against them."

This duty was now being enforced, but the headquarters of American whaling was not to abandon the industry. Instead, the Nantucketers had transferred one corner of it well inside the British tariff barrier.

Moreover, even while the group of Islanders were embarking for the move to Dartmouth, another plan for a migration was being advanced—one representing the most daring idea of all. William Rotch, Sr., had sailed for London in his ship *Maria*, accompanied by one of his sons, Benjamin Rotch.[14] They were bringing to the British government a proposal which was to make history on both sides of the Atlantic.

13. *P.A.N.S.*, Vol. 301, Doc. 78. Richard Cumberland to the Committee of Correspondence of the Nova Scotian Assembly, 25 February, 1786.

14. Logbook of the ship *Maria*, Nantucket Atheneum Library, Nantucket, Massachusetts.

4. The Pilgrimage of William Rotch

IT WAS late in the afternoon of July 29, 1785, when the coach from Dover to London reached the top of Shooters Hill in Greenwich, and the driver brought the horses to a halt for a breather. Among the passengers alighting to stretch their cramped legs were two men dressed in the sober brown garb of the Society of Friends. The elder man was William Rotch, Sr., of Nantucket, then fifty years of age; the younger was Benjamin Rotch, the second oldest of his three sons, who was then approaching his twenty-first birthday. They had reached England only a few days before.

William Rotch, Sr., had a mission—one without precedent in the history of whaling. He intended to approach the British government with a proposal that a colony of Nantucket whalemen be invited to settle in England, where they might establish a whaling center—bringing with them their ships, families, and artisans, as a complete colony. It was his firm belief that these Nantucketers, as the foremost exponents of the Southern Whale Fishery, should not be invited as individuals to take out the whaleships of London or Bristol, but should be given the opportunity to build their own lives as a community just as in Nantucket, bound together by the one industry which had nurtured them, and by which, as pioneers, they had developed. This Nantucket Quaker whaling merchant had a deep conviction that whaling was virtually a public necessity as it supplied oil and candles for the public good, and therefore was not the possession of any one nation but of the world. In this respect William Rotch was one of the first American internationalists.

As the two Americans stood on the commanding eminence, gazing at the brown waters of the Thames River as it stretched into the center of the great port of London, they were more than ever conscious of the significance of their mission. In later years, William Rotch, Sr., recalled

5. Captain Hezikiah Coffin, Master of the Brig *Beaver*, of Nantucket
The *Beaver* was one of the three vessels taking part in the
Boston Tea Party, December 1773
[Miniature in the collection of the Peter Foulger Museum, Nantucket]

vividly that moment of pause on Shooters Hill. In his own manner of writing he described it:

feeling the great distance which separated me from my family, myself a stranger in that land, the occasion that drew me there, and the uncertainty of its answering any valuable purpose, I was overwhelmed with sorrow, and my spirits so depressed that, in looking toward the great city no pleasant pictures were presented to my view. But I found I could not give way to despondency; reason resumed. . . . I was there and something must be attempted.[1]

In a successful career as a whaling merchant William Rotch had gained a wide knowledge of the industry. His experience with the London market, and association with the very merchants who had launched the British into the Southern Whale Fishery (once his customers but now his competitors), made him fully aware of the advantage they had gained.

1. William Rotch, Sr., *Memorandum Written by William Rotch In The Eightieth Year of His Age* (hereafter cited as *Memorandum*) (Boston, 1916), p. 37.

The migrations of his fellow-Islanders to Hudson and Nova Scotia—the latter he termed the "Halifax affair"—had brought about his decision to attempt his present course of action.

He did not believe that the Nova Scotia migration which was being deliberated when he sailed from Nantucket was a solution to the Island's problem. With prophetic insight William Rotch noted: "The main points necessary to be done to make it [the move to Nova Scotia] of any use to our inhabitants, is entirely out of Governor Parr's power, I fear. He seemed desirous that we should be helped but saw no way while we were a part of the United States."[2]

This was a point that Rotch was to stress in his negotiations to come—that Nantucket was a part of the British Empire until the Paris Treaty removed it from British territory. He regarded this action a political rather than an economic separation. He correctly surmised that the London whaling merchants would have the political power to curtail any competition on the part of Nova Scotia, despite the fact that the Dartmouth settlers were well within the tariff barrier. In this he was to be proven correct.

It was on July 4, 1785, that William and Benjamin Rotch had sailed from Nantucket in their fast sailing ship, the *Maria*, with Captain William Mooers, a favorite shipmaster, in command. After a voyage of twenty-one days they were in the English Channel. Anxious to get ashore to visit friends, the Rotches were put ashore at Dover on July 22, and the *Maria* continued on to Gravesend, then up the Thames to London.[3]

At Dover the Nantucketers took a coach for Rochester, where they visited Dr. William Cooper who, with his family, had sailed to England two years before on the *Maria*. Cooper agreed to accompany them on their journey to London. As the coach made its way along the Highway, William Rotch talked with Dr. Cooper about his plans. Conditions confronting Nantucket after the peace were self-evident, and Rotch believed no other alternative faced the Islanders than to recover the lucrative European market for their oil and candles—particularly the British market.

2. John M. Bullard, *The Rotches* (New Bedford, 1947), p. 214.
3. Logbook of the ship *Maria*, Nantucket Atheneum Library, Nantucket, Mass.

His own business future was a part of the plan to bring a colony of Nantucketers to Britain and, despite the loss of $60,000 in the Revolution, he was still willing to risk his entire fortune on the venture. Above all, his main desire was to bring a vital aid to his fellow Islanders.[4]

Among matters discussed by the company was that of the debts owed by the Americans to British merchants. Rotch shared John Adams' opinion that the young United States was evading this issue; that Congress must adhere to the agreements concerning these debts as had been provided by the Treaty of Paris. Rotch believed that to do otherwise was a "flagrant breach of a treaty of our own making."[5]

Arrived in London, William and Benjamin Rotch went to live in the home of their friend Thomas Wagstaff on Gracechurch Street. The elder Rotch had many friends in the city, one of them such a prominent figure as Robert Barclay, who later was to gain him an audience with Prime Minister Pitt. His business ties with London merchants extended over many years. The firm of Champion and Dickason had been his customers for numerous oil and candle cargoes; his brother, Francis Rotch, was now an influential figure in the London firm of George Hayley. Probably the very first place of business visited was the firm of Samuel Enderby and Sons on Upper Thames Street, where he found some letters from America awaiting his arrival.

The *Maria*, still under Captain Mooers, sailed from London within two weeks, bound for the Southern whaling grounds, "her crew yet remain intire" (as Rotch wrote), although attempts had been made to lure them away.[6]

The rapid advances of the London merchants in building their Southern whaling fleets were carefully noted. Writing to his son-in-law Samuel Rodman, the Nantucket merchant stated: "They can send out about twelve ships from here as good head-men (harpooners) as can be found anywhere, (mostly our former inhabitants), the others but in-

4. Rotch, *Memorandum*, pp. 36–37.
5. Bullard, *The Rotches*, p. 213.
6. William Rotch, Sr., to Samuel Rodman and William Rotch, Jr., London, August 2, 1785. Rotch Letters Collection, Old Dartmouth Historical Society, New Bedford, Mass. (hereafter cited as Rotch Collection).

different."[7] At this time (August 1785) there were some twenty-two whaleships representing the London fleet but not all, of course, were at sea at the same time.

In another letter to Nantucket, William Rotch reported: "The spirit of whaling seems almost running to a degree of madness; they intend if men can be got to send out 30 ships, but at present there appears no possibility of getting men."[8]

Of the shipmasters he listed, many were Nantucket commanders, such as Captain Henry Delano, Captain Thomas Brown, Captain Paul Pease, and Captain James Gwinn. The numbers of Islanders who had accepted berths on London ships were more numerous than even the astute William Rotch had realized. This was all the more reason why his plan of a colony of Nantucketers needed to be advanced as quickly as was possible.

But there was another Nantucket observer of the international whaling scene who, for quite a different reason, felt that all such migration to Britain should be promptly curtailed. Before William Rotch sailed, this man, Captain Alexander Coffin, an active supporter of the Revolutionary cause during the war, had written to Samuel Adams, under date of June 8, 1785:

Duty and inclination both point to the propriety of my giving the earliest information to those who have it in their power to prevent a transfer of that valuable branch of business, the whale fishery, one of the grand staples of the Commonwealth, as well as one of the principal nurseries of seamen. . . . I have it on good authority that the British ministry has it in contemplation to remove such inhabitants of this Island as are suitable for Masters, Mates, etc., of whaleships there, and pay them for all their property which cannot be removed—pay their passages to England and put them in as good a situation there as they left here. . . . Mr. Rotch is now bound to London in a ship of his own from this place, intending to fit her from London to the Brazils, and in my opinion, to close the contract with the Ministry in which he will save at least £700 sterling for every family he will pace down in England agreeable to contract; a very pretty fortune to himself at the expense or rather the

7. Rotch Collection, November 2, 1785.
8. Rotch Collection, August 2, 1785.

loss of the Fishery to this Commonwealth. He is now taking on board a double stock of materials . . . of which they have none in England. . . .[9]

Captain Coffin also reported that William Rotch intended establishing a whaling base in Bermuda. There is no evidence in the Rotch correspondence to bear this out. But the matter was brought to the attention of the state and a law was enacted prohibiting the exportation of necessary whaling materials. By this time, however, the *Maria* had arrived in England. A few months later, Captain Coffin himself had left Nantucket, to join the group of Islanders who had settled at Hudson, New York.

The Massachusetts Legislature was still trying to enact laws to aid the flagging whaling industry. Captain Daniel Coffin, arriving in London in the ship *Mercury*, reported the State had voted a bounty of £6 per ton on sperm oil brought into the Commonwealth by ships belonging to the Bay State.

William Rotch was skeptical. He wished the news were true but was doubtful of its effect if it were a fact. He wrote home:

How far this [bounty] will be adequate to the expense of the Fishery I cannot tell, not knowing what oil is worth in Boston . . . but if sufficient to satisfy our people, it will ease me of a great burden which I could wish to be clear of, and have only to conduct my own business.[10]

A further indication of how deeply William Rotch felt this obligation toward his fellow Nantucketers—his "great burden" he termed it—is revealed in his reaction to a letter he received from Timothy Folger, his close associate of the Nantucket crises on the Island during the Revolution. Folger and Samuel Starbuck were now well into their negotiations with Nova Scotia's Governor Parr for the removal to Dartmouth, across the harbor from Halifax, and Folger was confident of the success of the proposal. It was obvious that William Rotch had not changed his opinion

9. Capt. Alexander Coffin to Samuel Adams, *The Inquirer* (Nantucket, Mass.), October 15, 1822. Newspaper copy of original letter of June 8, 1785.
10. William Rotch to Samuel Rodman, Rotch Collection, August 9, 1785.

as to the practicality of the venture; and he considered the delay in receiving word from Folger a setback for his own plans.

Rotch still held out hopes for Nantucket being granted its petition for neutrality, but the Nova Scotia movement ended this dream. On September 5, 1785, he wrote to his family:

I suppose by these movements nothing was likely to be done as to setting our Island at Liberty, which if is the case I wish I could have known it in time, that I might have done something here for ourselves, which I avoided purely on account of the Island.[11]

Despite the fact that favorable consideration of the migration of the Nantucket whaling group to Dartmouth-Halifax would affect his own plans,[12] William Rotch retained a characteristic, unselfish attitude. In a letter to his son-in-law Samuel Rodman who, with William Rotch, Jr., was handling the business interests of the House of Rotch in Nantucket, he wrote: "If thou, my son Samuel, should find thy way clear to remove [to Dartmouth] I shall furnish thee with property sufficient to do business with."[13] This, as much as any of his deeds, reveals the dedication of William Rotch to the cause of preserving the Southern Whale Fishery through the activities of Nantucketers—no matter in what corner of the world they were to settle. It also shows the unshaken faith he had in the economic soundness of the industry.

Having put his affairs in London in as good an order as circumstances would permit, William Rotch completed plans for his pilgrimage—the search for a port where he might bring his colony of Nantucketers and establish a whaling center in England. Accompanied by Benjamin and an English friend named James Phillips, the departure from London on September 6, 1785 had a touch of drama, as a "violent wind from the southwest—a near hurricane—" marked the beginning of their journey.[14]

And what an unusual journey it was—mostly by sea, unique in the history of whaling—a tour of some nine hundred miles along the Channel

11. Rotch Collection, September 5, 1785.
12. Ibid.
13. Ibid., August 9, 1785.
14. Bullard, *The Rotches*, p. 209.

ports and west along the South coast of Britain, ranging from Southampton to Bristol. The heavy burden of his mission was always with William Rotch, but never more than during these weeks. He wrote:

Our long journey . . . I heartily wish was over, but I think shall be inexcusable to come thus far, and not be able to make some judgment of the propriety of removing to the country if necessity shall urge it (I mean respect'g the suitableness of the different places for the Fishery). I am much urged by all ranks to remove into this Kingdom, yet, I never wish to be compelled at this time of life to take such a Step, though I see no way at present for the Inhabitants to subsist at my native Spot.[15]

His observations during this extensive journey give further insight into his future plans. While Southampton had "a fine capatious harbor," it was also "a great resort of fashionable people from both City and Country . . . where too much gaity, dissipation abounds, unsuitable to bring a family in our way; nor are there any Friends [Quakers] there." Poole had a fine harbor, "but with a long crooked channel up to it"; Lyme and Bridgeport had harbors too small; Tingmouth had a "bar before it, of 16 feet at high water." Dartmouth's harbor was "of easy access," but the steepness of the town's streets, "and the town being surrounded with such very high hills is the greatest objection I have."

Plymouth impressed him, as he wrote, "having a noble harbor . . . the great objection remains that there is a large Dock Yard and Magazines for the Royal Navy, and a great din in time of war which would be very disagreeable to us."

With a practiced eye for the future potential he weighed all the advantages and handicaps of each of these time-honored ports. He found Looe had too shallow a harbor, while Fowey was geographically so situated like Dartmouth as to pose the same problems to his mind, although he was well aware of its fine, deep harbor.

We now come to . . . Falmouth, notwithstanding the land where the Town stands is higher than I would wish, yet there is such an extent of situation after entering this harbor that one is not confined to Falmouth Town but may have a choice of many miles, so that this place is

15. Rotch Collection, November 2, 1785.

sufficient to answer every purpose . . . and is rather preferable to the rest.[16]

In Bristol, William visited the grave of his brother, Joseph, who had died there eighteen years before (1767) during a business trip to England. At this point in his pilgrimage business called him back to London. He wrote his sons that he had "intended to have gone to Milford Haven," in Wales; "I suppose it is the best harbor in England or Wales; it is surrounded with a fine fertile country . . . but for my own part, I prefer one of the ports of the English Channel."[17] His comments on the location of Milford were not without a prophetic touch.

Late in October 1785 Rotch returned to London, where he found letters from Nantucket at Enderby's Counting House, having arrived on the ship *Diana*, of Boston, commanded by George Folger, of Nantucket. The *Diana*, with a cargo of whale oil, had arrived during the absence of the Rotch party and had sailed before their return to London.

In reply to his Nantucket letters, Rotch wrote on November 2, 1785, to Samuel Rodman and his son William Rotch, Jr., reporting a particularly important transaction connected with the arrival in London of the *Diana*. Accompanying Captain Folger was Thomas Boylston, the ship's owner. Realizing he could not pay the import duty on the cargo of oil brought by the *Diana*, Boylston proposed that the Enderby firm purchase it for £10 less than the price in London. Enderby petitioned Customs for such permission, on the ground that the oil—sperm—was not available from any other source at this time and was needed in London. The petition was denied, and the *Diana* sailed for France, where the oil was sold. This was a sharp reminder that Customs was determined to prevent any landing of American oil without payment of the high duty.[18]

Soon after this, the *Warwick*, one of the Rotch fleet of Nantucket, arrived in London with a cargo of sperm oil largely consigned to Enderby. On this occasion, the London firm successfully negotiated with Customs for the entry of the oil by declaring it was intended for immediate export

16. Bullard, *The Rotches*, pp. 209–211.
17. Rotch, *Memorandum*, p. 38.
18. *Board of Trade Records*, B.T. 6/95, p. 83.

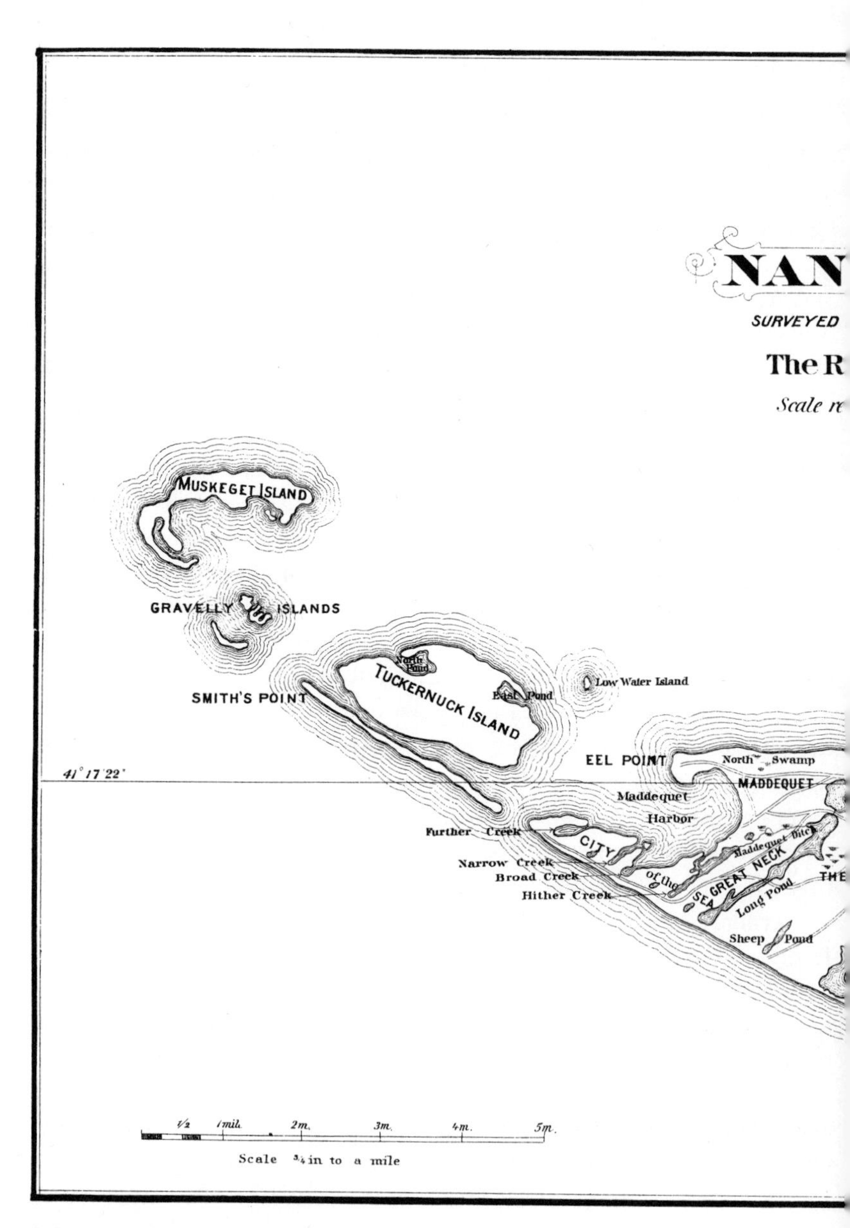

6. Map of Nantucket, 1869
"Historical Map" of Nantucket drawn by F. C. Ewer
[Courtesy the Nantucket Historical Association]

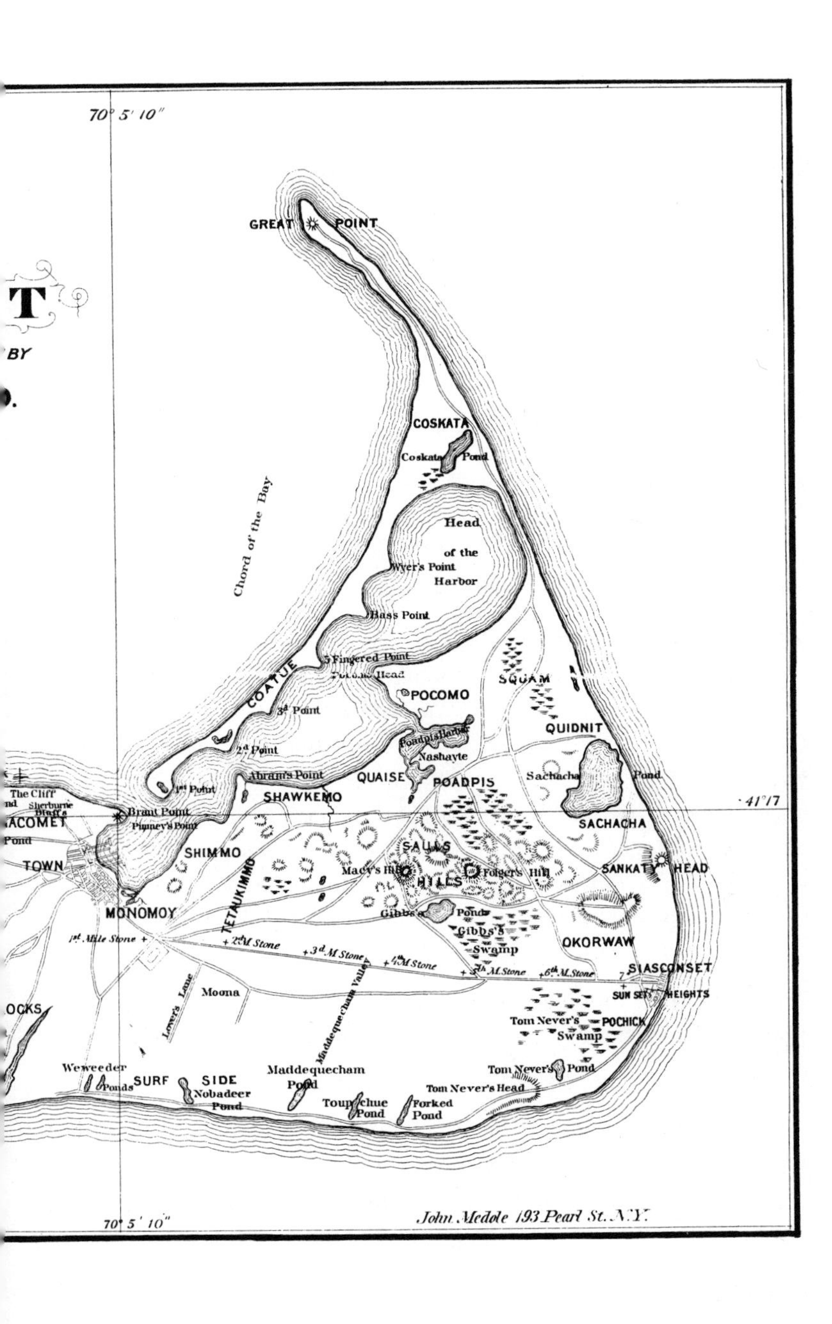

GREAT POINT
COSKATA
Coskata Pond
Chord of the Bay
Head of the Harbor
Wyer's Point
Bass Point
5 Fingered Point
COATUE
3d Point
2d Point
1st Point
Abram's Point
POCOMO
SQUAM
QUIDNIT
Nashayte
QUAISE
POADPIS
Sachacha Pond
SACHACHA
SHAWKEMO
The Cliff
Brant Point
Pinney's Point
TOWN
SHIMMO
TETAUKIMMO
SAULS HILLS
Macy's Hill
Folger's Hill
Gibbs's Pond
Gibbs's Swamp
SANKATY HEAD
OKORWAW
MONOMOY
1st Mile Stone
2d M Stone
3d M Stone
4th M Stone
5th M Stone
6th M Stone
SIASCONSET
SUN SET HEIGHTS
Moona
Lover's Lane
Maddequecham Valley
Tom Never's Swamp
POCHICK
Tom Never's Pond
Tom Never's Head
Weweeder Ponds
SURF SIDE
Nobadeer Pond
Maddequecham Pond
Toupachue Pond
Forked Pond
70° 5′ 10″
41°17
John Medole 193 Pearl St. N.Y.

and not for sale in Britain. William Rotch reported that his firm received £319 from Enderby, "who was glad of the oils." This transaction was an indication that the Nantucket merchant and his British customer were at this time still on friendly terms. While this personal feeling was to continue, the business relationship would soon undergo a drastic change.[19]

The *Warwick* had also brought some whaleboat boards (American cedar) which sold at a good profit. A return cargo of English manufactured goods and Russian hemp was put aboard the *Warwick* for her return to Nantucket. In his next letter, Rotch informed his firm that he "would not take any money of Enderby; therefore you may draw [on that firm] without any regard to me . . . I shall not need any money for my use, having enough here for the next twelve months to come, as I took the ship's [*Warwick*] canvas [extra suit of sails] off and sold it."[20]

His letters to his sons revealed thoughts not known to his associates, especially his reflections on his pilgrimage. He wrote:

the prospect of being obligated to abandon our native land, occasioned by the most ruinous system of conduct at present that perhaps ever befell any people . . . though I believe in judgement rather than mercy, and notwithstanding this is a fine country & I believe may have many encouragements, yet I desire heartily, if it may be consistent with Divine Providence, some way may be open'd to render it unnecessary to quit our habitations. . . . But, whatever we do, I hope it will be with a clear evidence that we are right . . . if not to bring an increase of the things of this World, yet an increase in peace of mind.[21]

Thus, the great merchant from the Quaker sea kingdom of Nantucket again revealed the courage of his convictions and his faith in the "rightness of his mission."

Writing to Nantucket from London in late October, William Rotch reported on the "hurry and bustle of the city" as being "amazing and far beyond my expectations; notwithstanding the almost total stagnation of American trade." He saw little prospect of any commercial treaty favor-

19. Bullard, *The Rotches*, p. 215.
20. Ibid., p. 212.
21. Ibid., p. 211.

able to America, "but I heartily wish there had been a more conciliatory disposition in our country at the conclusion of the War, and that the Treaty of Peace had been fully and amply satisfied and complied with on our part . . . but this by some might be deemed only the sentiment of a Tory, but I can say it is the sentiment of one who is a hearty well-wisher of both America and Britain."[22]

But the purpose of his coming to England was uppermost throughout all his letters. "If we [Nantucket] are placed in a Neutral State, . . . I fully believe something may be done." Then his conviction on the Dartmouth migration and its effect on the political leaders was expressed: "Those concerned in settling in Nova Scotia are powerfully making interest against us, but as I have had no opportunity of laying our situation before more than two of the Privy Council, I can not form much judgement." One of his former London business associates, the whaling firm of Samuel Enderby and Sons, surprised him with an offer to assist, "which, from some circumstances, I believe is real."[23]

The British Ministry had shown no indication of the fate of William Rotch's appeal, which he had placed in the hands of two of the Privy Council. Titled simply "Paper Received From Mr. Roach [*sic*]," it is altogether a remarkable document, written in his own hand. Starting out simply, and continuing in language typical of the man, it reads:

Nantucket is an Island of a barren soil, situated about thirty miles from the Continent of North America. The inhabitants are in number about four thousand five hundred, and are mostly of the People called Quakers. They derive their Subsistence almost wholly from the Sea by the Whale Fishery, in which before the late War they employed One Hundred and Fifty vessels in a year. Long experience and close application hath given them an acknowledged Superiority in this Business in which they are initiated from early youth.[24]

After recounting some of the devastating effects of the Revolution on Nantucket, forcing the virtual abandonment of whaling, Rotch de-

22. "Paper Received From Mr. Roach," *Board of Trade Records*, B.T. 6/95, pp. 51–53.
23. Ibid.
24. Ibid.

scribed the resumption of the industry during the last years of the war through permits obtained from the two British Admirals, Arbuthnot and Digby, whereby some twenty-four vessels finally sailed, and the oil "shipped to the British market, where it was admitted on the same footing as the oil imported by British subjects."

This was a point on which, in his later interview with Jenkinson, Rotch dealt at some length.

The proposal went on to offer the services of a colony of Nantucket whalemen, with ships and families and artisans for the trade. In his calculation of the costs involved in such a migration, Rotch estimated it would be fair to allow 100 pounds Sterling for the transportation of each family of five persons, and 100 pounds settlement—say £20,000 for one hundred families. The migration would involve the bringing of thirty whaleships, completely equipped for whaling, and Falmouth was the port preferred—certain landowners there having indicated a desire to encourage such a settlement.

The "Proposal" was to be placed in the hands of the recently appointed "Committee for the Consideration of all Matters Relating to Trade and Foreign Plantations," commonly known as the Board of Trade. Charles Jenkinson was the leader of this powerful agency of Britain's government under William Pitt the younger.

Jenkinson was to become William Rotch's chief adversary in the consideration of the Nantucket proposal. An experienced and gifted man, Jenkinson thought in the same terms as did Rotch, basing his conclusions on the economics of the situation. The ensuing duel of wits between these two men was to determine the fate of the Southern Whale Fishery not only as it affected Britain and Nantucket but Europe and America as well.

5. London Takes the Lead

IN THE city of London the prospect of the year 1786 was especially significant to that group of merchants who had suddenly found themselves deeply involved in this comparatively new (to Britain) phase of the whaling industry—the Southern Whale Fishery. The older branch of that business, the Greenland or Northern Fishery, had for years been heavily subsidized by bounties. Since the exclusion of American oil, the price of the Greenland, or right whale oil, had risen from £17 per ton to £21. The total number of ships in the whaling trade had risen to 126 on an average per year. The cost of subsidizing the fleet (increasing to £84,122 per year, averaging £600 per ship), was causing some bitter comment on the part of the Customs Commissioners.[1]

There was reason enough for the London merchants to seek a parity for their embryo Southern Whaling Fishery. They wanted bounties also. The sperm oil their ships brought home was worth twice as much as the black oil of the Greenland fleet. It had proved of greater value to Britain because it also served as a lucrative export to Europe.

But there was another important factor. The ships the Londoners sent to the South Atlantic, the Falklands, and the African coast, and to other areas where the Southern whalers cruised, manned almost exclusively by the imported Nantucketers, were now meeting ships sent out by the Nantucket whaling colony in Nova Scotia. They also learned of another Nantucket group ready to emigrate with William Rotch. Now came the question: As a check to such migrations, what role would government play in encouraging more of the Nantucket whalemen to come directly to London?

The Londoners knew they had powerful friends in government. Whatever private inquiries were advanced to the Privy Council were

1. *Board of Trade Records*, May 3, 1786, P.R.O., B.T. 5/3, pp. 457–469.

treated so sympathetically that the merchants were encouraged. On January 17, 1786, a well-prepared "Memorial" was submitted by Samuel Enderby, Sr., and Sons, Alexander Champion, and John St. Barbe.

From their opening statement the care in which these merchants had prepared their petition was obvious. They spoke for themselves and for a dozen other merchants who were represented by whaleships in the growing London fleet. However, as the leader in the petition, Samuel Enderby pointed out:

That your Memorialist began the Southern Whale Fishery in this country at the commencement of the War with America, from whence it was carried on before to a great extent; from the Island of Nantucket alone they sent from 120 to 140 sail of vessels on the Spermaceti Whale Fishery annually. It being a new Fishery here and very expensive, Government in the year 1776 granted the following encouragement for eleven years, £500 for the ship that should obtain the greatest Quantity of Oil, £400 for the next greatest Quantity, £300 for the next, £200 for the next, and £100 the next.[2]

Despite these premiums, there had been only a few ships fitted out because of the cost of outfitting for this type of whaling, the "Memorial" stated, and of these ventures even those who were fortunate enough to obtain a cargo, entitling them to the largest premiums, lost considerably. As a more effectual mode of encouragement the merchants proposed a system of bounties:

that 40 shillings per ton be allowed for three years on all British vessels (which are free according to the Act of Navigation and are solely owned in Great Britain) from 78 to 250 tons measurement and fitted out from thence for the Southern Whale Fishery after the 1st of February, 1786, and that no advantage of the Bounty may be taken by going to the Greenland Seas or Davis Streights, the Fishery to begin to the Southward of 59 degrees, 30 minutes,

2. *Memorial of Samuel Enderby & Sons, John St. Barbe, and Alexander Champion to the Board of Trade*, P.R.O., B.T. 6/93, p. 55. Also, *An Act For The Encouragement of the Southern Whale Fishery*, 15th Geo. III, Cap. 31, and 16th Geo. III, Cap. 47.

North Latitude, which are the limits of the Greenland Seas and Davis Streights. After 3 years the Bounty be reduced to 30 shillings per ton, no ship to be entitled to any Bounty without she brings home 50 tons of oil, all of her own catching and curing on that voyage taken to the Southward of the Greenland Seas and Davis Streights, or has been out 8 months, and that she being the whole of what she took during that voyage . . . [3]

The "Memorial" further stipulated that no time element be established for the departure of the ships, as they were fitted out at all seasons of the year, "either to the Coast of Brazils, Guinea, Falkland Islands, West Indies, or elsewhere, as it almost requires sailing at different seasons from each of the above named places."

The recognition of the serious competition which then existed in the Nova Scotian colony of Nantucket whalemen at Dartmouth was clearly demonstrated: "Your Memorialists have proposed the Bounty of 40 shillings per ton for three years as they know it will be an inducement to the Americans to settle in England, and it will be the means of establishing the Fishery here rather than in Nova Scotia." [4]

But another problem was presented—the danger of having American oil smuggled in by way of the Dartmouth colony vessels. The "Memorial" commented that the Nova Scotian competitor was so closely allied to Nantucket as to serve "as a cover to a great deal of American property and whaling vessels, besides the disadvantage of the Nursery of whalemen being in Nantucket, from whence it will be difficult to get them in case of War."

The "Memorialists" went one step further in their carefully arranged plan to lure the Nantucketers to London rather than to Nova Scotia:

We likewise request that the [import] Duty of 18 shillings 3 pence per ton on American and all Foreign oil may continue. As an inducement for the Nantucketers to settle here, and for the Encouragement of the Seamen to go into this Fishery, we recommend the Harpooners, Boat Steerers, and Line Managers be protected in case of Warr. [5]

3. *Memorial of Samuel Enderby & Sons*, et al., January 21, 1786; received by Board of Trade on February 4, 1786; B.T. 6/95.
4. Ibid.
5. Ibid.

It was carefully noted in the "Memorial" that the recruitment of American whalemen had been carefully planned, and the part played by the Enderby firm was significant: "Samuel Enderby, Jr., . . . lives with his father . . . has been over to Boston since the Peace expressly to get information about this Fishery, and to engage Nantucket men to come to England."[6]

Among the Nantucket shipmasters who went to London during and shortly after the Revolution were Captains Elisha Pinkham, Francis Macy, Robert Meader, Isaac Burnell, Ebenezer Calef, Tristram Pinkham, and Jonathan Coffin. Many of their fellow-islanders had now joined them, deciding to accept the opportunities to pursue their vocation in England, and these included mates, coopers, and harpooners, as well as shipmasters.

While the Lord Commissioners of the Board of Trade were examining this London "Memorial" they were also considering William Rotch's proposal—weighing one against the other. It was at this crucial moment that Charles Jenkinson was carefully but surely guiding the course of the destiny of Britain's future in the Southern Whale Fishery. His natural inclination was to favor the London merchants. But the sagacious Jenkinson, later to become President of the Board, was also consolidating his own ideas in regard to the method to be used. He resolved to pursue the matter so closely as to keep this highly important industry in the grasp of Britain through complete support of the London merchants.

There was another group of whaling merchants deeply interested in keeping the bounty on their ships—the Greenland ship owners. They had been enjoying that extensive period during the Revolution and afterwards in which American oil had been absent from the British market. Since the war's end, the increased price of their oil had been a welcomed circumstance. In one sense, the "Memorial" of the Southern Whale Fishery merchants provided aid as much for the Greenland fleet as for their own, as the 40-shilling-per-ton bounty worked for both. Another claim by the Greenland merchants was that their complement of some six thousand seamen represented a source of potential naval recruits, and

6. Ibid.

therefore government should continue to support them. This latter claim, however, was effectively disputed in Parliament by both Pitt and Jenkinson, who pointed out that only one-third of the Greenland crews were actually British subjects.

The Committee of the Board of Trade referred the "Memorial" to the "Commissioners of your Majesty's Customs." This body did not approve, reporting "they cannot under the present circumstances recommend the adopting of the Proposal of the Memorialists, nor can they even recommend the continuance of the Bounties already granted."

The Customs officials further declared that continuing the bounties was not warranted; that in 1785 the bounties paid the northern and southern whale fleets amounted to £84,122, or an average of £600 per ship; that the public treasury thus subsidized the ships to the extent of "$\frac{3}{5}$th of the value of the Produce, uncompensated by any Duty." While the whaling merchant received £21 per ton for his oil, with the bounties he actually received £34 per ton. (A British ton of oil then measured 252 wine gallons.) The high duty against American oil made the bounties doubly attractive and the Customs people predicted that if the bounty continued the number of whaleships would increase to such an extent that the total bounties paid would increase from £84,122 to £113,000—"and is an Expense of such Magnitude that, in their Judgement the Public ought to be relieved from it and the Trade left to stand upon its own Bottom."[7]

As a shrewd businessman, Jenkinson recognized the fairness of the Customs officials appraisal of the situation. However, he was also more aware that, considering the two types of whale fisheries involved, the Northern (or Greenland) and the Southern, the latter was the more valuable by far. In place of bounties, he recommended premiums to ships completing successful voyages.

Enderby, St. Barbe, Champion, and their colleagues advanced another argument, contending that the Southern Whale Fishery, if given Government support for a few years, would eventually build up such a fleet that the full 40 shillings bounty would not be necessary. That is why they proposed the decrease in bounty to 30 shillings per ton after three years.

7. *Report of the Commissioners of Customs to Board of Trade*, May 8, 1786, B.T. 6/95, pp. 59–61.

7. The Rotch Counting House, Nantucket
Erected in 1774 by William Rotch, Sr. Now the Pacific Club
[Photograph by the author]

From their experience with the Nantucket whalemen who had come into the London fleet the astute Enderby and Champion recognized the depressed state of the Island, and the drab outlook for the future of the once great center of the Southern Whale Fishery. Their feeling about the Nova Scotian competition was realistic, but they were depending on Jenkinson to check this threat. But they had great respect for William Rotch, and his proposed migration of an entire colony into Britain posed a definite competitive problem.

Cognizant of all the potential contestants for their hold on the industry, the London merchants were conducting their campaign armed with considerable knowledge of both their own and the American situation. These merchants made a strong point of a decisive factor they were planning as a part of the future of Britain's new industry:

We think we may be assisted by Parliament, to enable us to go on with the Fishery, and to induce the subjects of the United States [Nantucketers] giving it up from America; we cannot devise any so eligible [a mode] as the Bounty upon the tonnage.[8]

Like pawns in a maritime game the Quaker whalemen of Nantucket, recognized as the great specialists in the Southern Whale Fishery, were to be welcomed as individuals—but as a colony in another port, the potential posed too great a risk. As they were successful in obtaining these whalemen as individuals, the Londoners were anxious to maintain this policy.

William Rotch, at this time, was keeping close watch on proceedings in London. In August 1785, in a letter to his firm in Nantucket, he noted that six of the London whaleships had arrived, five commanded by former Nantucketers. He listed them, in typical fashion, by the names of the captains instead of the ships: "Delano [Captain Henry Delano in the *Kingston*] 85 tons; Bunker [Capt. Owen Bunker, *Brothers*] 85 tons; P. Pease [Capt. Paul Pease, *Kent*] 77 tons; Gage [Capt. Thos. Gage] 50 tons; Goldsmith [Capt. William Goldsmith, *Mentor*] 43 tons, the others considerable but not arrived."

He wrote that Captain Goldsmith had been in company for a time with the Rotch ship *Canton*, Captain James Whippey, and that the Nantucket ship had taken two large sperm whales in one day, only to lose both in a squall. "The ships parted about the 12th of 2nd month," the letter went on,

Whippey for the Islands [Falklands] and Goldsmith for home [London]. . . . Daniel Coffin in the Mercury *arrived from Boston, who informed me the General Court [of Massachusetts] have allowed us a Bounty on Sperm Oil of £6 pr. tun & abated the taxes, but his account is imperfect . . . how far this will be adequate to the expenses of the Fishery I cannot tell, not knowing what oil is*

8. *Memorial of Samuel Enderby & Sons*, et al., October 30, 1786, B.T. 6/95, pp. 67–68.

worth in Boston at this time, but if sufficient to satisfy our people [Nantucket] it will ease me of a great burden . . . and we will have only to conduct our own business.[9]

Rotch had reason to doubt the effectiveness of the Massachusetts' bounties, coming as they did too late to aid the Nantucketers. He continued to express the hope that the state would permit the Island to assume a position of neutrality, whereby it might effect an agreement with Britain to bring in the oil duty-free.

But Britain's entry into the Southern Whale Fishery had become so successful that such a hope was too remote for consideration. The Customs records gave the best indication of the Londoners' progress. During the year 1785 the whole product of the Southern Whale Fishery in the London ships was 650 tons, valued at £27,300, brought home by twelve of the eighteen ships engaged in the industry.[10]

Success was the key to the presentation of their "Memorial" so far as the London merchants were concerned. Their bold proposal that government should be willing to reward them for this success by the awarding of bounties had a strong appeal, and reflected the confidence they felt in the future of the industry out of London.

9. William Rotch, Sr., to Samuel Rodman, August 2, 1785. Rotch Collection.
10. *Paper Delivered In By The Petitioners, St. Barbe, Enderby & Sons,* March 22, 1786, *Board of Trade Records,* B.T. 5/3, pp. 111–112.

6. Quaker Merchant and Titled Lords

DURING his stay in England William Rotch had gained a knowledge of political as well as economic conditions in Britain. Having presented his "Proposals," he was aware that his cause must reach the leading figure in government, William Pitt, through a personal interview. To gain such an advantage he sought the help of his good friend Robert Barclay, who appealed to his associate Harry Beaufoy, a member of Parliament. Beaufoy was able to make the arrangements for the interview, which took place late in November 1785.[1]

Then in his twenty-seventh year, Prime Minister William Pitt was in the midst of consolidating his initial gains and forming his government. Tall and ascetic-looking, possessing an icy demeanor which often disconcerted those to whom he granted interviews, the brilliant young leader of post-war England formed a marked contrast to the dignified Quaker merchant, fifty-one years old, with his calm, almost deliberate manner, and sober dress.

In describing this fateful meeting, William Rotch carefully reported that Pitt received him "politely and heard me patiently," as he presented the situation of Nantucket. William Rotch went quickly to the nub of the matter:

When the War began we declared against taking any part of it, and strenuously adhered to this determination; thus placing ourselves as a Neutral Island. Nevertheless, you have taken away from us about Two Hundred sail of Vessels, valued at 200,000 pounds sterling, unjustly and illegally. Had that War been founded on a General Declaration against America, we would have been included in it, but it was predicated on a Rebellion, consequently none could have been in Rebellion but such as were in arms, or those that were aiding such. We have done neither. . . .[2]

1. Rotch, *Memorandum*, p. 38.
2. Ibid., p. 39.

At this point it is probable that Pitt found himself aware that this man was not an ordinary Loyalist pleading for remuneration of a lost fortune, but an American merchant whose logic was compelling—something that appealed to his own sense of orderliness and consistency in debate. Rotch continued:

You sent your Commissioners to restore peace, in which any province, county or town that should make submission, and would receive pardon, should be reinstated in their former situation. As we had no submission to make, nor pardon to ask—and it is certainly very hard if we do not stand on better ground than those who have offended—consequently we remained a part of your Dominions until separated by the Peace.[3]

The bold approach and forthrightness of the Quaker merchant impressed the Prime Minister. It was apparent that Pitt found the Nantucketer's statement intriguing, as he asked for further elaboration. Rotch reiterated his argument that most Nantucketers were aware that their distress during the war was "in great measure owing to their being considered by Great Britain as a part of America in Rebellion, and as this idea may yet have some operation to their disadvantage, it may not be improper to repeat that they would not in Justice be considered a part of the United States until it was so decided by the articles of Peace—and this being done without their knowledge or consent."[4] This argument was a cornerstone in William Rotch's presentation, and he gave it with a calm, convincing directness. It was a point not lost on Pitt, and he nodded his head in agreement. "Most undoubtedly you are right, sir" (he was quoted by Rotch). "Now, what can be done for you?"

Rotch continued:

I told him of the present state of things [on Nantucket]; that the principal part of our Inhabitants must leave the Island, some actually having already moved to the American continent [the Hudson, New York migration], and others wish to continue the Whale Fishery wherever it can be pursued to advantage [referring to the Nova Scotia colony]. Therefore, my chief business is to lay our distressed situation, and the cause of it, before this

3. Ibid., pp. 39–40.
4. *Paper Received From Mr. Roach, Board of Trade Records*, B.T. 6/95, pp. 51–53.

Nation, and to ascertain if the Fishery is considered an object worth giving such encouragement for the removal to England, as the subject deserves.[5]

The interview soon came to an end, and Rotch, with his new friend Beaufoy, withdrew. William Rotch thus placed his "Nantucket Proposal" into the hands of the Prime Minister, and Pitt as promptly referred it to the Board of Trade at Whitehall. Soon after, Rotch received word from Stephen Cotterel, one of the secretaries, that the Board would "sit at an early day, when they would hear what I had to offer." After awaiting that "early day" for a full month, Rotch inquired of Secretary Cotterel as to the delay, and reported: "The answer was that so much business lay before them that they had not been able to attend to it—but would soon."

At his lodgings in Gracechurch Street William Rotch anxiously awaited word from the titled Lords. The last month of the old year passed giving way to January 1786 with still not word.

Another inquiry of Secretary Cotterel resulted in the reply that the Board of Trade's Council was faced with so much business that it had not been able to attend to the "Proposal," but hoped to do so soon.[6] "Thus I waited," William Rotch recalled in his *Memorandum*, "not daring to leave Town lest I be called for. This state of things continued more than four months, during which time I received several what I called unmeaning court messages, 'that they were sorry they were not able to call me,' etc."[7]

There was an excellent reason for the delay. During the same period the Privy Council had received a "Memorial" from the London whaling merchants—Enderby, Champion, and St. Barbe—and this was being carefully weighed against the Rotch "Proposal." Of equal reason for delay was the sharp change in the character of the Board of Trade itself. Where its former function was to prepare drafts for the consideration of the Privy Council, in its new capacity it assumed a far more significant role. Now its deliberations would be reported directly to the Prime Minister, and for the years ahead its judgements would actually control the vital trade policies of the British Empire—policies which were to make possible her steady rise as a maritime power.

5. Rotch, *Memorandum*, pp. 40–41.
6. Ibid., pp. 41–42.
7. Ibid., p. 42.

At this time, the experienced and adroit Charles Jenkinson became the leading figure of the new Board in the post of President. Other members were Lord Grenville, a cousin of Pitt who became Vice President; Henry Dundas, the very knowledgeable Scot lawyer who was the head of the India Board; the Marquis of Carmarthen, Secretary for Foreign Affairs; and Lord Sydney, the Home Secretary.

Jenkinson brought new prestige to the Board's deliberations. He was highly respected in the business world. Through careful adherence to his duties, and application of his unusual mercantile abilities, he rapidly assumed a position of authority. His loyalty and unquestioned dedication to the Pitt government was to win for him soon the title of Lord Hawkesbury.

Another month passed and still Rotch waited. During this time, Jenkinson and the Board of Trade had received the London whaling merchants "Memorial" and granted them an audience. Enderby and his associates had now obtained the services of additional Nantucketers, and had already sent them to sea. The merchants were anxious to receive governmental support for their fleet through the continuance of bounties similar to those granted Greenland whalers.

Jenkinson and the Board weighed the "Memorial" against the "Paper Received From Mr. Roach [*sic*]." Here was the actual problem—whether to obtain the Nantucketers as individual whalemen for the Londoners or as a complete colony for Britain itself.

Another month passed, then another, and despite his frequent inquiries William Rotch found an audience with the Board still denied. At last, at the end of a five-month period of fruitless waiting, he requested an opportunity to meet one of the members, "that the matter might be brought to a close." This was done, but the man selected was Jenkinson, who by this time had become deeply interested in the proposals of the London merchants set forth in their "Memorial."

In the Board of Trade papers is a significant record, probably a memorandum prepared for Jenkinson by Secretary Cotterel:

A Mr. Roach [sic], a Quaker of Nantucket, is a person of the greatest proportion that Island, and has the greatest influence among the fishermen [whalemen], and wherever he goes it is supposed the greatest number of their fishermen will go also. This gentleman and his son are now in

London waiting to know the determination of Government. It seems he has visited most of the Western ports in England, with a view to fix on a Place to settle at in case Government should not prefer and adopt a plan for their settling in Nova Scotia. In that case it is said he has fixed on a place at Falmouth to purchase for that purpose; intends to build a Meeting House there and to fetch over his family, and all his effects, with fishermen and all kinds of workmen employed in the fishery and to carry it on from that Port—perhaps it would be proper to Encourage him in carrying such a scheme into Execution.[8]

This "Memo," together with additional information furnished by such London merchants as Samuel Enderby and Alexander Champion, provided both the Board and Jenkinson with whatever other facts that were necessary in making William Rotch better known. The respect such knowledge brought justified the careful observations that were noted in the "Memo."

The two men who were to become the great antagonists in the rivalry for control of the Southern Whale Fishery—William Rotch and Charles Jenkinson—met finally in March 1786 at the latter's office in Whitehall. At that time they were unaware of their roles in the history of that industry, but there is something in their reactions to one another which indicates a kind of instantaneous recognition of their future parts.

As men of the mercantile world their interests were similar. Both were shrewd appraisers of the market place; both had a sense of order and timing in business practice; and they shared a gift for looking well to the future. Where they differed was in their political careers. Jenkinson had become a successful figure in government, to which he applied his natural business ability. Rotch abhorred politicians, dealing with them only as a matter of necessity. As a highly respected figure in the Society of Friends, he applied to his daily life the tenets of that Society. In America, as the leading representative on committees from Nantucket dealing with both British and American authorities, he had been able to have the neutral position of the Island recognized. After succeeding in this unprecedented accomplishment in a time of war he had been bitterly disappointed

8. *Memorandum on The Southern Whale Fishery, Board of Trade Records,* B.T. 6/93.

when the same state was not granted in the coming of peace. This was the major factor which had determined his effort to salvage the Nantucketers' only industry by transferring it to Britain.[9]

Jenkinson ushered Rotch into his office. The two men were instantly on the defensive. One British historian commented:

The former had no love for New Englanders—haughty rebels and Britain's keenest competitors; and one can well understand the attitude of the trans-Atlantic Quaker, a disapproving stranger in the middle of London worldiness and ill at ease in the chilly presence of the English politician with the flickering eyes.[10]

The Rotch proposal, now quite familiar to Jenkinson, was immediately brought to the fore. The Quaker merchant, with his deliberate manner, presented his case quietly and carefully. Quickly, the cost of transportation of the Nantucket colony was brought up. Rotch stated his estimate of such a cost at £20,000 for one hundred families. Jenkinson demurred, also declaring it was a great sum, and came at a time when "we are endeavoring to economize in our expenditures." The Nantucketer bridled: "Thou mayest think it is a great sum for this Nation to pay. I think two-thirds of it a great sum for you to have taken from me as an Individual, unjustly and illegally."[11]

In this, Rotch was referring to the loss of nearly £60,000 during the War by his own firm of William Rotch and Sons. This was not including the loss sustained by his father, Joseph Rotch, when the British fleet raided the village of Bedford and burned out the waterfront and all of its shipping.

The major issue in the conference arose quickly—the request that the thirty Nantucket ships to be included in the proposal be entered as British bottoms and receive registers as such. To strengthen his request, Rotch reminded the President of the Board of Trade that this would save Britain a great deal of money, as the bringing in of of two Nantucket ships was equal to the cost of building one in Britain.

9. Rotch, *Memorandum*, pp. 36–43.
10. Vincent T. Harlow, *The Founding of The Second British Empire* (London, 1964), p. 297.
11. Rotch, *Memorandum*, pp. 42–43.

8. William Rotch (1734–1828)
Painted in 1825 by E. D. Marchant
[In the possession of Miss Caroline Snelling, South Lincoln, Massachusetts]

At this point, Jenkinson became somewhat incautious, saying: "Oh, we don't make merchantile calculations; it is seamen we want."

"Then two of our ships will answer your purpose better than one of yours," observed Rotch, quietly, "as they will double the number of seamen."

Recovering his composure, Jenkinson then made a calculation in which the cost of transporting one family from Nantucket to England was set at £87–10 shillings.

"I am about [drafting] a Fishery Bill, and I want to come to something that I may insert in it," announced Jenkinson. Rotch replied that the offer was not acceptable, and felt it should not be incorporated in the Bill. He then requested permission to withdraw.

"Well, Mr. Rotch, you'll call on me in two or three days."

"I see no necessity for it."

"But, I desire you would."

"If it is thy desire, perhaps I may call." [12]

Upon meeting again it was soon evident that the old impasse was to prevail, and the two men again felt the antipathy each held for the other. Rotch then informed his host that he had learned of a movement in France whereby "Nantucket had agreed to furnish France with a quantity of oil." In his "Memorandum," Rotch described Jenkinson's studied calm, then his stepping across the room to his bureau [secretary], taking from it a file of papers, "and pretending to read an entire contradiction, although I was satisfied there was not a line there on the subject."

The Nantucketer knew that the prospects of favorable action by France had a strong basis in fact, as preliminary negotiations he had advanced had been well received. His next statement contained the crux of his determination to bring a colony of Nantucketers to Europe.

"If there is any contract sufficient to retain us at Nantucket," said William Rotch, "neither you nor any other nation shall have us. If the French situation is insufficient, I will enlarge it."

He spoke so quietly, as was his custom, that Jenkinson once again failed to recognize the strength of the man's character.

"What?" he remarked, with a trace of mockery in his voice. "Quakers go to France?"

12. Ibid., p. 44.

"Yes—but with regret."

Thus ended the meeting. The two men were not to meet personally again—but it was not the last confrontation of their respective concepts for the control of the Southern Whale Fishery. Despite their dislike for each other, they shared a common bond—the practicalities of the industry.

William Rotch's appraisal of Charles Jenkinson was surprisingly bitter for one whose Quaker principles made him charitable towards his fellow man: "A greater enemy of America, I believe, could not be found in that body [the Board of Trade] nor hardly in the Nation."[13] In this opinion William Rotch was not alone, as John Adams, then in London, reached a similar view a short time later.

Despite Charles Jenkinson's contemptuous regard for the new American nation, and Americans in general, it must be recognized that he was holding fast to a rigid policy he had established for himself long before. It was a policy which was to prove effective in Britain's commerce, adding vitality to her trade. Combined with Pitt's realistic fiscal policies, aided by Dundas' management of affairs in India, and the control of the West India trade, it was a course of action that so strengthened Britain's maritime activities as to place her foremost among the world's trading nations for years to come. And what was to be an important part of this policy was the way Jenkinson manipulated events to bring about Britain's supremacy in the Southern Whale Fishery.

An illustration of Jenkinson's calculated planning was his close study of the Rotch proposal before he granted the Nantucketer an interview. First, he requested a thorough review of the "Paper" by one of his trusted assistants, Grey Elliott, Under Secretary of the Plantation Bureaus. A month before Rotch was granted his interview, Secretary Elliott's observations were in the hands of Jenkinson, and through him placed before the Board of Trade. The influence of Elliott's report may be judged by perusal of some pertinent sections:

Considering the settlement of the Fishermen from Nantucket in Nova Scotia to be out of the question, I have not touched upon that part of the proposal; and the advantages that may be derived from their settlement in Great Britain being obvious and acknowledged, I have confined my

13. Ibid., p. 42.

observations to what, I conceive, are the conditions under which they mean to become British subjects. You may perhaps think I entertain too harsh an opinion of them. I know the people of New England too well to imagine that any proposition of theirs can be without some latent view. . . . I may consider their proposals with too cautious and may be Jealous Eyes. At all wants, it may perhaps not be improper to act on this occasion as if they meant what I suspect they do. And to guard against imposition, the different applications they have made to get the oil they say is the produce of their Fishery, imported free from the Foreign duty, prior to this proposal.[14]

Secretary Elliott then commented that the Nantucketers' emigration "is not only voluntary but in search of certain and considered advantages, which must be deemed equivalent to whatever property they leave behind and which their present situation denies them . . . if they be permitted to take 35 ships from the United States, of what size they think fit to be made free, they fitting them out in the first instance from thence and manning them with Americans will certainly be the means of covering many frauds, and the proposal of their manning them in the future with half Americans [as crews] is equally improper."

The astute Jenkinson, realizing the desperation of the Nantucketers, saw the opportunity of getting these whalemen as individuals, rather than as a unit which would constitute a competition to the London merchants. In the Board of Trade records is a report titled "Observations on Mr. Rotch's Proposals," with the following recommendations:

That they be allowed to transport themselves to a certain port in Great Britain in such vessels as are adaptable to the Fishery, and actually and bona fide their *property; such vessels, or such as they may fit out for the Fishery, not to exceed a certain number, to be made free.*

That upon their arrival the sum of £10 be paid as a bounty for the passage of every master of a family, owner, whaler or mechanic connected with the Fishery, and becoming a British subject—£7, 11s. as a bounty for the passage of every Boy above the age of 15 and under the age of 18 years; and £5 as a bounty for the passage of any Female or

14. *Observations On Mr. Rotch's Proposals,* Grey Elliott, January 24, 1786, Board of Trade Records, op. cit., 6/95, p. 40.

Boy under the age of 15, and that £70 to £100 be allowed toward erecting Houses, etc.

Upon this plan the expenses will be:

100 heads of families, etc. at £10	*£1,000*
150 Boys, etc. at 7.10s.	*1,125*
250 Females & Boys at 5	*1,250*
	3,375
100 Houses, etc., at £85	*8,500*
	11,875
Contingencies	*1,125*
Total	*£13,000*

An act will be necessary not only to make the vessels free but to naturalize the Emigrants, and will doubtless be taken by the provisions of it that no frauds be committed, either as to property of the Vessels, the Navigation, or the future produce of the Fishery.[15]

From this statement it was apparent that the Rotch proposal was far from being shelved. Such careful preparation for the possible advent of the colony received a further observation in a memorandum dated April 3, 1786, and headed:

Additional Offers Made to Mr. Roach [*sic*]*: "that 30 vessels shall be made free if 500 (total number of colony) come. The vessels to fit out the first voyage at Nantucket, and the oil to be allowed to be sold, the high duty to be retained, and if the owners become settlers, be returned to them. Some allowances to be particularly made to Roach and his Family."*[16]

If these terms were complied with an agreement was to be entered into with merchants at Poole and Dartmouth (in the English Channel) to provide "a settlement and a capital."

This decision on the part of the Board of Trade came after the "Memorial" of the London merchants had been strongly supported and tacitly approved. Jenkinson's first consideration had, most naturally, been the

15. *Recommendations on Mr. Rotch's Proposals, Board of Trade Records*, B.T. 6/95, pp. 46–47.
16. Ibid., April 3, 1786, p. 47.

sponsoring of these British merchants and their plans for increasing their fleet in the Southern Whale Fishery.

His patience worn thin by Jenkinson's delay, William Rotch finally decided to journey to France. Soon after this decision was made, he received word that Jenkinson again wished to see him. It was not to be, as the Quaker merchant had parted from his adversary for the last time.

The President of the Board was surprised to learn that William Rotch was no longer available for another conference. When Jenkinson's messenger arrived at 33 Gracechurch Street he found that the Nantucket merchant, together with his son Benjamin, had boarded a packet in the Thames bound for France. A tradition in Nantucket recounts that the messenger managed to get word to the vessel before it sailed, but that he was informed: "If Mr. Jenkinson wishes to see William Rotch he will find him on board." Whether this may or may not ever be substantiated it is known that Harry Beaufoy (who had aided Rotch in obtaining his interview with Pitt) had received an inquiry from Jenkinson as to the whereabouts of the Nantucketer and had been told he had gone to France. Alexander Champion, a longtime business associate of Rotch, later wrote to him in Paris that Jenkinson was now ready to discuss provisions governing the proposal for the Nantucket colony's settlement in Britain.

Under the date of April 8, 1786, William Rotch wrote to his sons in Nantucket.

We shall proceed pretty direct for Paris, as we think there is a prospect of introducing our oils there and yet remain where we are. . . . I think it is best not to mention particulars, especially as my most hearty desire is not to be obliged to quit the Island.[17]

Thus, through too close practice of political expediency, the British had lost a rare opportunity to obtain, in one fell swoop, a colony of whalemen from the very headquarters of the Southern Whale Fishery—Nantucket.

Charles Jenkinson was not to forget this experience, as later developments showed. When such an opportunity presented itself again he was to adopt tactics that, more than anything he might have said, reveal how much he realized what he had lost by underestimating William Rotch.

17. Bullard, *The Rotches*, p. 222.

7. Politics and Whale Oil

WHILE William Rotch and Charles Jenkinson were meeting for their final conference, the London merchants were already pressing the Board of Trade for action. Their "Memorial," having been referred to the London Customs Commissioners, had received a highly critical analysis. On March 13, 1786, the merchants learned the "Memorial" was to be further discussed and the Board had asked the whaling group to appear at Whitehall for questioning.[1] Some of the merchants' statements had been challenged by the Customs people. But the merchants in turn protested certain statistics advanced by the Commissioners.

Of particular concern to the merchants was the statement by the Commissioners that £16,000 had been paid out in bounties to the ships in the Southern Whale Fishery from 1777 through March 4, 1786. The range of payments was reported as: £2,400 in 1777; £1,500 in 1778; £1,400 in 1781 and 1782; none in 1783; £3,600 in 1784 (the year the London merchants plunged heavily into that industry); £1,700 in 1785; and, for the first two months in 1786, some £1,500 had been allotted.[2]

Quickly challenging this record by Customs, the London merchants asserted that "some mistake must have arisen . . . as the firm of Sam'l Enderby themselves have brought into this kingdom oil which has sold for £77,600 and they, who have been entitled to many of the premiums for nine years past, have received only £7,700 [in bounties]. This shows your Lordships that the account you have before you must be erroneous."

According to their statement, the merchants reported some 650 tons of oil had been brought into London from the Southern Whale Fishery

1. *Letter to Merchants of London Engaged In The Southern Whale Fishery*, March 11, 1786, P.R.O., B.T. 5/3, p. 111.
2. *An Account of Monies Paid For Bounties on The Southern Whale Fishery From The Commencement of The Said Bounties To The Present Time*, P.R.O., B.T. 6/93.

ENGRAVED FOR THE SENATOR.
Charles Lord Hawkesbury
Drawn from Life by C. Benazech, & Engrav'd by C Warren
Publish'd according to Act of Parliament, Jan.y 7th 1791.

9. Charles Jenkinson, Lord Hawkesbury
[Courtesy the Nantucket Historical Association]

during the previous year (1785), and had been sold for £27,300. This was the product of some fifteen ships which sailed that year, owned by six firms.[3] The total number of whaleships in the London fleet in that year had grown to twenty-two, according to the observation of William Rotch, who wrote thus to Samuel Rodman in Nantucket in November 1785.

Thus, the British Southern Whale Fishery had been gradually developed over the decade following the Enderby's fitting out their first ship in 1775. In 1776 there were ten whaleships owned in London; twelve in 1777; thirteen in 1778 (Adams had been accurate in his report to the Continental Congress that year, his informant being a Nantucket shipmaster); dropping to four in 1779 and seven in 1780, but rising again to ten in 1781, with Liverpool being represented by one vessel, Poole with two, and London with seven; six in 1782, of which total London had four, Bristol and Cowes (Isle of Wight) one each; fifteen in 1784, all from London; and eighteen early in 1785, fourteen from London, two from Bristol, and two from other ports.

The merchants continued:

We have considered your Lordships' requisition of adopting some other mode . . . but pray you to recommend it to Parliament to grant us a Bounty on the tonnage. We will be content with 10% on each ton for 3 years, 3% per ton for 5 years, and after that 2% of a ton upon all ships fitten from Great Britain & returning with their cargoes directly here.[4]

Being fully convinced of the necessity of attracting as many Nantucket whalemen as possible, the petitioners carefully stated:

The Committee wish that the whole of the crews (Common sailors as well as Harpooners, Line Managers or Boatsteerers) of the ships employed in the Fishery may be protected from being impressed, and they conceive that this protection will be an inducement to many of the Fishermen remaining at Nantucket to come and settle in Great Britain.[5]

3. *Draft Report of The Lords of The Committee For Trade Upon The Memorial of The Merchants and Others Concerned In The Southern Whale Fishery*, P.R.O., Privy Council Registers, P.C. 2, Vol. 131, pp. 251–257.
4. *Memorial of Samuel Enderby & Sons*, et al., P.R.O., B.T. 6/93.
5. Ibid.

The London merchants went into further detail concerning this important matter, as their continuing thoughts clearly showed:

The Master, Mates, officers and half the common seamen may be subjects of the United States—the other half subjects of Great Britain. Should the Master attempt to import any oil but such as may be taken by the crew of the ship he commands, he shall be subject to a penalty of 500 pounds and imprisoned until it is paid, with a reward of 100 out of the 500 to any one or more of the crew who shall inform against him, provided the information be laid within 30 days after the ship shall be reported at the Custom House.

The merchants were particularly concerned with the problem of American oil from Nantucket ships. There were two ways in which this importation had been attempted; first, efforts to sell in a British port a cargo of oil in an American ship by transferring it to a British firm for export only; second, attempts to use a British registered vessel in a manner as described in the memorandum already noted as concerned Francis Rotch and the ship *United States.*

Fully aware of their advantage over the Americans, the British took no chances in allowing any importation of American oil in British vessels by trans-shipment at sea. The Nantucketers in the Dartmouth, Nova Scotia, fleet were suspected of this operation, but there was no direct proof. The affair of the *United States*, and other Falkland Islands voyages, gave the Londoners a case in point. It was much easier to deal with the *Maria* of Nantucket, William Rotch's ship, which arrived in England in 1786 from the Falklands. She was not able to get the oil sold without paying the high duty and so went to France.

Preventing the smuggling of Nantucket oil by way of the Nova Scotian imports (which came in under the small Colonial duty) was another problem the London merchants were anxious to have government investigate. They suggested:

We know many tricks will be played, therefore we would recommend it to your Lordships to give every encouragement to induce all people to carry on the Fishery from this Kingdom, for by that means you will have all the crews (except the officers, who we would wish to be protected) at your command, in case of a capture with any other power—and which will not be the case if it is carried on from Nova Scotia. For whatever Fishery is carried on in that part of the world will be managed by the people from Nantucket, who, to a man, will return to their own country, particularly should the difference arise between Great Britain and the United States. That will not be the case if they with their families come to reside in England, for in a short time we shall become one people of one interest.[6]

These straightforward opinions, so close to those of William Rotch, indicate how intimate a relationship existed between that Quaker merchant and his London associates, Enderby and Sons, and Champion and Dickason.

Continuing, the memorialists pointed out:

Should a Fishery be established in Port Roseway [Nova Scotia] we are certain but few families would remove from Nantucket, but every season the men, when returned from the Fishery, would go home for a month or two to visit their families. This, we have every reason to believe, a business the Americans would wish to bring about which, if they cannot accomplish, we are given to understand they mean to reside in England. And you will then not only deprive the Americans of many seamen but keep them in Great Britain.[7]

Recruiting the Nantucketers, the great specialists in the Southern Whale Fishery, was a procedure already taken for granted. The question was how to get them into Britain proper, instead of allowing them to build up the competitive potential in the Dartmouth colony. As has been noted it was the influence of Jenkinson that was the determining factor in bringing about the initial rejection of William Rotch. He had to support, also, the stand of the Customs Commissioners in that the Nantucket ships should not be allowed to enter as British registered vessels and thus be eligible for bounties. Jenkinson gained the confidence of both the London whaling merchants and Customs by adopting the arguments of both, consolidating them into a policy which the Board of Trade would recommend to Pitt and the Parliament.

6. Ibid.
7. Ibid.

By far the most momentous proposal in the London merchants' "Memorial" of February 1786 was contained in the request that they be permitted to send their ships "round the Cape of Good Hope, where they are credibly informed there are great numbers of whales."[8]

This was in the domain of the powerful East India Company, and if permitted would crack the once impregnable monopoly it exercised over British shipping in these seas. To one as interested in all phases of commercial prospects as Jenkinson this posed a fascinating series of possibilities. He had pressed the point with Prime Minister Pitt and they had called in Admiral Sir Hugh Palliser, whose experience had included a first-hand knowledge of the Whale Fishery in the Americas. Palliser strongly favored encouraging the whalers in their search for new grounds noting that "distant Seas and Coasts, now little known may be explored and be better known, which may hereafter be used in other respects."[9]

The potential in such an extension of these whaling voyages appealed equally to the quick imagination of Pitt, to Jenkinson's mercantile mind, to the experience of Admiral Palliser, and to the knowledgeable Henry Dundas, whose India men had no doubt reported the great numbers of whales east of the Cape of Good Hope.

In following Jenkinson's lead and recommending that the "corporations" of the East India Company and the South Seas Company grant the London merchants permission to round Good Hope and Cape Horn, the Board of Trade was supplying the key which was to unlock the door to political support for such a development. The fine hand of Jenkinson appears in the wording of one particular paragraph: "This may tend to encourage the Spirit of Adventure, to promote Navigation, and to render this Fishery still more extensive and beneficial to the country."[10]

The "Whalefishery Bill," as it was called, came before Parliament in April 1786, and in the House of Commons. Chairman Rose of the special

8. Ibid.
9. Sir Hugh Palliser to Lord Hawkesbury, November 3, 1786. P.R.O., B.T. 6/93.
10. *Report of The Lords of The Committee For Trade, Proposing Certain Premiums, Etc., For The Encouragement of The Southern Whale Fishery* (hereafter cited as *Report of The Lords of The Committee For Trade*), P.R.O., B.T. 5/3, pp. 457–469.

committee on the bill recognized Charles Jenkinson, the Honorable Member of the Board of Trade. Jenkinson spoke at some length on the proposals for encouraging this industry and pointed out that the Southern Whale Fishery was now the most important branch of it. He recommended that 30 shillings per ton be granted as a bounty, and that at least three-quarters of the crews be British subjects. Only two members spoke in opposition, and when a motion containing these recommendations was made and seconded it was "carried without dimunition."

The formal report of the Board of Trade on the "Memorial" was drawn up on May 3, 1786, and two days later it was submitted to the Privy Council for approval. Present at this meeting were Prime Minister Pitt, the Marquis of Carmarthen, Earl of Denbigh, Lord Amherst, Earl of Cartown, Admiral Viscount Howe, Lord Sydney, and Henry Dundas, Esq.

The report had been drawn with close attention to historical sequences concerning the Southern Whale Fishery both in Britain and other countries of the world, with especial detailed references to the "Memorial" of the London merchants, and comment on having "personally interviewed the several persons who subscribed to the Memorial before mentioned."

It noted further that the Southern Whale Fishery was carried on in Europe only in "Portugal, and in North America in the countries now belonging to the United States of America, and particularly the Island of Nantucket." As for the Portuguese, went on the report, the success of their Brazil fishery was small, and the "Americans, and particularly the Island of Nantucket, carried on the Fishery to a great extent, and with great success before the War," and their oil having been excluded by the high duty since the end of the war, "the people of Nantucket appear disposed to remove from the said Island to some other country where they may be able to carry on their Fishery to more advantage."[11]

Further mention of these Nantucketers was carefully combined with reference to the inauguration of the Southern Whale Fishery out of Britain by their coming to settle, and the adoption of their custom of paying shares instead of wages—the "lay" system.

It appeared also to the Committee that the men who have hitherto been employed in ships carrying on this Fishery from Great Britain have

11. Ibid.

been Americans, some of whom have been in the Service of the Memorialists before mentioned at least ten years; and this Fishery is carried on by such men upon Shares, which, in the judgment of the Committee, is the most beneficial.[12]

The careful course which Jenkinson was steering between the London merchants and the Customs Commissioners was noteworthy. The report noted that the latter did not recommend "the continuance of the Bounties already granted . . . and those proposed by the Memorialists will be a much heavier charge on the Public than those at present," and that the Board of Trade was also in sympathy with the Commissioners' argument that the Greenland Fishery should not be supported with further bounties.

In keeping with the policy of compromise, however, and at the same time seeking a stimulus for the Southern Whale Fishery, the Report recommended that premiums be offered in lieu of bounties to vessels sailing south of 7 degrees north latitude. Added to these, special premiums were suggested, ranging from £700 to £300, to be offered to the five whaleships bringing home the largest cargoes of oil—provided they took this oil beyond 30 degrees south latitude and had been absent not less than eighteen months.[13]

The Privy Council studied the Report with not a little surprise, especially that section which recommended that whaleships be permitted to sail around the Cape of Good Hope and Cape Horn. This came at a time when the directors of both the East India Company and the South Seas Company were still trying to collect their respective breaths. When first apprised of the fact such requests were to be made, these two companies, long protected by the British government, were dismayed by the prospect.

They agreed such an arrangement might lead to a clandestine trade that would spread through the domains so long controlled by the two companies. It was felt that the whalers, being a far-ranging breed, might so circumvent the regulations as to enter "other ports" not permitted under the authority of the respective companies, and they were suspicious of "some other object in view," besides whaling.

12. *Privy Council Registers*, Vol. 131, p. 256.
13. *Report of the Lords of The Committee For Trade*, B.T. 5/3, p. 465.

But when Mr. Pitt supported the Board of Trade's Report the Honorable East India Company gave ground. Yet, the concession was that only a few whaleships should be provided with such a license to cruise in waters *beyond Good Hope*, and these for one voyage only.

Jenkinson was not one to lose this slight advantage. If one voyage was to be permitted, why not a dozen voyages? He spoke privately to Mr. Pitt and the Prime Minister arranged for a meeting with a committee representing the East India Company. Representing the Privy Council would be Pitt and Dundas, while Jenkinson would be the spokesman for the Board of Trade.[14]

On May 7, 1786, when the London merchants were asked if they intended to apply to the Honorable Company for licenses to sail into the seas then under the Company's jurisdiction, they quickly replied in the affirmative. Two days later, Jenkinson had arranged for the East India Company to draw up a definite statement in which the whaleships would be granted licenses to sail into the areas around Cape Horn and east of the Cape of Good Hope, provided certain regulations as regarded geographical limitations were concerned and agreements as to trading in certain ports were met.

The regulations which the Honorable Company imposed included:

1. The ships proceeding to fish beyond Cape Horn shall be confined to and shall not proceed beyond the limits hereinafter mentioned, to wit, in Latitude as far as, but not to the northward of the Line [equator], and in Longitude not exceeding 100 Leagues from the Coast of America.

2. The ships proceeding beyond the Cape of Good Hope are not to go to the northward of 30° Latitude and not more than 15° of Longitude eastward from the Cape.[15]

Violators of these stipulations were to pay the Honorable Company, as a forfeit, £4 per ton of the tonnage of the ship. If the whaleship proceeded "to any port in the East Indies or China, or was found at any port

14. *Collections With Regard To The South Sea Company*, P.R.O., B.T. 6/93. *Minutes of The South Sea Company, To The Board Of Trade, May 1–3, 1786.* P.R.O., B.T. 5/3, pp. 379 ff.
15. Resolution of the Directors, United East India Company, May 10, 1786, P.R.O., B.T. 5/3, p. 379.

10. Captain William Mooers of Nantucket
In command of the *Bedford* on February 7, 1783, hoisted the first United States flag ever displayed before the port of London
[Miniature in the collection of the Peter Foulger Museum, Nantucket]

or place beyond the Limits herebefore described for their Fishery, it shall be lawful for the East India Company or their Servants to seize or cause such ships, with their tackle, furnishings, and appurtenances, to be seized and the same shall be confiscated to the use by the Company."[16]

The whaleships intending to proceed beyond the Capes were required to receive a license from the Company, and their voyages into the Indian or Pacific oceans were not to exceed three years.

This was, indeed, a great victory for Jenkinson and the Board of Trade—a breakthrough in the monopoly of the "Honorable John Company"—a full scope of the triumph not to be realized for several years to come. Not since the days of the illustrious Cook would there be such voyaging as was to take place during the next decade and beyond. It was to give England a new era of maritime adventure and to lead to the further glory of her seafarers.

But Pitt and his far-seeing colleagues, Jenkinson and Dundas, did not leave the problem half solved. The next group to yield to the Board's supported demands of the whalers was the South Sea Company, which had also been for years receiving Parliamentary protection, and thus enjoying a monopoly. Jenkinson had assigned his trusted Chief Secretary, George Chalmers, to investigate this Company's activities, and Chalmers produced a remarkably well-documented report. It was shown that the Company, from its inception in 1711, had done little to justify its existence as a promoter of South Seas commerce; in fact, it had not engaged in foreign commerce since 1750. Chalmers' caustic criticisms included the fact that after eight years' effort to promote the Greenland Fishery it had abandoned it as a "losing adventure," and that since 1753 the "Company has been merely a money corporation."[17] In fact, Chalmers believed that an Act passed some years before virtually repealed the establishment of a South Sea Trade by the company.[18]

In the Report of the Board of Trade the nub of the policy of Jenkinson, now close to fulfillment, was contained in the following statement:

16. Ibid.
17. Memorandum By George Chalmers, *Collection As To The South Sea Company*, May, 1786, P.R.O., B.T. 6/93.
18. *Minutes of The South Sea Company*, At a Court of Directors, May 4, 1786, p. 143.

This trade . . . is of great Importance in the present Moment, when the American Fishery is declining and its doubtful to what countries the People heretofore employed in it may resort and whether Great Britain, or any other Foreign Country, may get possession of it, that a proper effort should be now made to secure to this Country the advantages of a Fishery which was once so lucrative to America.[19]

It was in May 1786, through the sponsorship of William Pitt, that Charles Jenkinson was appointed President of the Board of Trade and then elevated to a seat in the House of Lords as Lord Hawkesbury. In writing to his mother, the Dowager Lady Chatham, Pitt had this comment: "Mr. Jenkinson is to preside [over the Board of Trade], with the honour of a peerage. This, I think, will sound strange at a distance, and with reference to former ideas; but he has really fairly earned it and attained it at my hand."[20]

The "Fishery Bill," which Jenkinson prepared, was passed by Parliament on June 7, 1786, becoming 26 Geo. III, Cap. 50, an act for the "Encouragement of the Southern Whale Fishery." In a very large sense it was a personal triumph for Jenkinson—now Lord Hawkesbury—who, in his seat in the House of Lords was able to provide strategic support in that body while Prime Minister Pitt gave equal support in the House of Commons.

In drawing up the Bill, Jenkinson had steered a careful course between the "Memorials" of the London merchants and the Customs Commissioners' demands for a reduction in bounties. His consistent advocacy of the London whaling merchants' cause was based on the logical proposals that group had advanced for the development of the industry. In the "Fishery Bill" the Customs Commissioners, who had opportunity to review it, had insisted that only British ships be eligible for the whaling bounties. Jenkinson supported this stand, and this completely frustrated any plans William Rotch may have had for bringing Nantucket whaleships into the Kingdom.

19. *Report of The Lords of The Committee For Trade*, B.T. 5/3, May 3, 1786.
20. William Pitt to the Dowager Lady Chatham, July 13, 1786. Quoted by *Harlow, Second British Empire*, p. 240.

Several months later, in a letter to William Pitt, this fact was pointed out by James Phillips, an English Quaker merchant who had accompanied William Rotch in his pilgrimage along the English Channel coast, seeking a port for his Nantucket Colony. Phillips wrote:

I shall proceed to inform thee that I have lately received undoubted information that the Nantucket fishermen are now about to settle in considerable numbers at Dunkirk—I need not inform thee that some time since a Bill was passed to encourage them to settle in England and it would certainly have had that effect if it had not undergone a short but very important alteration after it was shown and approved by William Rotch, who was here in behalf of himself and other inhabitants of Nantucket. This small alteration which, I have heard passed, both Houses unnoticed consisted of (I think) these words—'except the bounties' —the insertion of which totalling defeated the intentions of the bill, and the Nantucketers immediately found themselves under a necessity of accepting the terms offered them by France. I have never heard what reason induced ministry to make the alteration after the bill had been settled with Rotch. It has been conjectured that they entertained a suspicion that a part of them purposed going to France and so to play a double game. I can vouch however for William Rotch that he and his connexions, who form the greatest part of the inhabitants, were warmly decided in favor of England, nor have I heard of a person amongst them who does not prefer England if they could exist there.[21]

Having secured the strong framework for Britain's whaling future Jenkinson once again turned his attention to the possibilities of acquiring the group of Nantucket whalemen under William Rotch. From strong evidence it is apparent that he drew up a measure to this effect. But there was one vital factor in his proposal—the Nantucket whaleships would not be granted British registries, and thus were not eligible for the whaling bounties.

It was at this stage of the whaling "chess game" (September 1786) that William Rotch returned to London after his success in Paris. He found a message awaiting him from George Rose, then Secretary of the Treasury, requesting him to call. Rotch did so and, after a cordial reception, was asked by Rose if he had entered into a contract with the French

21. James Phillips to William Pitt, *Chatham Papers*, P.R.O, G.D.8, May 5, 1788.

government. "I told him no," reported Rotch, "that I did not come to make contracts—propositions were the extent of my business."[22] Secretary Rose came quickly to the point: "You are then at Liberty to agree with us, I am authorized by Mr. Pitt to tell you that you shall make your own terms."

One may well imagine the thoughts which were now coursing through the mind of William Rotch. He recounted the episode in his "Memorandum," written many years later, and his words still ring with the dignity of this unusual man:

I told him it was too late. I made my very moderate proposals to you but could not obtain anything worth my notice. I went to France, sent forward my proposals, which were doubly advantageous to what I had offered your Government. They considered them but a short time, and on my arrival in Paris were ready to act. I had a separate interview with all the Ministers of State necessary to the subject, who all agreed to, and granted my demands. This was effected in five hours, when I waited to be called by your Privy Council more than four months.[23]

Rotch went on to state that Lord Sheffield also sent for him, and that the offer advanced by Rose was repeated. But Rotch's reply was the same: "It was too late." The Nantucket merchant recalled that despite his distaste for Jenkinson he refused to furnish evidence requested by certain members of the British minority party which might have been used to attack the President of the Board of Trade.

Some time later, in reviewing his last attempt to negotiate with William Rotch, Jenkinson wrote in a letter to Charles Greville (1792), that he had seen "Mr. Rotch, and opened the propositions to him;" then went on to state:

Mr. Roach was so far from finding fault with the terms offered that he acknowledged they were very handsome, but he declined accepting them and from his conversation Lord H. had then Reason to conclude that he had already entered into some Engagement with the Government of France.[24]

22. Rotch, *Memorandum*, p. 50.
23. Ibid.

It is possible that, in the stress of the years between, Jenkinson may have become confused, as Rotch, in his "Memorandum," carefully noted that he had never seen Jenkinson after their last conference in Whitehall in March 1786. Subsequent exchanges in negotiations were made through Rose and others.

Jenkinson may have regretted losing the opportunity to acquire, through the Nantucket Colony, a virtual center which would insure Britain's superiority in the Southern Whale Fishery. If these thoughts did occur, coupled with them was the satisfaction that, having turned the Nantucketers away as a colony, he provided the London Whaling merchants the opportunity to secure the Island whalemen as individuals, thus gaining a similar result without creating a rival center in England.

What had now developed did not escape this remarkable man. Having turned away from the opportunity offered by William Rotch he was allowing France an opening whereby that nation might gain a competitive edge? The answer was not long in coming.

24. *Memorandum* of Lord Hawkesbury to Charles Greville, September 29, 1792. *Official and Private Correspondence and Papers of the Earl of Liverpool*, British Museum Manuscripts, Addit. Mss. 38,310, f. 81 (hereafter cited as *Liverpool Papers*).

8. Britain Consolidates an Industry

WHILE the passage of the "Act For The Encouragement of the Southern Whale Fishery" was a greater victory for Britain's new industry than was at first realized, another move by Government also had a salutory effect in strengthening London's control. Prior to the passage of the Act, the Privy Council had taken action to check the growth of the fleet of whalers out of the little port of Dartmouth, Nova Scotia.

Upset by the fact that this little offshoot of Nantucket had not only completed its first whaling voyages but had sold their oil in Britain at a profit (after paying the small Colonial duty), the London combine recognized the challenge was not apt to diminish. The only hope was to appeal to the Board of Trade for some way to resolve the problem. Again, Jenkinson's careful touch became the decisive factor in checkmate. The knight on the chess board chosen to make the move was Lord Sydney, Secretary for the Home Department.

Late in the winter of 1785 Governor Parr of Nova Scotia had written Lord Sydney, reporting his great satisfaction on the marked progress of the Dartmouth whalers. After six months, during which time there were no adverse observations from London, there came from Sydney a letter that both shocked and saddened the Governor. The Home Secretary stated that further immigration of Nantucket whalemen into Nova Scotia must stop.

In a terse and unequivocal paragraph, Parr was informed:

It is the present Determination of Government not to encourage the Southern Whale Fishery that may be carried on by Persons who may have removed from Nantucket and other places within the American States excepting they shall exercise the Fishery directly from Great Britain.[1]

1. Lord Sydney to Governor Parr, April 20, 1786. *P.A.N.S.*, Vol. 33, Doc. 26.

The consternation of the Nova Scotian Governor may well be imagined. The growth and enterprise of the Nantucket colony during those few months at Dartmouth had both cheered and encouraged Parr, whose political gamble showed every indication of success. In February 1786 Richard Cumberland, the Agent for the Province in London, had expressed "great satisfaction" on the progress of the Dartmouth Fishery, and he hoped there would be no diligence spared in "prosecuting this great source of wealth." In that communication, Cumberland had written: "I have no reason to fear but Government here will take Every prudent Measure for the prosperity of your Trade."[2]

Lord Sydney's terse statement, therefore, was a rude awakening for Governor Parr. It reflected clearly the political success of the London whaling merchants and the support of their opinion, so well expressed in their "Memorial" to the Board of Trade:

If the Nantucket men have not the privileges granted in Nova Scotia, there seems to be no doubt but they will come to England and carry it on in conjunction with our merchants in England, who in that case would soon have the whale fishery in their own hands which in a short time, will be an immense acquisition of wealth and increase in shipping in this country.[3]

At Dartmouth, the Nantucket colony accepted this news with typical Quaker determination. They now numbered forty families and, shortly after Sydney's letter came and its peremptory contents were noted, their fleet of six whaleships returned from the Brazils—May and June, 1786—with full cargoes valued at £14,000.

Now, despite this success, and their willingness to build their future outside their home Island, they found themselves deprived of their greatest opportunity for insuring their future—the prospect of being joined by fellow Nantucketers and increasing their fleet of whaleships.

The Government's action had another and deeper impact on the two leaders of the Dartmouth colony—Samuel Starbuck, Sr., and Timothy Folger—an effect that was to undermine their original strong belief in

2. Richard Cumberland to The Nova Scotia Assembly, Committee of Correspondence, February 23, 1786, *P.A.N.S.*, Vol. 301, Doc. 78.
3. *Memorial of Samuel Enderby & Sons*, et al.

the future of Dartmouth; a fact which was to surface in the crisis that was not yet apparent within the colony.

There can be little doubt that had the British government not checked the close association of Nantucket and Dartmouth, the latter colony would have literally become "New Nantucket," and the London branch of whaling could not have met the formidable challenge presented by such a colony.

In London, the merchants planned to expedite the advantage provided by the new act. With the concessions obtained from the East India and South Sea companies, the great opportunity now lay ahead. To round the Cape of Good Hope to Africa's east coast, and to enter the Pacific by way of Cape Horn or the Straits of Magellan, placed the Londoners in a commanding position. With the failure in England of the plans of William Rotch, and the curtailment of the Nova Scotia development, many Nantucket whalemen accepted the offers of the London merchants and went directly to that port. For Lord Hawkesbury, matters were falling neatly into place.

The success of the merchants in their political adventures was duplicated by the success of the ships in the London fleet. The Enderby firm's fleet had made excellent voyages, notably the *Kent*, Captain Paul Pease, and *Friendship*, Captain Abishai Delano, both Nantucket masters, as did Alexander and Benjamin Champion's fleet, one of which was the *Adventure*, under the veteran Nantucketer Captain Elisha Pinkham.[4] Seven other London firms were represented in the Southern Whale Fishery with two to four vessels each.

Returning on November 14, 1787, the ship *Triumph*, under the Nantucket captain Daniel Coffin, brought to Alexander and Benjamin Champion a full cargo of oil and bone which sold at £47 per ton for the spermaceti body oil and £57 for the "head matter," and Brazil right whale oil at £19 per ton and 1.7 shillings for bone per pound. The *Triumph* was the first London whaleship to sail after the passage of the Parliamentary Act in June 1786. In direct contrast to the £47 per ton received for the sperm oil of the Southern Fishery, the Greenland Fishery oil—the black oil from the northern right whale—fell to £19 per ton,

4. *Minutes of the Board of Trade*, P.R.O., B.T. 6/230 and B.T. 6/95. Listing of Southern Whale Fishery Ships, p. 160.

the same figure which the right whale oil from Brazil Banks voyages commanded. But the whale bone from the Greenland whale sold for ten times that of the Brazil right whale, being longer and heavier. A total of £2,400 in premiums was earned by the Southern Fishery, as well.[5]

In 1787 the London fleet increased to fifty ships, totalling 11,555 tons and employing 988 men. The price of sperm oil this year went up to an unprecedented £55 per ton, with "head matter" at the startling figure of £65. Right whale oil, on the other hand, fell to £16 for the Brazil and £18 for the Greenland. Whalebone stayed at the same figure. A new by-product of the Southern Whaling, that of fur seal pelts, was also being placed on the market, bringing 3 shillings, 6 pence per pelt—a low figure, indeed. Added to the return for the oil and bone—totalling £40,949—the ship owners received £5,500 in premiums, as guaranteed under the Act of May 1786.[6]

While the fleet owned by the London merchants for the five-year period from 1786 to 1790 averaged forty-eight ships each year, this does not mean that all of these vessels sailed in a single year. It must be remembered that while some ships sailed others would be fitted out in the Thames, and still others would be unloading their cargoes at the wharves. For example, in 1788, out of the total of fifty ships owned in London, not more than one half had sailed within that previous fall (1787) and the remainder had gone out early in 1786.

In the complete listing of British whaleships in the Southern Fishery sailing in 1787, at least half were from the port of Dartmouth, Nova Scotia, and hence rivals of the Londoners. These Dartmouth ships were:

Ship *Romulus*, 148 tons, Captain Latham Chase
Brig *Rachel*, 161 tons, Captain Obed Barnard
Ship *Lively*, 184 tons, Captain Jonathan Chadwick
Ship *Parr*, 174 tons, Captain Tristram Folger
Brigantine *Sally*, 145 tons, Captain Paul Worth
Sloop *Watson*, 124 tons, Captain Daniel Ray
Brig *Argo*, 142 tons, Captain Daniel Kelley
Brig *Lucretia*, 120 tons, Captain Jonathan Coffin
Brig *Somerset*, 122 tons, Captain Stephen Gardner

5. Ibid., p. 100–110.
6. Ibid.

11. "Mill Cove" (upper left), Dartmouth, Nova Scotia
The Nantucket whaling colony built their first wharf here
[Courtesy Dartmouth Heritage Museum, Dartmouth, Nova Scotia]

Brig *Industry*, 130 tons, Captain William Chadwick
Schooner *Hero*, 62 tons, Captain Valentine Pease
Brig *Hibernia*, 124 tons, Captain Francis Coffin
Schooner *Jasper*, 83 tons, Captain William Pinkham
Sloop *Peggy*, 93 tons, Captain Silas Paddock[7]

A British merchant who had been at Halifax in 1787, wrote that the Dartmouth fleet was active in whaling on the Brazil Banks, and that the oil produced was all exported to Britain. The Quaker colony was prosperous despite the duty of 15 shillings, 5 pence per ton it had to pay as Colonial oil exported to Britain, and, also, not being eligible for the premiums paid London, Bristol, and Liverpool whalers.[8]

During this crucial period for American whaling—1785–1786—only eight ships sailed from Nantucket, the former headquarters of the Southern Whale Fishery; barely a dozen more cleared from the old Boston district and the Bedford (Dartmouth) area; Hudson and Sag Harbor in New York State sent out two or three each. Britain, at this same time, fitted out sixty-five vessels in the Southern Fishery alone, a definite indication of the ascendancy of the London branch. Nantucketers in command of British whalers now numbered thirty-four—another remarkable illustration of the ready demand for the services of these experienced whaling masters. Added to these were the numerous harpooners, officers, and seamen from Nantucket serving in British vessels, thus bringing the total figure up to some 450 men in the London fleet alone. It was to increase with the next five years and by 1793 the Nantucketers, especially the shipmasters, were to give the London Fishery a leadership unrivaled on either side of the Atlantic.

The Enderby firm continued its success. Of the original 1784–1785 fleet, the *Hero* (originally built in America), was commanded by Captain William Folger; the *Rasper*, by Captain Matthew Gage; the *Atlantic*, by Captain Henry Delano; the *Greenwich*, by Captain John Lock; the *Friendship*, by Captain Abishai Delano—all were Nantucket men. There was a combination of Nantucket and British crews. These vessels averaged

7. *Folios of Mediterranean Passes* (whaleships), *P.A.N.S.*, 1785–1791. Files of *The Nova Scotian*, June 15, 1825.
8. *Letter* of John Turner to Board of Trade, January 10, 1788. P.R.O., B.T. 5/5, p. 15.

some ninety feet in length, the largest being the *Hero*, which was 104 feet, and some 250 tons. From Bristol merchant Sydenham Teast, the Enderbys had purchased the *Kent* and placed her under the command of Captain Paul Pease, another Nantucketer. The new ship *Sandwich* was given to Captain James Shields, who was to make whaling history on his very next voyage when he took command of the famous Enderby ship *Emilia*.[9]

Other London merchants sending out ships in the 1786–1787 fleet were Lucas and Spencer, with five vessels: the *Spencer*, Captain Owen Bunker; *Ranger*, Captain Matthew Swain; *Fox*, Captain Ransom Jones; *Waterford Packet*, Captain Francis Barrett; *Lucas*, Captain Paul Coffin—all Nantucket masters. Curtis and Co. sent out the *Dolphin*, Captain William Swain, *Ann Delicia*, Captain Timothy Fitch, *London*, Captain Joshua Coffin—also Nantucket captains.

Alexander and Benjamin Champion, who shared with the Enderbys in the inauguration of the Fishery from London, had their own trusted Nantucket shipmasters commanding their ships. The *Venus*, a new ship, was dispatched in 1788 under Captain Daniel Coffin—perhaps the veteran of the Nantucketers. Captain Coffin had been captured during the Revolution and taken into New York by a British frigate. He was liberated only upon signing on for service in the packet *Lady Gage*, sailing several voyages between New York and London.

In 1777, the packet was captured by an American privateer and Captain Coffin was put aboard a prison hulk in Boston harbor, despite his proof of Nantucket residency. After several months he was released, and came back to Nantucket to resume his whaling career. At the Treaty of Peace he returned to London and entered the employ of the Champions. He was the first of the British whalers to approach the Madagascar whaling area of the Mozambique Channel.[10]

Other Nantucketers were most successful as whaling masters and shareholders in ships they sailed, including Captain Elisha Pinkham, who, as already cited, took out Alexander Champion's new ship the *Adventure* in May 1787, and Captain Thomas Delano, who made a record voyage in the *Lord Hawkesbury*.

9. Listing of Whaleships, B.T. 6/230.
10. Diary of Keziah Coffin, *Historic Nantucket*, N.H.S.

London merchants who were newcomers in the whaling industry included Timothy and William Curtis—the latter to enter politics and eventually become Lord Mayor of London—who also had experience in the Greenland Fishery; Daniel Bennett, a blacksmith who invested wisely; and James Mellish, an outfitter and victualler, who was a successful provisioner in beef and pork for the Royal Navy.

Along the curving waterside of the Thames, just below the Tower of London, was a market known as "Doctors' Commons," a maritime exchange, where ships which were overdue were often placed for sale on speculation. On one of these occasions, Messrs. Bennett and Mellish had purchased the risk on a whaleship that had been overdue for some time. After another prolonged period, in which no word had been received on the missing vessel, Mellish's son met Bennett and, in a bantering tone, inquired as to the ship. Bennett quickly asked if the elder Mellish was becoming "sick of the bargain," and offered to buy out his share. Soon after, Mellish senior agreed to sell his risk. An agreement was made whereby Bennett purchased the other's share for £1,000, giving him complete possession of the whaleship should she ever return.

Whether or not the shrewd blacksmith had received knowledge of the vessel's whereabouts may only be conjectured, but less than a week later, like a ghost, the long overdue whaler was reported off Gravesend, with a full cargo.[11]

With the money gleaned from this gamble, Bennett invested in two more ships, buying them cheaply enough as surplus craft in the Danish Royal Navy. One of these he named the *Africa*, and sent her out under command of Captain Ransom Jones, the Nantucketer who had previously sailed for Lucas.[12] Thirteen months later the ship was home with a full cargo of oil and seven thousand seal skins. Bennett added steadily to his fleet, eventually becoming one of the wealthiest shipowners in London. A number of years into the next century, one of his captains named an island in the Pacific "Bennett Island" after this enterprising shipowner.[13]

11. *Memorandum and Papers* of Frederick C. Sanford, Nantucket Atheneum Library, Nantucket, Mass.
12. *Whaling Long Ago*, Frederick C. Sanford Papers, Nantucket Atheneum Library, Nantucket, Mass.
13. Log of the Whaleship *Resource*, National Maritime Museum, Library, Greenwich, England.

Of considerable interest is a report from the voyage of the *Lucas*, under command of Captain Aiken. Sailing from the Downs on August 25, 1787, she was reported at Trinidad in May 1789, homeward bound. She took 160 tons of sperm whale oil on the voyage, but also brought back 100 tons of spermaceti transshipped from the *United States*, of Nantucket, placed on board the *Lucas* at the Falkland Islands. As Francis Rotch, then with the Hayley firm in London, was one of the owners of the *United States*, he claimed the oil. This was a practice which the London whaling merchants feared—Nantucket oil attempting to be entered free of duty. They complained bitterly, and the Customs people were alerted.

When the ship *Sappho*, owned by Ogle and Co., returned from a voyage to the Falkland grounds and arrived in London in March 1788, she reported the following whaleships in the South Atlantic grounds:

SHIP	MASTER	OWNER	PORT	TONNAGE
Good Intent	R. Brown	W. Sims	London	250
Ranger	M. Swain*	Lucas & Co.	London	150
Rasper	M. Gage*	Enderby & Sons	London	200
Lucas	J. Aiken*	Lucas & Co.	London	200
Tiger	H. Barton	J. Hall	London	250
Fox	R. Jones*	Lucas & Co.	London	175
Active	D. Ferguson	D. Bennett	London	178
Experiment	T. Gage*	T. Guillaume	London	205
New Hope	R. Hillman*	T. Yorke	London	200
Liberty	T. Clark*	Lucas & Co.	London	276
Adventure	E. Pinkham*	A. Champion	London	231
Queen	F. Bolton	G. Thornton	London	198
Quaker	J. Hopper	T. Gilbert	London	170

* Nantucket captains.[14]

The list of these owners is of interest as to their location and their business establishments:

W. Sims, Ropemaker, Sun Tavern, Shadwell.
Isaac Lucas and Co., Oil Merchants, 73 Holdburn Bridge.
Samuel Enderby and Sons, Oil Merchants, Paul's Wharf, Thames St.
J. Hall, Ship Chandler, 265 Wapping Wall.

14. Lists from *Board of Trade Records*, B.T. 6/95.

Daniel Bennett, Brazier, Wapping.
Thomas Guillaume, Ship-Builder, Limehouse Bridge.
Thomas Yorke, Oil Merchant, Soho.
Alex. and Benj. Champion, Oil Merchants, 71 Old Broad Street.
G. Thornton and Sons, Russia Goods, 25 Austin Friars.
T. Gilbert, Sail-Maker, 222 Wapping.

Other firms entering the business were Robert Curling and Co., whose ship *Dolphin* was commanded by Captain Andrew Swain, and whose *Experiment* was commanded by Captain Gage, both from Nantucket; John Thompson, whose ship *Stormont* had Captain Reuben Ellis of Nantucket in command; LeMesurier and Secretan, listing a ship named *Guernsey Lily* (apparently they were natives of that island) with Captain Henry Folger of Nantucket as her master; John Hall, who owned the *Attempt*, Captain David McCormick, *Audacious*, Captain John Lovejoy, and *Saucy Ben*, Captain William Raven, the latter to become well known from his subsequent voyagings.

John St. Barbe was represented by the *Aurora*, Captain Peleg Long of Nantucket; *Liberty*, Tristram Clark, also of Nantucket, as master, and *Southampton*, Captain William Akin, another American.

James Montgomery sent out the ship *Nimble*, with Captain Francis Gardner of Nantucket, master. An unusual entry was that of the ship *New Hope*, owned by Thomas Yorke, an American Loyalist from Philadelphia who had moved to Bristol. Her master was Captain Jethro Daggett of Nantucket, succeeding Captain Hillman.[15]

Thus, out of the port of London alone, the British Southern Whale Fishery was showing a vigor and growth that was a prophetic indication of what was to develop as the "golden age" of that nation's newest maritime industry.

15. *Protections*, Southern Whale Fishery, 1777–1793, Public Record Office, Adm. 7/389. London Directory, *Board of Trade Records*, B.T. 6/95.

9. The Seaport With the 'Back Door'

WITH the advent of spring 1786 William Rotch had completed the preliminary negotiations with the French government's ministers leading to the transfer of his fleet and other Nantucket ships to France and the port of Dunkirk. Early in the year, having heard nothing from Lord Hawkesbury and the Board of Trade, Rotch had crossed the Channel with his son Benjamin, to be joined at Dunkirk by his brother Francis. Proceeding to Paris for a further meeting with the French authorities, the Nantucketers were accompanied by a Frenchman bearing a name then well known on the Island—Francois Coffyn, a native of France—and by Captain Shubael Gardner, of Nantucket.

These two personages deserve to be better known. Francois Coffyn had become deeply interested in American commercial affairs and had gained the confidence of both Franklin and Jefferson in Paris. He had been in correspondence with William Rotch, and had served as an interpreter for Francis Rotch, who was in Dunkirk before his brother William arrived. Coffyn had become intrigued with the idea of procuring for the Nantucket whalemen "an asylum in my own country." Captain Shubael Gardner was a veteran master mariner with several years experience in selling oil and candles in the European market at London, Amsterdam, and Dunkirk, and had become a trusted messenger of William Rotch and Sons.

Coffyn had been an invaluable aid to the Rotches over a period of months, having first met Francis Rotch in Dunkirk. In March 1786 Coffyn had written a letter of invitation to Nantucket whalemen, sending it directly to the Island in the hands of Captain Shubael Gardner. As an emissary of William Rotch, Captain Gardner had gone to Dunkirk from London in November 1785, bearing a list of proposals which William Rotch had written as an application to the French government similar to those which the British had failed to act upon. These proposals had been also advanced by Coffyn.

Captain Gardner took the Coffyn letter, dated March 10, 1786, directly to Nantucket, where in May he presented it to the Board of Selectmen. It is altogether both a remarkable document and a strong support of William Rotch's plans. Declaring his motives were to offer "a friendly hand to a set of people who, in my opinion, greatly deserve the assistance of all men," Coffyn wrote his object was

to make an application to this Government in their behalf, and to make use of all the Interest with the King's Ministers I was capable of, to make such proposals as were penned by our Friend William Rotch and which you delivered to me in his name on your arrival from London in November last [1785] . . . every article which I thought capable of contributing to the welfare and happiness of those Inhabitants who would chuse to remove to the Town of Dunkerque . . . you may also communicate to the Selectmen the negotiations commenced with our friend William Rotch. . . .[1]

While the Selectmen were passing along to Nantucketers the contents of this document of invitation, William Rotch had again arrived in France. Having rejected the last-minute offer of Lord Hawkesbury (which added to his distrust of the British leader's policies rather than encouraged him to reconsider them), William Rotch was now fully committed to establishing a whaling base at Dunkirk. He wrote home: "I took my leave, giving notice of my intention of going to France, that no just censure of my taking any secret step should be placed on us as a Society."[2] By the latter, he meant the Society of Friends.

Immediately upon arrival at Dunkirk, William Rotch, Francis Rotch, and Benjamin Rotch were joined by Coffyn and proceeded to Paris. First, at Versailles came an interview with M. Necker, Minister of Finance, one of the most capable of the men on whom King Louis XVI depended. Later, the Nantucketers met the aged Vergennes, Minister of Foreign Affairs, who had been close to Benjamin Franklin; Marshall de Castre, Minister of Marine; and M. de Calonne. The reception was cordial, with a swift reappraisal of the Rotch proposition, and an immediate decision

1. Francois Coffyn to Captain Shubael Gardner, "Letter of Invitation for Nantucketers to Remove to France," March 10, 1786. Manuscript Collection of the Nantucket Historical Association.
2. William Rotch, Letter of June 6, 1786, in Bullard, *The Rotches*, p. 228.

12. Hive of Nantucket Whaling Masters
Coffin-Paddack House, oldest on Nantucket (1686)
[Photograph by Cortlandt V. D. Hubbard]
[Courtesy Historic American Buildings Survey]

to accept the terms for the Nantucket whalemen as to their coming to Dunkirk as a colony to establish the Southern Whale Fishery.

One can imagine the reaction in the breast of the Quaker advocate. In reporting the historic transaction to his son William, Jr., and his son-in-law Samuel Rodman at Nantucket, he wrote:

We were received with all the civility and politeness that the policy of this nation dictates . . . Our business was well done and signed . . . I most gladly set my face toward this country, with the hope that in all our transactions there the truth professed by us had not suffered.[3]

The terms William Rotch had successfully negotiated were similar to those contained in his "Proposal" to the British Ministry, and included land and waterfront areas at Dunkirk; dwellings for families; free exercise in their trades by the coopers, shipsmiths, sailmakers, and cordage makers; exemption from all duties for the importation of whaling stores

3. Ibid.

and gear; a bounty of 43 livres per ton on whale oil brought in the fleet; import duty on other oils being imported, so as to protect their fleet; exemption of the whalemen from military or naval service; and last, but in the forefront of the stipulations, the guarantee that the colony would be permitted to conduct their religious affairs as members of the Society of Friends.[4]

Of major importance was the extra concession which Rotch was able to obtain from the French Ministers—the privilege of bringing into Dunkirk, duty-free, 250 tons of Nantucket whale oil per annum, exclusive of the regular fleet. This was actually more than a mere concession; it was the creation of a mode of business that would enable Rotch to help his fellow merchants in Nantucket to ship their oil into the Market of Europe. Apparently he intended to take swift advantage of this opportunity as he authorized his firm in Nantucket to purchase oil there at £50/5 per ton, so that, in Dunkirk, he could "keep pace with the London market." It was evident that he intended his ships, as both catchers of oil and conveyors of oil taken by other Nantucket ships, to bring the oil to the French market by way of his counting house in Dunkirk, stating that Islanders could "reap the advantage and *not disclose that part for the present*. If I live to return, I will open up the whole business."[5]

Rotch directed the initial steps to be taken at the Nantucket end. The ship *Warwick*, which had arrived in London during his stay, had been ordered to return to Nantucket with a general cargo including Russian hemp, needed for cordage. When the *Maria* had sailed from London, August 18, 1785, Rotch had ordered Captain Mooers to return to England, as he hoped to complete his proposal with Pitt, and he had given similar orders for other vessels in the Rotch fleet then in the Falklands. Now, the returning Rotch ships found orders awaiting them in England and Nantucket to sail for Dunkirk and France.

First of this fleet to appear was the *Canton*, Captain James Whippey, which arrived at Falmouth in May 1786. Rotch was now authorized to get her cargo in duty-free—hence she was ordered to Dunkirk, where that privilege had been granted. Rotch, in a letter home, advised the Nan-

4. Rotch, *Memorandum*, pp. 46–47.
5. Rotch to Rodman, quoted in Bullard, *The Rotches*, p. 227. Italics are the author's.

tucket vessels in the Southern Fishery to "go after Right Whales if they are more plenty than Sperm, which will do best in France," and ordered all his own ships "that are fitted for Whaling into that business."[6]

That his plans were well advanced is shown by his order that his ships return immediately to Nantucket from "the Brazils," there to have the sperm oil separated—the "head-matter" from the body—have it refined, and then brought directly to Dunkirk. This was important, as there was not at that time the proper equipment for the refining process at the French port. He requested that frames for spermaceti presses be shipped, together with screws, there being none in Dunkirk, as the manufacture of sperm candles was still a secret process, known only to a few in New England. The ship *Bedford*, which had already made history by unfurling the flag of the new United States before London three years previously, was ordered to Dunkirk with stores, oil, and candles.

In keeping with his awareness of the mercantile situation, he wrote his sons: "When you ship our oil, write Champion & Dickason for insurance &c. from Nantucket to Dunkirk, with liberty to touch at an adjacent Harbour to fill up, & write them to insure it clear of risque of the Algerians & other Barbary Cruisers, which will lower the premium as the risque on that Head is very little." Thus, the Londoners would be knowingly sharing in William Rotch's Dunkirk whaling venture—by sharing in the risks!

The reputation of William Rotch and Sons had been well established in France over a period of years before the Revolution. At Dunkirk, the Rotch firm had shipped oil and candles to Europe through the firm of Louis DeBauque and Bros., and Nantucket oil had found good customers at L'Orient and Bordeaux.

Francis Rotch, who had preceded his brother William to Dunkirk, had finally dissociated himself from the home office of the Hayley firm in London, after being Madame Hayley's chief emissary for more than seven years. Upon his arrival in England in 1775, and after his unsuccessful efforts to recoup his fortunes following his Falkland Islands venture, Francis assumed a leading post with the Hayley Company. Madame Hayley, who had continued the business after her husband's death in 1777, found Francis Rotch's knowledge of whaling and whale products

6. Ibid., pp. 224–225.

of considerable value to the firm. More than this, she accepted the Nantucket man as her constant companion. He was then thirty-five years of age, while the lady was thirty years older. Madame Hayley was the sister of the famous British radical John Wilkes; she possessed a quick mind, and maintained one of the social centers near London, where she entertained frequently.

When Francis Rotch sailed for America in November 1784, Madame Hayley accompanied him so that she might negotiate for the transfer of some property she owned in this country. One of the members of the party was a young twenty-one-year-old clerk named Patrick Jaffrey, who was given the task of carrying out the business arrangements in Boston, while Francis Rotch journeyed on to Bedford and Nantucket, where the will of his deceased father, Joseph Rotch, was being probated. Shortly after this it was found necessary that he return to London, to handle some pressing business for the Hayley firm. Soon after he arrived in England, Francis received word that Madame Hayley and the clerk, Jaffrey, had married. Thus, Francis Rotch removed himself from the London scene and went to Dunkirk.[7]

However, he did not entirely free himself from the business affairs of Madame Hayley. Her firm owned a majority of shares in the whaleship *United States*, and when that ship returned to England from her long and memorable voyage to the Falkland Islands, and found her oil not clearing British customs without paying the high tariff, Francis Rotch and his brother William ordered the ship to go to Dunkirk. Here the oil was sold at a handsome profit. The Hayley shares in the vessel were sold to the firm of De Bauque Freres, at Dunkirk. She was then loaded with French goods, together with some hemp and iron (both in heavy demand) and dispatched to America, her cargo consigned to Christopher Champlin at Providence, Rhode Island.

In a letter from Dunkirk dated October 9, 1786, Francis Rotch advised Champlin that the *United States* had been placed under the command of Captain William Hayden. Upon arrival at Providence she was to be fitted out for the Northern or Greenland Fishery, but if circumstances prevented, Champlin was to notify the firm of William Rotch and Sons in Nantucket "or to notify the former captain of the ship, Benjamin Hussey."

7. Bullard, *The Rotches*, pp. 53–57.

From this letter it was apparent that after leaving the Falkland Islands the *United States* first returned to Nantucket, where Captain Hussey went ashore, and then continued on to England under Captain Uriah Swain. This pattern of Rotch-managed whaleships demonstrates the careful planning and arrangements for the marketing of the oil.

Further, as an example of the close relationship that continued between the Rotch firm and his former London business associates, the Francis Rotch letter carefully directed:

Should any accident happen to make an application to the underwriters, you will . . . have the vouchers well arranged and authenticated and sent by two conveyances to Messrs. Alexander and Benjamin Champion, New Lloyds, London, who are the agents for my friends, Messrs. De Bauque, and who have done the insurance, ship and cargo, to the amount of £3500—£2500 on the ship and £1000 on the cargo, in the office of the London Assurance Co.[8]

Upon returning briefly to London, Francis Rotch again wrote to Champlin (November 1, 1786) that the ship was now called *Le Dauphin*, no doubt due to the investment by the De Bauque firm. He mentioned a commercial treaty between England and France appeared to be passed favorably by Parliament.

The *Dauphin* (former *United States*) had a stormy passage across the Western Ocean and arrived in a "wrecked condition." The De Bauque firm, in answer to Champlin's letter announcing this, ordered that the proposed voyage to Greenland be cancelled, and that the ship be loaded "with all kinds of staves" and sail for Dunkirk. A letter from Francis Rotch corroborated these orders, and the ship was directed to go first to a port in Virginia (probably Norfolk) for the cargo of white oak staves, and to "guard against disappointment in some degree by taking aboard tobacco or freight to the Farmers General in France."

The cost of repairs on the *Dauphin* amounted to £1,351, while the sale of her cargo totalled £1,804. Rotch wrote that he did not doubt but that the insurance for damages could be adjusted "with the Underwriters amicably." The *Dauphin* returned to Dunkirk under Captain Hayden,

8. Francis Rotch to Christopher Champlin, October 9, 1786. Whetmore Collection, *Commerce of Rhode Island*, Vol. II, pp. 292–293.

and then sailed to the Brazil Banks under her former master, Captain Uriah Swain.

At Dunkirk, William Rotch and his son Benjamin were now busying themselves in arrangements for the transfer of their ships and the Nantucket families who planned to remove to the French port. The cordial reception by the French officials in Paris was heartening, and the Nantucketers, accompanied by Francois Coffyn, met such personages as the aged Count Vergennes, the Prime Minister; Carnot, Minister of Marine; and many other dignitaries who gave them "liberal encouragement." In all instances, as William Rotch wrote, the simple dignity of the Quakers was respected:

We appeared [at Versailles] in our usual way with our heads covered [the Quaker custom at formal affairs] which being explained, policy directed their apparent full approbation, & though a singular appearance yet every mark of respect shown us . . . I most gladly set my face toward this country . . . with a degree of thankfulness in being discharged from a burthen respecting our business which has lain so heavily on my shoulders since leaving home.[9]

With the continued help of Francois Coffyn, "our friend and assistant" (and also interpreter), the Rotches completed their preliminary plans at Dunkirk. The first two ships to be fitted out from the French port were the *Canton*, still under the veteran Captain Whippey, and the *Mary*, Captain David Starbuck. The *Young States* would sail later in the year from Nantucket, under Captain Thaddeus Coffin, to join them at Dunkirk.

Soon after his arrival in France William Rotch recognized the need of direct communication between Nantucket and Dunkirk. At this time the firm of William Rotch and Sons were developing a branch in Bedford (later New Bedford) which was eventually to become the firm's headquarters. The elder Rotch wrote to Nantucket:

We cannot carry on our business without a vessel to run between Dunkirk and America regularly, therefore thou must look out for purchase of a carrying Brig of 150 tons, try to get one built of white oak. If she will

9. William Rotch to Samuel Rodman and William Rotch, Jr., June 7, 1786. Quoted by Bullard, *The Rotches*, p. 229.

make no money she will save some. Take out a new register for the Ann *in thine and my name [as] of Nantucket.*[10]

William Rotch now made plans to return home as quickly as possible, so that the arrangements at the Nantucket end would be coordinated through the Rotch firm, so ably represented on the Island by his eldest son, William, Jr., and his son-in-law Samuel Rodman. In late September 1786 he crossed the Channel to Dover and took the stage to London. One wonders as to his thoughts as, once again on Shooters Hill, he looked down at the great city and recalled that first visit he made just over a year before.

With characteristic energy, William Rotch put his short stay in London to good advantage. Arrangements for the insurance of his vessels out of Dunkirk and Nantucket were completed with the London firm of Champion and Dickason. At the counting house of the Enderby firm on Lower Thames Street he found letters awaiting him from Nantucket. He was pleased to learn of the safe arrival of the *Warwick* of Nantucket, and that his letters home had afforded his sons "that satisfaction which we mutually enjoy at receiving the account of each others welfare." In his replies he advised as to business matters, and in a revealing sentence, requested that his Nantucket firm "refuse, at any rate, if offered," the paper currency being issued by the State of Massachusetts.

He repeated instructions as to the ships returning from the Southern Fishery, reiterating that they should first return to Nantucket to refine the oil, thence to Dunkirk. "If I can find a cheap American ship here, fit for whaling, I intend buying one." At Rotherhithe, below the Tower of London, he found a suitable ship and purchased her. She was probably the American ship for which he was looking, and he named her the *Penelope*, placing her under the command of his favorite shipmaster, Captain William Mooers.

Finally, on October 14, 1786, William Rotch and son Benjamin embarked. The *Penelope* slipped away down the Thames to Gravesend, and the two Nantucketers began their voyage home.

It was on the first day of the new year, 1787, that the anxious pair at last reached their destination of Nantucket, having been gone eighteen months. During this absence from Sherborn they had tested the whaling

10. Ibid.

potentials of two nations, and negotiated for the pursuit of an industry that had been the peculiar property of this tiny kingdom of Nantucket—now representing the youngest of the world's nations, the United States of America.

In the history of mercantile affairs it is doubtful if there had ever been a more logical plan for conducting such an industry as whaling as that conceived by William Rotch. Far ahead of his time in the concept of international trade, he considered this whaling industry as a world public utility—lighting the lamps and candles of nations and lubricating the machines which were to launch new eras in world manufacturing and trade—and held that it should not be the prerogative of any one nation. Herein, in his island home, he gained new inspiration for carrying out his ideas, because the success of this enterprise would guarantee the continued existence of Nantucket as a whaling center.

Recognizing that Britain had now taken over the leadership of the Southern Whale Fishery through the energetic application of American ideas by the London merchants, William Rotch intended that Dunkirk would not only become the rival to London in this industry but that Nantucket whalemen would be diverted from England to France and be more free to conduct his basic plan—getting Nantucket whale oil through the "back-door of Europe."

It was in this plan—called by one historian the "coup de whale"—that William Rotch saw the great opportunity to challenge the supremacy of Britain. By trans-shipping the oil to Dunkirk from whaleships actually sailing from Nantucket, the total augmented by that brought in by his own fleet from Dunkirk itself, could supply the French demand and compete with Britain for that valuable market. By such an arrangement Nantucket would be greatly aided in regaining its former position as the headquarters of the Southern Whale Fishery. The Quaker merchant knew that London had gained a sizable advantage and must be challenged. The very daring of this plan was based on William Rotch's knowledge of whaling as an industry and of Nantucket as the chief exponent of that industry.[11]

As he had written so many times in his letters, his chief hope was "to remain in his native place." Now, he was returned home, and in a

11. Stackpole, *The Sea Hunters*, p. 143.

13. Whaleship *Edward* of Dunkirk, Captain Micajah Gardner
A rare drawing of one of the Rotch fleet out of the French port
[Courtesy Mrs. Christel Mitchell]

position to bring his exceptional business talents to the task of recovering both his fortunes and those of his fellow Islanders.

Benjamin Rotch had another reason, aside from aiding his father, in his return to Sherborn. During his absence he had continued wooing by letter his childhood sweetheart, Elizabeth Barker, and had now received her consent to become his wife. They were married in the old Friends Meeting House in Upper Town on March 29, 1787. A brief but characteristic glimpse of the wedding was later written from a family reminiscence:

The bride was dressed in a pale peach colored silk gown, with the skirt wide open in front, showing through a transparent apron a quilted satin petticoat of light blue color. A light drab satin cloak, lined with white reached nearly to her ankles, and was wide enough to make in after years two mantles, one for each of her daughters.[12]

12. Eliza Farrar, *Memorials of The Life of Elizabeth Rotch By Her Daughter* (Springfield, Mass., 1861), p. 24.

This may seem slightly unusual for a Quaker bride's costume but it does present a picture of the day, when some of the worldliness of the Nantucketers was another indication of their independence of spirit. After all, the gown was covered by a "drab satin cloak . . . reached nearly to her ankles!"

Six vessels were fitted out from Nantucket to serve with other ships of the Dunkirk fleet. These were all Rotch ships—the *Warwick*, Captain Christopher Mitchell; the *Maria*, Captain Coffin Whippey; the *Ospray*, Captain Benjamin Paddock; the *Young States*, Captain Thaddeus Coffin; the *Ann*, Captain Prince Coleman; and the *Penelope*, Captain Elisha Folger—the latter being the same ship purchased in London which had brought William and Benjamin home.

In Dunkirk, Francis Rotch and Francois Coffyn had continued to carry out the plans for the colony from Nantucket. Such matters as wharf property and customs registers had been expedited. A former rope or cordage factory was acquired as a warehouse and refinery, and a sperm candle manufacture was established on the Isle of Jeanty in the harbor. Several dwellings were engaged for families removing from Nantucket, and other arrangements completed.

Most important for the removal of the colony to France, several decrees had been passed by the French government that guaranteed the importation of the American oil duty-free. This decree continued:

Such of the Inhabitants of the Island of Nantucket as are desirous to settle in France, on purpose to engage in the Whale Fishery, are hereby permitted to transport themselves, with all their vessels, boats, tackle &c necessary for the said Fishery; and they shall also be permitted to enjoy all the advantages of the French Flag and all other privileges hitherto granted to the Nantucket Fishermen in the Ports of France, provided the vessels so imported are destined solely for the said Fishery.[13]

Among the closest observers of the novel situation was Thomas Jefferson, recently the successor to Franklin as the American Ambassador to France. In a letter to John Jay, written from Paris, Jefferson noted:

She [Britain] fears no rivals in the whale-fishery but America; or, rather, it is the whale-fishery of America she is endeavoring to possess

13. *Boston Gazette*, November 4, 1791.

herself. It is for this object she is making the present extraordinary effort, by bounties and other encouragements, and her success so far is very flattering. Before the War, she had not one hundred vessels in the whole trade, while America employed 309. . . . Now they have changed places. England has gained what America has lost. France, by her ports and markets, holds the balance between the two countries.

Writing further to Jay, he observed:

The French government had not been inattentive to the news of the British, nor insensible to the crisis. They saw the danger of permitting five or six thousand of the best seamen existing, to be transferred by a single stroke to the marine strength of their enemy, and to carry over with them an art which they possessed almost exclusively. The counterplan, which they set on foot, was to tempt the Nantucketois by high offer, to come and settle in France. . . . The British, however, had in their favor, a sameness of language, religion, laws habits and kindred. Nine families only, of 33 persons in all, came to Dunkirk, so that the project was not likely to prevent their imigration to England if nothing else had happened. . . . It became tolerably evident that the terms offered the Nantucketois would not produce their emigration to Dunkirk; and that it would be safest in every event, to offer some other alternative. . . . The obvious one was to open the ports of France to their oils so that they might exercise their fishery by remaining in their native country. . . .[14]

The astute Jefferson was fully aware of the strong start the Rotch fleet had made in establishing a whaling industry at Dunkirk. He knew that it was ostensibly a French industry but appreciated the potential importation of Nantucket oil. It was apparent that he believed that the Rotches, having a virtual monopoly of the whale oil market, would oppose freeing the importation to allow all American oils. The effect of the new French and British commercial treaty, permitting the entry of British oil, was soon to be felt, but at this time William Rotch, in Nantucket, was confident he could match prices with the British imports.

14. Thomas Jefferson to John Jay, *Memoirs, Correspondence and Miscellanies From The Papers of Thomas Jefferson.* Edited by Thomas Jefferson Randolph (Boston, 1830), pp. 392–393.

But the competition was stronger than supposed. The De Bauque firm at Dunkirk, the principal purchaser and distributor of the oil furnished by the Rotch fleet, was concerned. Realizing that the Rotch interests needed a representative in France who was familiar with conditions on both sides of the Atlantic, Benjamin decided to return to Dunkirk. He was the logical person to fill the role, but the prospect of separation from his wife haunted the twenty-four-year-old merchant. A son, named Francis after the uncle, had been born in January 1788, and this made his decision to leave Nantucket all the more difficult. But the strong character of the young man was never more evident, and he made arrangements to leave. The ship *Bedford* was loaded with whaling supplies and, under Captain Abishai Hayden, sailed for Dunkirk on June 6, 1788, first going to Bedford for some additional material. Thirty-eight days later the ship arrived in France.[15]

Benjamin found the affairs at Dunkirk were progressing in good fashion. During the balance of that year and well into 1789 the Rotch fleet grew steadily, with the Southern fleet bringing in full cargoes. To balance the importation, the ships *Penelope*, Captain Tristram Gardner, and *Maria*, Captain William Mooers, were sent into the Northern fishery for the black oil of the Arctic whale.

The log of the *Penelope* recounts some of the aspects of such cruisings:

May 23d—Thick snow storm. Latter part saw whales but could not get at them, some ships in sight.
June 5. Saw two wrecks, one was the ship London, *of London, the other belonging to Whitby, men saved, ship lost in last gale.* [*Whitby ship named* Chance].
June 8. Killed a ten-ft. bone whale [*measurement of baleen*]. *Mated with Capt. Mooers* [*in* Maria] *and took one. Several ships in sight.*[16]

In Nantucket the success of the Rotch firm's management at Dunkirk was reflected in the rapid increase in the island's fleet. From only seven vessels sailing in 1786, the number of the Nantucket-based fleet increased to twenty in 1789. By the end of the latter year the Dunkirk fleet numbered fourteen ships, so that the combined Southern Whaling fleets

15. Bullard, *The Rotches*, p. 112.
16. "Leaves From The Log of the *Penelope*," *The Inquirer and Mirror*, issue of April 15, 1859.

(Nantucket-Bedford and Dunkirk) numbered thirty-four. While this was a remarkable and substantial gain it was still well under the London fleet which had increased at this same time to total well over sixty vessels.

The outstanding feature of the situation lay in the fact that, of the whaling fleets engaged in the Southern Whale Fishery, sailing from three countries, at least three-quarters of the shipmasters and officers manning the ships were Nantucket men. Also to be remembered, as an economic factor, the trades allied to whaling were being similarly aided. As one merchant who was familiar with some of the men involved once wrote:

Not all who participated in the migration to Dunkirk were shipmasters or seamen. There were numbers of islanders who were employed from time to time in construction of boats, manufacture of lines, harpoons, and other instruments. The same was true of Le Havre.[17]

As an example of the value of the cargoes brought into Dunkirk by the Rotch whaleships the returns of the voyage of the *Maria* are of interest. After her arrival in London in July 1785, the ship sailed for the Falkland grounds on August 18, under Captain William Mooers, and came back to England with a full ship. Being prevented from selling the oil in Britain, the *Maria* was ordered to France, and fitted out at Dunkirk for her first voyage from this port. As Captain Mooers was needed for transfer operations from Nantucket to Dunkirk, the *Maria* was placed under the command of Captain Bartlett Coffin, and sailed in the fall of 1786.

In April 1788 the *Maria* returned to Dunkirk from a successful voyage to the South Atlantic. Her cargo was promptly sold, bringing £3,162/10 10d, of which amount the master received $\frac{1}{15}$ lay after the expenses of the ship were deducted, and also a percentage of the bounty. The De Bauque firm purchased the oil and bone, which was gauged as follows:

Whale Oil (81 tons, 6 bbls., 24 gals.)	*£2182 10s 09p.*
Sperm Oil (1649 "hoops")	*203 11s 11p*
Head Matter (432 "hoops")	*76 3s 09p*
Bone (6302 lbs)	*700 4s 05p*
	£3162 10s. 10p[18]

17. Ibid.
18. Rotch vs. Coffin, Nantucket Court Records, Nantucket, Mass. Book II, pp. 42–43.

There was a dispute between the owners of the ship and Captain Coffin as to the proportion alloted him, and the case was decided in Nantucket, where three referees, appointed by the Court, recommended the Rotch owners receive the judgment in their favor.

The *Maria* became a famous ship. Built in the North River at Pembroke, Massachusetts in 1782, she sailed from Nantucket, London, Dunkirk, New Bedford, and Talcahuano, Chile, before being finally laid up in a cove on Vancouver Island on the Northwest Coast of the North American continent—ninety years old. But it is doubtful if in her long career she had a more unusual experience than during the period from 1785 to 1788, when she sailed from three different nations, one in the New and two in the Old World.

It is of more than ordinary interest to note that at the time William Rotch had returned to Nantucket to consolidate his plans, his leading London rival in the Southern Whale Fishery, Samuel Enderby and Sons, still held him and Nantucket in high regard. In a letter written by Samuel Enderby, Jr., to George Chalmers, one of the trusted secretaries of Lord Hawkesbury and the Board of Trade, under date of December 2, 1788, appears this remarkable paragraph:

I have received two or three letters from Mr. Rotch at Nantucket, by which it appears that the French bounties [at Dunkirk] are very large and most of the money received. I now dispair of getting that family to this country. We have been apply'd to by some of the Southern Whale Fishery owners to join in an application for a heavier duty or a prohibition being laid on oil caught by the subjects of the American States; we have declined it, and if such application is made, I have no doubt, but we can convince Government of the impropriety of such a step.[19]

19. Samuel Enderby, Jr., to George Chalmers, Dec. 2, 1788, Chalmers Collection A322, Mitchell Library, Sydney, Australia.

10. 'Round the Capes: and New Worlds

THE success of the British whaleships in the Southern Whale Fishery was now becoming measured by the rich rewards in their oily cargoes. By 1788 the price of sperm oil had risen to the unprecedented price of £60 per ton, with the prized head matter (spermaceti) reaching an all-time high of £68 per ton. Fifty whaleships sailed that year, forty-three from London, bringing a total catch valued at £72,931—more than twice that of the American fleet. The total tonnage caught by the rival fleets was even more in contrast—3,414 tons of oil for the British and 496 for the United States. The premiums granted the former added considerably more to the profits of the owners and in 1789 the total value of the British Fishery totalled £77,805—the product of fifty-two vessels.[1]

While the Greenland, or northern whalers, could remain on the whaling grounds for only four months at best, the Southern fleet had a much longer period, their voyages now nearing eighteen months in duration. The range of their cruisings increased during the 1785–1789 period to an extent never anticipated. After proceeding to the South Atlantic, the ships cruised on the "Brazil Banks," then to the south and the "River Plate" grounds, and on to the Falklands and the coast of Patagonia. Following the example previously established by the Nantucketers another pattern for voyages through the South Atlantic took the whalers to the west coast of Africa and south to Walfish Bay, then down the coast to Cape Town. From here their voyages proceeded around the shoulder of Table Mountain into the Indian Ocean, and up the east coast to Delagoa Bay, following that great migratory route of the whale.

The several published accounts of the voyages of the great Captain James Cook had given valuable information as to the presence of vast

1. Samuel Enderby to Lord Hawkesbury, *Board of Trade Records*, B.T. 6/95, pp. 95–99.

numbers of seals and sea elephants on that remote mountainous island of South Georgia, beyond the Falklands and southeast of Cape Horn. Although the great navigator carefully noted the rocky shores and bays of this huge island, it remained for the whalers and sealers to actually explore and inhabit it. While the seal pelts were well worth the taking, it was the sea elephant who was chiefly hunted at this time. On the flinty shores of the island, the ponderous creature was easy prey for the hunters who stripped his thick blubber from his body just as the whale's was removed, and then boiled it in shore-side cauldrons for its rich oil.

From a report presented to the Board of Trade in 1788, it was recorded that the London whaler *Lucas*, under Captain Thomas Smith, was the first to go to South Georgia, reaching that island in 1787 and returning home the same year with a cargo of sea elephant oil and seal skins.[2] One of Alexander Champion's ships, the *Lord Hawkesbury*, had a most unusual catch—ambergris. Captain Joshua Coffin, the Nantucket master, returned in late December 1790 from a voyage of fifteen months (sailing in October 1789) and had a full ship—seventy-six tons of sperm oil and head matter. In reporting his ship's cargo, Alexander Champion wrote:

Captain Coffin . . . brought home 360 ounces of Amergris [sic], which the Captain took from the body of a female Spermaceti whale on the coast of Guinea [Africa]. I have sold it at a private sale at the high price of 19/6 per ounce. I understand it was bought principally for exportation to Holland, France & Turkey, where it is used as a perfume and is in great estimation. [Actually, it is used as an additive to enable the perfume to retain its scent far beyond the usual length of time.] This circumstance establishes what naturalists have always expressed doubts about respecting the origin of ambergris. I thought it would give your Lordship pleasure to hear of it, & when I have next the honor of waiting on you with Mr. Enderby, I will bring some of it with me for your Lordship's inspection.[3]

While this was in the nature of an incident during a whaling voyage, it is important in that it related a circumstance concerning the "first parcel [of ambergris] to be brought home by a British whaler."

2. Ibid.
3. Alexander Champion to Lord Hawkesbury, *Board of Trade Records*, January 2, 1791. B.T. 6/95, p. 187.

Ambergris was not new to Captain Joshua Coffin, commanding the *Lord Hawkesbury*. This peculiar substance, found generally in the intestines of a sick sperm whale, was known to have been first brought to the attention of scientific societies as early as 1724, when Thomas Boylston of Boston wrote to the Royal Society of London about this material having been brought in by some Nantucket whalemen who had "three or four years past made the discovery."[4]

Prices for the coveted sperm oil and head matter continued high. The 1789–1790 value of the Southern Whale Fishery reached the unprecedented figure of £84,493, the product of fifty-one vessels, and for the first time this surpassed the Greenland Fishery's total of £66,662. The relative value of the respective kinds of oil may be better appreciated when it is noted that the Greenland whaling fleet was represented by some 150 ships (English and Scotch), while the Southern fleet numbered but one-third of this total.[5]

Finding themselves on a high tide of success, the enterprising Londoners decided to press for a further advantage. In February 1787 the Board of Trade received another "Memorial" from Samuel Enderby and his associates which read, in part:

That the Southern Whale Fishery, under the present encouragement, is likely to become very extensive and valuable to this country, but cramped by the limited time fixed for the sailing and returning of their ships, it is recommended that all ships fitted for the 20/ Premiums granted by an Act passed in the 26th year of his present Majesty, may be permitted to sail any time on or after the 1st of May and on or before the 31st of October, and those entitled to 15/ in premiums to return on or before the 31st of August in the following year; those entitled to 15/ premiums to return as the Act specifies within 28 months from May 1 in the year they sailed, and not until 8 months from the time they sailed. . . . Some method requested that they may not suffer from any error by not taking out Licenses from the South Sea Company for fishing on the coast of Brazil, which they now understand is within the limits of the South Sea

4. Letter of Thomas Boylston to the Royal Society, London, *Philosophical Transactions*, Vol. 33, p. 193.
5. Report of The Whale Fishery, B.T. 6/95, p. 160.

Company's charter—Sections 14, 15, 16 and 17 of the Act permits ships to go to the eastward of the Cape of Good Hope and westward of Cape Horn or through the Streights of Magellan. . . . No ship has as yet sailed to the westward beyond Cape Horn for the purpose of whaling. . . .[6]

In responding quickly to this appeal Charles Jenkinson, now Lord Hawkesbury, continuing his open support of the whalemen, resorted to his customary political chess game. Instead of passing the petition directly into the hands of the Honorable East India Company, Hawkesbury sent it first to his colleagues, comprising the India Board, which had Lord Sydney as President, William Grenville (Pitt's cousin), and Henry Dundas. The latter was not only the most knowledgeable member of the Board as regarded affairs in India, but was Pitt's confidant and Hawkesbury's closest associate on the Board of Trade. The India Board promptly voted that in its opinion the whalers should be allowed to sail around the Cape of Good Hope "as far northward as the Equator and as far Eastward as 54° east from London . . . as there cannot be any reasonable objection on the part of the East India Company to comply with this request."[7]

As expected, the Honorable Company demurred. But Hawkesbury was too strong for them and in March 1788 he proposed that the whaling merchants be supported by a new statute for their further encouragement, and the Board of Trade promptly agreed.[8]

The East India Company, however, feeling now more clearly the threat to its trade by the wide-ranging whalers, remained adamant as to the extent of ocean to be allowed their use east of the Cape of Good Hope, and held firm for keeping the eastward limits at 51 degrees East Longitude. To retain this, however, they had to concede in another area—that west of Cape Horn—and the whaleships were permitted to extend their voyages around Cape Horn into the southwestern Pacific ranging northward as far as the equator and as far as 180 degrees West Longitude. These provisions were incorporated in a new Act for the encouragement of the Southern Whale Fishery—28 Geo. III, Cap. 20—which amended the Act of two years before.

6. *Memorial of Samuel Enderby & Sons*, et al., B.T. 6/95, p. 55.
7. William Broderick, Secretary of the East India Company, to W. Fawlkner, Secretary of the Board of Trade, London. B.T. 3/4, p. 377.
8. Minutes of the Board of Trade, March 7, 1788. B.T. 5/5, p. 63.

14. Captain John Hawes (1768–1824)
Master of whaleships from Dunkirk and New Bedford
From *Old Dartmouth Historical Sketches*, no. 22
[Courtesy the Old Dartmouth Historical Society]

Now that the great sea road into the Pacific was open to the whaling fleet the coming year was to witness the beginning of a new era not only in Britain's whaling industry but in the history of the Southern Whale Fishery itself. Insofar as Britain was concerned, it opened new and far-reaching opportunities in its maritime activities—and more than ever gave another purpose to its Royal Navy, its life-line.

The South Sea Company, in yielding to political pressure, had begun issuing licenses to the whalers with its meetings beginning October 26, 1786, and the first to receive such a permit to sail into the newly opened areas were the ships *Prince of Wales*, Captain Samuel Moore, and *Triumph*, Captain Daniel Coffin, both owned by Alexander and Benjamin Champion. The latter vessel had already sailed, however, on September 1, 1786, so as to reach the whaling grounds in season, and returned on November 14, 1787, with a full cargo, her Nantucket master cruising mainly "on Brazil Banks."[9]

Next to obtain licenses were ships owned by the firm of Joseph Lucas and Christopher Spencer: the *Astrea*, Captain Horne; the *Lucas*, Captain J. Aikin; the *Ranger*, Captain Andrew Swain; and the *Spencer*, Captain Owen Bunker—the latter three all Nantucket whaling masters.[10] It is to be noted that British whalers often changed masters on successive voyages.

It was at this time that the whaleships participated in adventures that were not anticipated. In 1787 the British government began its settlement of the new continent called New Holland (Australia), planning to use the Botany Bay area of the east coast as a place to establish a penal colony. The British jails were crowded, and the Ministry decided to remove the most unwanted of these unfortunate people to the remote shores of this little known colony of New South Wales.

In March 1787 the firm of Timothy and William Curtis petitioned for a license to sail within the territorial domain of the South Sea Company, having "fitted out the ship *Lady Penryhn*, under Captain William C. Sever, for a voyage to Botany Bay, and intended to proceed from thence

9. South Sea Company, Minutes, *Liverpool Papers*, Addit. Mss., 25,521, No. 38.
10. *Liverpool Papers*, Addit. Mss. 55,222, March 8, 1787, f. 6.

on the Southern Whale Fishery." Joining the fleet under Captain Arthur Phillip, with their living cargo of convicts, the *Lady Penryhn* sailed with them on May 13, 1787. As a companion vessel, the *Prince of Wales*, another ship with a license for whaling, was also one of this so-called "convict fleet." The *Lady Penryhn* arrived at Sydney on January 26, 1788 (having previously stopped at Botany Bay), and the *Prince of Wales* arrived with the others.[11]

The question arises as to whether either of these ships did any whaling on their return voyage home. The *Lady Penryhn*, under Captain Sever and with Lieutenant Watts on board, sailed on May 5, 1788 for Tahiti, and then for China, arriving at Macao on October 19. The *Prince of Wales* was one of a fleet of four vessels which left Sydney on July 14, all under command of Lieutenant John Shortland, bound for Endeavor Straits on the voyage home. But the fleet was scattered and the *Prince of Wales* and a companion vessel, the *Borrowdale*, came back to England by the southern passage.

Despite the lack of any evidence that the two ships did any whaling on these voyages, their voyages should be recognized as of a pioneering nature, as the *Lady Penryhn* is credited with a discovery of reporting Tongareva Island. The voyages of these two vessels actually gave the whalemen an opportunity to learn something of the seas which the Honorable East India Company still controlled by an exclusive privilege, as during their return to London they were able to observe numbers of whales in the Indian Ocean. The firm of Alexander and Benjamin Champion, owners of the *Prince of Wales*, thus obtained valuable information about whales in the area near Madagascar, and the astute Curtis firm received similar information from their *Lady Penryhn*. Oddly enough, neither of these vessels noted any whales off the coast of New Holland (Australia) or Tasmania—this area to be the discovery of whaleships which followed.

Another of the London whalers licensed to enter these seas was the *Adventure*, owned by Alexander and Benjamin Champion, which sailed in June 1787 under Captain Elisha Pinkham, the veteran Nantucket master. She returned in November 1789. The Enderby ship *Kent*, under

11. Shipping Arrivals at Sydney, 1788–1825, by J. S. Cumpston (Canberra, Australia, 1963), p. 24.

the Nantucketer, Captain Paul Pease, left London at the same time. In May 1787 there sailed the ship *Intrepid*, owned by the new whaling firm of Hall and Downing, and commanded by Captain John Leard, a retired Royal Navy officer, and the ship *Duke of York*, under Captain John Wolfe, owned by Richard Cadman Etches. All were licensed to round Cape Horn. These ships were to make history.[12]

Britain's ministers, especially Lord Hawkesbury, were impressed by the progress of the British Southern Whale Fishery. It was a triumph of private enterprise strongly supported by government subsidy—the willingness of the merchants to invest their capital and the agreeableness of government to supply the bounties. But a third factor actually made the success possible, and this was the daring and skill of the adventuresome whaling masters and their crews.

Early in the development of the industry both the Board of Trade and the merchants recognized that the service of the whaleships as explorers in little known seas was of tremendous value to Britain. As a maritime nation, the pioneering efforts of the whalers was very quickly appreciated. Naturally, any efforts by a rival power to interfere with her Southern Whale Fishery was interpreted by Britain as a challenge.

One such incident occurred in 1787–1788. Captain John Leard, in the London whaleship *Intrepid*, returned in July 1788 with a cargo of oil and six thousand seal pelts. Captain Leard had found himself in the rather doubtful position of taking seals along a coastline in Patagonia, South America, controlled by Spain. His voyage had been an outstanding success and he was well aware of the potential in voyages for seal skins, but his training as a naval officer made him appreciate that he was operating in Spanish waters. After discussing the matter he was advised to write a memorandum for the attention of Lord Hawkesbury and the Board of Trade. A significant portion of this memorandum reads:

From my having made a voyage to the Coast of Patagonia for the purpose of carrying on the Seal Fishery independent of the Whale Fishery, which voyage is the very first that has been made on this plan, I therefore think it is my duty (as well as by the desire of the Earl of Stanhope) to send your Lordship a copy of my observations on the Seal

12. Liverpool Papers, Addit. Mss. 29,522.

Fishery; and as I mean to continue the carrying on of this Fishery, provided that it is not offensive to the Spaniards. I should therefore be very thankful to be informed what some of the Spaniards have thought of the English carrying on the Fishery on the Southern coast of Patagonia, Tierra del Fuego, and Staten Land, which are not inhabited by the Spaniards, and beg leave to observe that, altho the aforementioned coasts are described by Anson to be barren and of little consequence—yet I am persuaded that if the Fisherys are properly managed . . . that the Seal Fishery would produce a neat import of at least £60,000 per annum, besides the advantage of employing many seamen.[13]

Copies of the memorandum were sent to Prime Minister Pitt, Marquis Carmarthen, and Lord Hood. Captain Leard had further noted he did not wish to "carry on the Fishery, or visit any place which would give offense to the Spaniards—which would also hurt myself and my connexions."

The voyage of the *Intrepid*, while not the first of its kind (as Leard believed), as two other London ships had engaged in taking sea elephant oil and seal skins from South Georgia before the *Intrepid* sailed, was important because of Leard's memorandum. His success led to quick participation by other London merchants who recognized the advantages of a combined sealing and whaling voyage.

The *Intrepid* found both seal and sea elephant numerous; Leard reported that in a coastal area of Patagonia not exceeding nine miles his crew had killed and brought on board the "fatt and skins of 11,800 seals in the course of three months."[14] The value of the seal skins on the China market had become known from Captain Cook's journals.

Insofar as sealing is concerned, however, the ubiquitous Nantucketers had actually preceded the *Intrepid*'s experience. Two years before Captain Leard left London, the ship *Maria* had sailed to the Falkland Islands on a whaling voyage and there taken hundreds of seal pelts. Francis Rotch's ship, *United States*, had gone to the Falklands under Captain Uriah Swain in 1786, and where she spent the winter is still called "States Harbor." When the Nantucket whaleship *Canton* arrived at Falmouth, England in June 1786, she brought four thousand seal skins

13. Captain John N. Leard to the Board of Trade, *Memorandum*, B.T. 6/95.
14. Ibid.

as well as whale oil from the Falklands.[15] The *Intrepid*, however, had ranged the Patagonia coast to procure her cargo and so was probably the British pioneer in this particular section.

As for the question raised by Captain Leard concerning the legality of a British vessel utilizing this coast, the government was not prepared to answer specifically, other than that the coast of a deserted land was not "off limits" to a vessel belonging to his Majesty's kingdom. But it was quite another matter when a ship had to enter a foreign port, and especially in a remote part of the world such as the Patagonian coast. (Such a state of affairs was bound to take place.)

In March 1788 the ship *Sappho*, under Captain Thomas Middleton, sailed from London for the Falkland Islands region on a whaling and sealing voyage. A few months later (October 1788) the *Elizabeth and Margaret*, one of Spencer and Lucas' ships left the Thames on a similar voyage, with Captain James Hopper in command. In April 1789 the two vessels sailing in company went into Port Desire, on the Patagonian coast, for water and supplies. While they were in this primitive port—hardly more than a watering place—crews were sent to Penguin Island, some four leagues away, on a sealing expedition. A Spanish frigate and two consorts subsequently came in and ordered the two British whaleships away from the coast, confiscating seven thousand seal skins.

Upon returning home the two British captains entered their protests, and the owners, supported by Enderby, St. Barbe, Alexander Champion and others, presented a memorial to the Board of Trade in October 1789. Lord Hawkesbury, for the Board, prepared a formal statement for the use by the British Government to the Spanish Court at Madrid.

Both England and Spain had been quarreling over the ownership of the Falkland Islands and the former had steadily refused to acknowledge Spanish sovereignty over the Islands. But the South American coast was quite another matter. Phrasing his representation carefully, Hawkesbury declared that Madrid should not support the Spanish authorities in their action of prohibiting British vessels from fishing in these waters as the British had no intention of establishing settlements in this coast, nor were they planning to engage in illicit trade on "the desert shores of that

15. William Rotch to Samuel Rodman, June 7, 1786. Bullard, *The Rotches*, p. 223.

part of the Continent where as yet no settlements have been made"; that the spermaceti whale was usually caught ten leagues from the coast and the black whale five leagues; and the crews of the vessels should be allowed to land on the shore for refreshment, as they had "no intention to form a settlement," but to catch seals.[16]

The London merchants had stated their case well and the firm support of Hawkesbury was a vigorous stimulant. Not to let any opportunity pass to further their case of "improving and extending the Trade and Navigation of His Majesty's Dominions," they informed the Board of Trade:

We are not competent to say how nigh the Spanish Coast, westward of Cape Horn, it will be necessary to carry on our Spermacetti Whale Fishery to be attended with success, but from all the information we can get, we support the soundings to be almost close to shore. We think if we had liberty to fish within 5 leagues of the coast it would be as near as we shall want. Many Spermaceti Whales have been seen about the Island of Juan Fernandez. We suppose no objection will be made close to any islands, so long as no attempt is made to trade. There is an Island near Juan Fernandez in the South Seas called Massafuero on which are many seals. If uninhabited may we send boats to kills the seals?[17]

Thus, both sides of the South American continent were under consideration by the merchants—and this before Enderby's *Emilia* had returned from her pioneer voyage around Cape Horn.

It was at this time that the great chapter in the history of the whaling industry began—rounding Cape Horn! It was a chapter which, as it enlarged page by page, would contain such a series of adventures as to be unparalleled in its own time and unmatched since. This chapter opened with the sailing from London on August 7, 1788, of the whaleship *Emilia*, owned by Samuel Enderby and Sons.

In a letter, quoted by Dakin's *Whalemen Adventurers*, the Enderbys wrote to George Chalmers (Lord Hawkesbury's confidential secretary) that they had "purchased and fitted out a very fine ship at a great expense to go around Cape Horn; she is now ready to sail; we are the only

16. Minutes of the Board of Trade, B.T. 5/5, pp. 408–424.
17. Letter of Samuel Enderby and Sons, A. and B. Champion, J. Lucas and Co., to the Board of Trade, November 5, 1789. B.T. 6/95, p. 187.

15. Captain Matthew Jones (1775–1818)
Portrait painted in France [Courtesy the Nantucket Historical Association]

owners intending to send a ship on that branch of the Fishery." Continuing, the letter stated that sperm whales in the Pacific were "to the northward of our limits," according to the general idea of "most people acquainted with those seas." The Enderbys then made a request which was entirely reasonable, and couched it in simple and effective prose.

As we appear to be the only adventurers willing to risk their property at such a great distance for the exploring of a Fishery, we humbly request their Lordships (if there is no impropriety in the request) that in case our ship should not meet with spermaceti whales to the Southward of the Line [the equator] in the South Pacific Ocean, that she, or any other vessel employed in the Fishery may have permission to sail as far as 24 degrees of North Latitude, as it is generally allowed that there are spermaceti whales on the Spanish Coast but uncertain whether to the Southward or Northward of the Line.

The concluding sentence epitomized the courage and purpose of that outstanding pioneer London whaling firm:

On the success of our ship depends the Establishment of the Fishery in the South Pacific Ocean, as many owners have declared they shall wait till they hear whether our ship is likely to succeed there. If she is successful a large Branch of the Fishery will be carried on in those seas; if unsuccessful we shall pay for the knowledge.[18]

When the *Emilia*'s lines were cast off from Paul's Wharf on that August morning in 1788, and she dropped swiftly down the Thames with the tide, she attracted no more attention than the average ship. Her commander was Captain James Shields and in her crew of twenty-one men there numbered several Nantucket men, headed by the young First Mate, Archaelus Hammond. The *Emilia*, according to the London Custom House register, was of 270 tons, built in Bristol in 1782, and registered in London in July 1788. She was 102 feet long, with a beam of 26.6 feet, and had the carved bust of a woman as a figurehead.

Although the logbook of the *Emilia* does not appear to have been saved a letter written by Captain James Shields to the owners, Samuel Enderby

18. Samuel Enderby to George Chalmers, Chalmers Collection, Public Library of New South Wales, Sydney, Australia (Mitchell Library).

and Sons, has fortunately survived. It is a letter of dramatic importance to whaling history as it was written from Deal on March 6, 1790, while the ship awaited a favorable tide so that she might ascend the Thames for her berth at St. Paul's Wharf, London, on the last bit of her return home. Captain Shields' letter presented a graphic picture of the historic voyage:

This is to inform you of the ship Emilia*'s arrival after a long, fatiguing voyage. I never had it in my power to give you any account of my voyage till this present time. On Jan. 13, 1789, I made the land, Terra del Fuego, being then in the Lat. 53° 18′ So. I saw nothing worth observing until I opened the Straights of La Maire, where I saw a good many black whales. On Jan. 18th I took my departure from Staten Land, where I found a current setting to the eastward at the rate of five miles an hour. After I got into 57° So. Lat. I found no current at all. I sailed no further south than the Lat. of 59°. Upon my passage I fell in with a great many Islands of Ice.*[19]

Rounding Cape Horn, and meeting the contrary winds and ocean currents, was a typical experience for a sailing ship, and the *Emilia*, on this pioneer passage into the Pacific, was beginning a pattern to be followed by the hundreds of whaleships that were to come after her. While sailing north toward the warmer climes of the Pacific, Captain Shields was continuously dogged by bad weather, noting "hard gales of wind for near 21 days from NW to WSW." He did not seek a harbor along the west coast of South America "without hazarding both the ship and lives, [as] it blowed a gale of Wind the whole time."

The first place into which the *Emilia* sailed for a long-awaited landing was the Island of Massafuero, a few miles west of the more famous island of Juan Fernandez, and in the latitude of Valparaiso, Chili. Here the men found fur seals "lying as thick upon the beach as they could be clear of each other," and many goats. They caught as many fish in two hours time "as lasted the crew 4 days." After an unsuccessful attempt to pull one of the whaleboats around the Island's shores, to look for fresh water, Captain Shields sailed to the eastward, standing "in for the mainland to look for whales."

19. Captain James Shields to Samuel Enderby, Jr., March 6, 1790. Chalmers Collection, Public Library of New South Wales, Sydney, Australia.

On March 1st, 1789, sperm whales were sighted running to the southeast, the ship then being "40 Leagues from the Land in the Lat. of 32° 5′ South." Boats were lowered and the whales were chased but they were soon too far away to pursue further. Two days later, March 3, Captain Shields reported the following historic incident—the taking of the first whales in the Pacific Ocean by a whaleship:

I fell in with a very large school of Sperma Coeti Whales in the Lat. of 31° 20′ South, out of which I killed 5, saved 4; the ground very lively, having blowing, blustery weather. I thought it necessary to go further to the Northward.

Thus did the *Emilia* inaugurate whaling in the greatest ocean of them all. The first man to harpoon and take the first sperm whale in this ocean was the Mate of the London whaleship, Archaelus Hammond of Nantucket. A contemporary of Hammond's on the Island, Frederick C. Sanford, noted this fact on several occasions as did other whalemen who knew of the event from Hammond himself.[20]

Sailing to the north, in latitude 27 degrees south Captain Shields took a large sperm whale, and, as he wrote, they "should have got more but it set in very foggy, the boats obliged to come on board. I found the ground very lively. . . . I sailed to the northward 'til I got good weather." The whaling grounds discovered by the *Emilia* were afterwards known as the "Coast of Peru" and were probably a section of the migratory route of the whales following the ocean current that carried their food.

In May 1789 the *Emilia* was at the Island of Lobos, where vast numbers of fur seals were found, and Shields wrote: "Any man may load a ship of one thousand tons with fur seals [pelts] if they please." Quantities of fish caught at the Island's shores were welcomed by the whalemen. On June 2, they sailed from Lobos, again headed to the north and within sight of the rugged coastline. A few days later a Spanish vessel bound from Guayaquil to Lima was sighted. Captain Shields hove to and, when the stranger followed suit, the Londoner sent a boat to the Spaniard with letters which he requested be forwarded. But the latter refused to take them, stating they would not be so honored at Lima. "He was very

20. Frederick C. Sanford, *Proceedings* of The American Antiquarian Society, Worcester, Mass., October 21, 1871. No. 57, footnote, p. 28ff.

cross," reported Captain Shields, "and would not believe that I came from London into the South Seas!"

Sailing to the north as far as 17 degrees south latitude the *Emilia* continued to take sperm whales. Finding a strong current setting him toward the coast, Captain Shields felt it safer "to leave the land and stand to the south and west 'till I got into the Lat. of 23° South; then I was 300 leagues from the Land." But once again he turned to the north and reached 8 degrees 25 minutes south latitute where he took sperm whales 110 leagues from the coast.

His next paragraph in his letter also becomes a valuable part of the history of whaling:

Aug. 15. Being in the Lat. of 15° So. I took my departure from the Coast of Peru with 147½ tons of Sperm Oil, the scurvy having made its appearance among the crew a month before this time. The coast from 32° to 14° South is a bold one. I ran in with the land in a great many different places as near to the shore as two miles and never got soundings but twice; that was in the Lat. of 14° and 15° South. I never saw so many large Sperma Coeti Whales all the time I have been in the business as I have this voyage. I never saw but two schools of small whales. . . . It is necessary for all Ships which go to the Coast of Peru to be coppered up to the Bends, and ships that will be fast.[21]

Upon leaving the Pacific coast of South America on Sept. 5, 1789, the *Emilia* had six men down with the scurvy. After having been "8 months and 2 days to the westward of Cape Horn," the ship rounded that famous landfall on September 22, and a month later (October 17) sailed into Rio de Janeiro with sixteen men down with the scurvy. Remaining in port for nearly two months, Captain Shields sailed for home on December 10, 1789, arriving in the English Channel the first week of March, 1790. In his letter from Deal he mentioned that the *Emilia* sailed remarkably fast, "and is one of the finest ships in a hard gale of wind that I ever had my feet on board of . . . and is one of the compleatest ships for this Whale Fishery that was ever built."

21. Captain James Shields to Samuel Enderby, Jr., March 6, 1790. Chalmers Collection, Public Library of New South Wales, Sydney, Australia.

The London firms promptly followed up the advantage provided by the voyage of the *Emilia*. When Enderby's ship again sailed from the Thames for Cape Horn she was followed by the *Atlantic*, Captain John Bassett; *Kitty*, Captain Matthew Swain; and *Greenwich*, Captain John Lock—all Enderby ships; also by the London ships *Prince William Henry*, Capt. Benjamin Swift; *Fanny*, Captain William Barney; *Nereus*, Captain Joshua Coffin; *Kingston*, Captain Henry Delano; *Liberty*, Captain David Baxter; *New Hope*, Captain Tristram Bunker; and *Yorke*, Captain Thomas Daggett—the latter two sailing from Falmouth, and owned by the American Thomas Yorke, who had come to England from Philadelphia. All these whaleships but one was under the command of a Nantucket shipmaster.

Captain James Shields, in his second voyage in the pioneer Enderby ship *Emilia*, made another successful venture into the Pacific. John Nicol, a young Scot who had signed on as a cooper, wrote an interesting account of the voyage. For some reason he thought the ship bound for Botany Bay, as he sewed money into his clothes, intending to buy his wife's freedom at the penal colony. The *Emilia* sailed in July 1790 and Nichol wrote of the ship stopping at St. Jago in the Cape Verdes, then crossing the Southern Atlantic for the Falklands, where they again landed for provisions and water. Before doubling Cape Horn, Captain Shields also stopped at Staten Land, off Tierra del Fuego's southeastern shores, and then rounded the Horn into the South Pacific.

Nicol states the *Emilia* cruised for whales between 18 degrees south latitude and the equator. At Lobos Islands, off the Peruvian Coast, boat crews landed and took thirty thousand seals. During one landing (Nicol does not say where) an enraged sea lion caught the second mate, Parker, by the arm and nearly drowned him before he was rescued. The whaleship cruised as far north as the port of Paita, Peru, where she put in for supplies, becoming the first whaleship to touch at that South American port along its western coast.[22]

The occasions when these pioneer whaleships were cruising off the west coast of South America and sighted one another must have provided some interesting "gams." On her way through the South Atlantic to-

22. "The Adventures of John Nicol, Mariner," *The Sea, The Ship and The Sailor* (Marine Research Society, Salem, Mass., 1925), pp. 141–145.

ward Cape Horn and the Pacific, the *Rebecca*, of Bedford (which was the first American whaler to return from a Pacific voyage), under Captain Joseph Kersey, "spoke" several whaleships which were bound to their home ports after successful cruisings in the South Pacific. The *Hope*, of Dunkirk, with 1,500 barrels, was sighted first, and Captain Prince Coleman informed them of his experience. The *Atlantic*, of London, was spoken, and Captain John Bassett informed them "that between the latitudes 30° and 8° south there is whales enough." They were also told that Captain John Lock, in the *Greenwich*, had 1,600 barrels and Captain Henry Delano, in the *Kingston*, had 400 barrels. Two of the three British whalers were under American masters, Lock and Delano being Nantucketers, as was Coleman of the Dunkirk whaler.[23]

The *Rebecca* made a safe passage around Cape Horn, to enter the Pacific early in February. Here she spoke three more London whalers, as well as the *Lydia*, Captain Benjamin Clark, out of Dunkirk, and learned that the whaleship *Beaver*, of Nantucket, had been the first American whaler to enter the Pacific, having sailed in August 1791, being followed by the *Washington*, Captain George Bunker, the *Hector*, Captain Thomas Brock, and the *Rebecca* (of Nantucket), Captain Seth Folger. Captain Bunker, in the *Washington*, during his voyage, put in at Valparaiso, and became the first American shipmaster to hoist the flag of the United States in a Spanish port in the Pacific. The *Beaver* returned home March 25, 1793, but was preceded to the American coast by the *Rebecca*, which arrived at Bedford a month earlier, February 21, 1793.

A journal kept by young Elijah Durfey on board the *Rebecca* has been fortunately preserved. During that ships' 1791–1793 voyage Durfey recorded that there were forty whaleships in the Pacific—twenty-two from England, eight from Dunkirk, seven out of Nantucket, with one each listed from Bedford, Hudson and Boston. However, thirty out of the forty whaleships listed were under the command of Nantucket men.[24]

Of the three nations involved, though, Britain had the largest number, sailing out of London, with an equal number of ships in the South Atlantic and at Woolwich (Walfish) Bay and Delegoa Bay, both sides of

23. Elijah Durfey, Journal of the *Rebecca*, Marine Historical Association, G. W Blunt White Library, Mystic, Connecticut.
24. Ibid.

the African coast. London still maintained her lead in the market place of products from the Southern Whale Fishery.

Of equal importance with their whaling success, the London whaling merchants had also gained something that could come only by the way of experience—the need of training whalemen to replace the veterans. With confidence and self-assurance these merchants continued seeking the staunch support of Lord Hawkesbury and his Board of Trade. Early in 1791, the following petition was presented to the Board by Samuel Enderby and Sons, Alexander and Benjamin Champion, Curtis and Co., and other whaling merchants:

We likewise very earnestly request your Lordships' attention to have the apprentices in the Southern Whale Fishery protected, they being the instruments by which the Fishery must become an English instead of an American Fishery. Many boys of very creditable parents are indentured and brought up for officers in the Fishery, and if [they are] suffered to be impressed we never shall get any more lads of creditable parents to be indentured in the Fishery.[25]

25. Letter of S. Enderby and Sons, A. and B. Champion, Curtis and Co., and others to Board of Trade, March 31, 1791. B.T. 6/95, p. 257.

11. Whale Oil Diplomacy and Adventure

IN THE spring of 1788 the commercial treaty between France and England, recently enacted, made possible the sale of whale oil from London at a price that, despite the import duty, enabled the British to undersell the Dunkirk product and overstock the market. Francis Rotch, at Dunkirk, wrote to Minister Lambert in Paris, declaring that the future of whaling from France was endangered. He applied for a loan of two million livres, and requested the privileges extended the Dunkirk fleet be granted to all other ports in France. Lambert refused to consider any loan, and stated that the favors asked had been already conveyed.[1]

But the protest from Dunkirk, where the Rotch fleet had increased to ten whaleships, was strongly supported by the French Minister Luzerne, who was in a position to be heard. An "Arret," or government act, was proposed, whereby all European oils would be excluded from the French market thus providing a monopoly for the French (Dunkirk) and American oils. The "Arret" was so drawn, but just before the Ministry voted it into law, the word "European" was struck out, so that the phrase "all oils" remained. This, of course, meant that American oil would be excluded with others from France. This "Arret" became effective in September 1788.

In Paris, Thomas Jefferson, now well into his term as Ambassador, became alarmed and wrote to the French Minister M. deMontmorin in protest against the "Arret," and arranged a conference with the Ministry. It was at this time he drew up his remarkable review of the whaling industry, which he called his *Observations*. In writing to John Jay on the subject, he stated that he "suspected" the striking out of the words in the

1. M. Lambert to Francis Rotch, June 12, 1788. Archives Affaires Extraordinaires, Correspondences, Politique, E. U. XXXIII. Transcript in Library of Congress.

16. A View in Port Jackson, New South Wales (1793)
[Courtesy the Nantucket Historical Association]

prohibiting phrase of the "Arret" was the work of M. de Count of Luzerne, minister of the Marine, but could not "affirm it positively." His letter to Jay contained the following comment:

I found them [the French ministers] possessed by the partial information of the Dunkirk fishermen; and therefore found it necessary to give them a view of the whole subject in writing, which I did [his Observations*] . . . and inclose you a printed copy . . . I was led to them by other objects. The most important was to disgust M. Necker, as an economist, against their [the Dunkirk] new fishery by letting him foresee its expense. . . . At a conference in the presence of M. Lambert, on the 16th [Nov. 1788] where I was ably abetted by the Marquis de Lafayette, as I have been through the whole business, it was agreed to except us from the prohibition.*[2]

But there was another angle to the Paris Conferences. Benjamin Rotch had been in Dunkirk for several weeks, following his return from

2. Thomas Jefferson to John Jay, Paris, November 19, 1788. *Papers of Thomas Jefferson,* edited by Julian Boyd (Princeton University Press, Princeton, N. J., 1958), Vol. 14, pp. 213–214.

Nantucket with his family, and had been invited to attend the conferences as the representative of the Dunkirk fishery. William Rotch made a report of his son's experience in a letter written to Nantucket:

Benjamin had a meeting at Paris with the Minister of State, the Marquis de Lafayette, and Jefferson, the American Ambassador. Part of the conversation was upon the petition from Dunkirk. . . . Benjamin insisted at that time on American oil being admitted [duty-free], but the Minister [Lucerne] was determined it should not, and, dashed out the words "American oil" with his own hand, upon which the general prohibition took place.[3]

In his letter to Jay, written soon after the conferences, Jefferson had declared his suspicion that Luzerne had changed the wording of the "Arret." But William Rotch, in a letter home, reported that Jefferson had accused Benjamin as being the "author of the prohibition of American oil," which the latter promptly denied. When the Ambassador continued to look doubtful, Benjamin remarked: "The original petition from Dunkirk is here. It contained a clause requesting American oil be brought in without duty, but this was dashed out by one of the ministers now present."

In William Rotch's account of the affair he described Jefferson as then requesting the identity of the Minister referred to, but Benjamin then replying that, as the gentleman was present, he could affirm the truth of the matter. Jefferson then acknowledged he believed Benjamin spoke the truth. It was at this time, the elder Rotch stated, that Jefferson drew up a petition of his own, his *Observations*, "to endeavor to influence the Ministry." Rotch observed that this action "was warrantable as it was his duty to his country in the best way he could."

In commenting further on the Jefferson petition, which the Ambassador had sent also to Washington, Jay, and Adams requesting the document be kept secret, William Rotch wrote:

If the Memorial contained nothing but the truth, which I will not say it did, I justify him [Jefferson]. But if secrecy was intended, this or some other cause disgusted the Minister and, instead of any privacy, he sent it,

3. William Rotch to Samuel Rodman, October 12, 1790. Bullard, *The Rotches*, p. 231.

I believe, with the letter also, direct to De Bauque and Benjamin's lodgings for their perusal and answer, which they did and so effectually refuted all his arguments that the petition suffered the fate of many others.[4]

The memorial or petition to which the elder Rotch referred was a summary of Jefferson's research on the whaling industry and one of the Virginian's remarkable studies. Titled *Observations on the Whale Fishery*, it was prepared for Necker, the French Minister, and was printed by the Paris firm of Jacques Clousier. It resulted from Jefferson's conversations and correspondence with many people. Only a few copies were printed.

The French Ministry had received rumors that the Rotch fleet at Dunkirk was bringing in oil actually taken by ships out of the Nova Scotian (Nantucket) whaling community at Dartmouth (or Halifax, as the rumors conveyed), and transferred. This was entirely probable, and the oil was entered as French oil at Dunkirk. Jefferson noted: "They [the French] will require vigorous assurance that the oils coming under our [American] name are really of our fishery. They fear we shall cover the introduction of the English oil from Halifax. . . ."[5]

Jefferson felt that, with the readmission of American oil, "this branch of commerce will be on a better footing than ever, as enjoying jointly with the French oil a monopoly of the markets." He believed the continuance of this monopoly depended on the growth of the French whaling at Dunkirk. He wrote to Jay:

Whenever they [the French] become able to supply their own wants, it is very possible they may refuse to take our oils. But I do not believe it possible for them to raise their Fishery to that, unless they can continue to draw our fishermen from us. Their 17 ships this year had 150 of our sailors on board. I do not know how many the English have got into their service.[6]

Writing to John Adams in London, on December 5, 1788, Jefferson gave a resume of the episode, which was resolved to his satisfaction when

4. Ibid. Also p. 232.
5. Jefferson to Jay, in Boyd, *Papers*, p. 214.
6. Ibid.

"an immediate order was given for the present admission of our oils," and "the draught of an Arret was communicated to me which re-established that of December 29 [1787]." He went on to report that the Consuls in the ports were to certify as to the admission of American oil, and so prevent the importation of Nova Scotian oil.

As a part of the check on the introduction of fraudulent importations, the French inspectors were ordered to examine the logbooks of the whaleships, and carefully to question all French sailors who were members of the crews. Jefferson felt it urgent that this be done so that any trans-shipments of English oil taken by Nantucketers at Nova Scotia would not jeopardize the American oil imports. After a time, when no evidence of illegal cargoes was discovered, the French ministers were convinced the rumors could not be supported. However, it is to be noted that Chardon, the supervisor at Dunkirk, was himself an investor in the Rotch fleet, and was directly under the control of Luzerne, Minister of Marine.

It is clearly apparent that Jefferson's idealism and William Rotch's practical application of liberty of action would not merge. As one American historian has written:

> *The American whalefishery was conducted by a clannish, inter-related, resourceful people, who were singleminded in their pursuit of the whale. The Rotches, the Starbucks, the Coffins, and others, who engaged in the whalefishery, were a sort of Atlantic cartel, devoted to commerce and rising above the political storms of the day.*[7]

There was every reason why the Nantucketers at Dunkirk could not accept Jefferson's reasoning. Why should they be agreeable to competition from any source? Had they not, by their own initiative and daring, without American governmental help, established an industry at Dunkirk which was to the advantage of France as well as themselves? Rotch had found a "back door" for the entry of American oil into Europe without political maneuverings.

Jefferson believed he had won a diplomatic victory by securing the exemption of American oil from the prohibition of the "Arret" of September 1788. As his eminent biographer, Julian Boyd, pointed out:

7. Boyd, *Papers*, p. 221.

This was correct but far from comprehensive, and the immediate context that Jefferson had to deal with in achieving this diplomatic victory was not only less congenial to him but much more formidable than the larger matters of policy. . . . In the conjectures and analyses of the immediate context [of his Observations] *Jefferson was pursuing the right trail, but it was a trail that had been confused and some times deliberately concealed.*[8]

If the controversy had no other result so far as the Ambassador was concerned, a major accomplishment was the production of Jefferson's *Observations on the Whale Fishery*, which, as noted, he had prepared for the perusal of the French Ministry. This pamphlet, printed in Paris, had been written after research in a combination of correspondence and conversation with a number of merchants in the whaling industry, including John Brown Cutting, Nathaniel Barrett, Francis Rotch, Daniel Parker, Thomas Boylston, Tourtille Sangrain, and several others. As Boyd notes: "Most remarkable of all, the information Jefferson sought . . . included some data of an authoritative nature from some of the most experienced and successful whalemen in the world—the officially favored group of Americans known as 'les Nantuckois.' "[9]

Francis Rotch, whose experience in whaling placed him in a unique position as regards knowledge of the present condition of the industry, was interviewed by none other than the Marquis de Lafayette. The interview was after a style of "inquiries" with Rotch furnishing the replies. Francis, then thirty-eight years old, after a life of varying fortunes was now deeply involved with the Dunkirk whale fishery. His opening statement reveals his confidence and caution:

I wish not to be considered as the entire representative of the fishery at Dunkirk, but as an individual much interested therein, whose influence may be supposed to extend to his family connections. In order to ascertain this degree of actual and probable representation, I will state the different interests in the fishery as it now exists both out of, and in, the port of Dunkirk, under the convention; super-adding what yet remains in actual service of M. de Calonne's Company ships.

8. Ibid., p. 220.
9. Ibid.

The listing of the Dunkirk fleet is important, and Francis Rotch reported the following:

5 ships of my own to the South Seas	*1485 tons*
2 ships of my brother's ,, ,,	*500 tons*
2 ships of my brother's to Greenland	*700 tons*
3 others ,, ,, to South Seas	*660 tons*
12 out of port (Dunkirk)	*3445 tons*
2 others in port getting ready	*255 tons*
	3700 tons
3 of the Company's ships (Calonne's)	*800 tons*
	4500 tons[10]

Francis reported that these seventeen ships had 150 Nantucket seamen—the agreement with the Ministry providing that the balance of the crews be completed with French sailors; that three Nantucket families were already residing at Dunkirk and that two former London whaleship masters from Nantucket had removed to France with their families; that three more Nantucket families were expected soon; and the total by the year's end (counting his own) would be nine Nantucket families and thirty-three individuals.

Jefferson had reported to John Adams in London (December 5, 1788): "If they [Dunkirk] succeed in drawing over as many of our Nantucket men as would supply their demands for oil, we might then fear an exclusion [of American oil]. But the present Arret as it shall be passed will, I hope, place us in safety."[11] A few weeks later he admitted that Necker, the French Minister of Finance, had written to him that "whenever the National [French] fishery shall be able to supply their demand for whale oil we must expect a repeal of the Arret." This, he believed, was unlikely as "their navigation being the most expensive in Europe." He continued: "I am informed there will be fewer French adventurers the next year . . . so if there be an apparent increase in the whale fishery it will be by drawing over more of our fishermen."[12] Writing to Jay at this time the

10. Francis Rotch to Marquis de Lafayette, in Boyd, *Papers*, p. 234.
11. Jefferson to Adams, December 5, 1788. Boyd, *Papers*, p. 335.

Ambassador stated "For the rivalship of the English is the only one we have to fear."

Francis Rotch told Lafayette that if a further emigration from Nantucket should appear inevitable there were many reasons for the Islanders to prefer England to France. In October 1788 Montmorin had inquired of Necker if England was offering further inducements for Nantucket whalemen to migrate there, and Necker assured him the King Louis was determined to keep in France, with every means possible, "des individus aussi utiles, et a en propager meme l'establishment."[13] Also, that the King had charged the Minister of the Marine, Luzerne to use every persuasion to induce the whalemen to remain in France. Despite the financial crisis in France the 50 livres per ton bounty for the Dunkirk ships was retained—another indication of the value placed thereon by the French government.

When Lafayette asked Francis Rotch if it would not be wiser to ascertain whether the French fishery could supply French needs for whale oil before a prohibition on all other oils was imposed, the experienced Nantucketer replied:

A free trade with America as far as it relates to the whale fishery holds out, I fear, a prospect more than experience hereafter will be found to justify. The use of oil is extremely diffuse. The greatest purpose is to give light. It enters more or less into every manufacture, but it is hardly ratable per centum in the capital of any. Leather, and perhaps soap are exceptions. I have known it extremely dear in England; but none of their manufacturers or the purposes of commerce have been obstructed on that account. It offers no staple for manufacture except in that of spermaceti candles, which is a branch of trade already too well established in America to leave room for a manufacturer in France to hope for a supply of head matter from that quarter. It appears to me a greater opening of benefit to the merchant than to the manufacturer, who may have more reliance upon a national than a foreign fishery for supply of a raw material to make spermaceti candles.[14]

12. Boyd, *Papers*, p. 225.
13. Francis Rotch to Marquis de Lafayette, Boyd, *Papers*, p. 237.
14. Jefferson to George Washington, December 4, 1787. Boyd, *Papers*, p. 331–332.

In reply to Lafayette's question as to whether a combined American and French whaling fleet would be an advantage to the Kingdom, Francis Rotch stated that if the Americans could be permitted to build, outfit, and navigate their whaleships directly from their home ports, "the advantages which they now enjoy under the convention, by touching at Dunkirk to change their papers and sail from thence, and the navigators left at perfect liberty, the two fisheries would interweave immediately."

Still strongly believing that the "Arret" of November 1788 which permitted American oil to be imported duty-free into France, and excluded all other foreign oils, would build up the American fishery, Ambassador Jefferson wrote to George Washington: "This will put 100 English whaleships immediately out of employ."[15] Ambassador Jefferson was wrong. The loss of the French market had none of this effect on the British fleet. At this time the London fleet had increased to sixty vessels, the bounties provided by Jenkinson and the Board of Trade were having a salutary effect, and the masters of the British whaleships were on the threshold of opening an era of the greatest expansion in the Southern Whale Fishery.

With his efforts to curtail the Dunkirk fishery Jefferson was sincere in his belief that it would encourage the recovery of the American fishery. But the heart of the latter fishery was at Nantucket, and her efforts had been curtailed by the loss of her markets. At this time, as Francis Rotch explained to Lafayette, from the port of Dunkirk there was a greater number of tons of ships in the whale fishery being sent than there was from the entire American fleet, and the Rotch management was the controlling factor.

Jefferson's respect for the whaling prowess of "les Nantuckois" was equalled by his determination to prevent the Dunkirk whaling venture from gaining more advantages over the hoped-for importation of American oil. On the other hand, William Rotch, Sr., very quietly set about his plans to increase the Dunkirk enterprise, commenting in a letter to Samuel Rodman, "I wish our great men may consider they have not sufficient power to overturn us in this branch, but it is probable the war may do it for them if it takes place."[16]

15. Rotch to Rodman, Bullard, *The Rotches*, p. 232.
16. Ibid.

Sukey of Nantucket. G. W. Gardner's Master. off the North head of Albemarl

17. Whaleship *Sukey* of Nantucket at the Galapagos Islands, 1808
Painted in reverse on glass [Author's Collection]

The Quaker merchant was a prophet.

The years 1788–1790 found the Rotch firm at Dunkirk still experiencing continued success. The fleet was increased by the addition of the *Ville de Paris*, with the veteran Captain Shubael Gardner in command; the *Penelope*, Captain John Worth; *Falkland*, Captain Obed Fitch; *Edward*, Captain Micajah Gardner; *Favorite*, Captain David Folger; *Ann*, Captain Prince Coleman; and the *Hope*, under Captain Obed Paddock, formerly in command of the *Falkland*.

But the political news from France was disquieting. The deep rumblings of the growing Revolution sent tremors throughout all of Europe. Following King Louis' authorization early in 1789 for the convening of the Estates General, that meeting of nobility, clergy, and bourgeoisie (which had not met in over 150 years), and the formation of its successor, the National Assembly, the affairs of France had worsened. The capture of the Bastille in July 1789 was but the prelude to the terrible years still to come.

William Rotch in Nantucket was aware of the dangers in the offing and decided to sail for France. On July 29, 1790, he embarked with his family on board the new ship *Maria and Eliza*, Captain Hayden. Accompanying him were his wife Elizabeth Barney Rotch, his two daughters, Lydia and Mary, and his son Benjamin's wife, Elizabeth Barker Rotch, with the young son Francis. After a passage lasting one month and one week the ship reached Dunkirk (September 6, 1790).

It was a voyage Elizabeth Barker Rotch, Benjamin's wife, was not to forget. From the moment she left Nantucket she suffered so severely from seasickness that she never could face another ocean voyage. Because of this, her parents, the Josiah Barkers, were bitter in condemning the Rotches for taking their daughter from her Island home. This breach between the two families was to remain during their respective lifetimes, notwithstanding William's and Benjamin's efforts at reconciliation. The worst that the elder Rotch could bring himself to write concerning the Barkers was in a letter to his son-in-law, Samuel Rodman: "Josiah, Jun's wife, is an amiable woman, I think worthy of a more equal partner."[17]

The return of William Rotch, Sr., to Dunkirk during the first week in September 1790 was most timely. Archaelus Hammond, former mate on the *Emilia*, of London, had already arrived in France bringing details of that British whaler's pioneer voyage around Cape Horn. Hammond's arrival at Dunkirk marked the beginning of a new era in French whaling, as guided by the Rotches, who planned to fit out two ships for the Pacific.

In London, the *Emilia* was readied for her second voyage around Cape Horn. With justifiable pride, the Enderbys reported to Lord Hawkesbury the prospects of the new whaling grounds in the Pacific, where large numbers of sperm could be found. The Enderby firm promptly fitted out six ships for Cape Horn voyages, while their associates followed suit. Two months before either of the Dunkirk vessels sailed, and ten months after the first Nantucket ship set out for the Pacific by way of Cape Horn, a dozen whaleships had already departed from London for that distant sea.

One of the strangest developments in whaling fortunes took place when the *Emilia* was homeward bound from her famous first voyage around Cape Horn. This incident was reported to William Rotch in Dunkirk by Archaelus Hammond, the mate of that ship. Somewhere in the South Atlantic, the *Emilia* sighted the whaler *Hope*, outward bound from Dunkirk, under Captain Thaddeus Swain, then forty-six days from

17. Rotch to Rodman, October 19, 1790. Rotch Collection.

France. Captain Shields was anxious to heave to for a "gam," knowing his news from round Cape Horn was certain to startle the skipper and men of the other vessel. But the *Hope*, unsure of the signals, did not at first heave to, it being some time before she finally "hauled her wind," then only long enough to hail the English whaler. The answering hail: "The ship *Emilia*, of London, from the Pacific Ocean by way of Cape Horn with 150 tons of sperm," came across the water. Captain Swain quickly hauled his yards bringing the *Hope* on a tack to draw nearer the *Emilia*, anxious to know more. But Captain Shields, angered by this tardy acknowledgment of his first signals, promptly put his ship before the wind and sailed away—despite his anxiety to learn from the *Hope* if France and England were at war.

In a letter home, William Rotch added a postscript to this affair: "Nor could the importunity of Mate Hammond influence Shields' mulish decision not to stop, even to get the intelligence which might have—if a war—been of great advantage to him. . . . I hope he [Captain Swain] clearly understood that he [Captain Shields] had been round Cape Horn, so as to have taken the same route." Unfortunately, the *Hope* did not take advantage of the news.

With the return of his newest ships, the *Falkland* and the *Harmony*, William Rotch immediately fitted them out for the Pacific, with the former, under Captain Obed Paddock, sailing on November 12, 1790, and the latter, with Captain David Starbuck in command, leaving Dunkirk several days later. A letter to Nantucket was promptly dispatched, giving details of the sperm whales 'round Cape Horn.

The first to return to Dunkirk was the *Falkland*, Captain Obed Paddock, arriving February 7, 1792, and reporting the extent of the whaling fleet in the Pacific, together with other particulars. Ten days later the *Harmony* reached Dunkirk and Captain David Starbuck brought other facts. Rotch, in the interim, had fitted out and dispatched nine other ships and they entered the Pacific within a few months of one another. To fit out these ships in such rapid succession, and get them safely into that distant sea in such a short space of time, was not only a tribute to the enterprise of the Rotch firm but an indication of the skill and ability of their shipmasters. In outfitting the fleet, Rotch had the great advantage of the background of Nantucket, so that experienced men—riggers, outfitters, ship chandlers, warehousemen—were available.

During this period, the Nantucket-Dunkirk fleet was operating so consistently that the Island's recovery was the best proof of their success. In true American style, vessels were owned by a company of several men, each purchasing a share or shares, with the largest shareholder usually acting as the agent. The Rotch firm thus controlled several vessels in this combination of home port and Dunkirk fleet, and no doubt had "double registers," that is, French and American, as follows: *Maria*, Captain William Mooers; *Canton*, Captain Coffin Whippey; *Penelope*, Captain John Worth; *Hope*, Captain Thaddeus Coffin; *Falkland*, Captain Obed Paddock; *Diana*, Captain Uriah Swain; *Bedford*, Captain Laban Coffin; and *Mary*, Captain George Whippey.

Also shuttling between Nantucket and French ports were vessels operated by the firm of Shubael Coffin and Co., such as the *America*, Captain Tristram Gardner; *Favorite*, Captain Valentine Swain, and *Swan*, Captain Silas Jones. Out of Bedford, trading with the Rotches at Dunkirk, were the *Hudson*, Captain Micajah Gardner; *Bedford*, Captain Mooers, and *Dauphin* (still owned in part by the De Bauques), with Captain Paul Coffin in command.

Thomas Boylston, of Boston, was shipping oil to Bordeaux in a vessel commanded by the veteran Captain George Folger. An interesting development was revealed by an application to sell American whale oil in France sent from Boston by Patrick Jaffrey, the young clerk who had married the Widow Hayley. Francis Rotch, her confidant for several years, who had sailed with her to Boston from London, apparently retained no bitterness as in one of his letters to America he requested that a Westphalia ham he had shipped to Boston be forwarded to Mrs. Jaffray.

The Nantucket whaling firms who were trading directly with France during this time, as well as those selling oil through the Rotch firm in Dunkirk, welcomed the open market for American oils. Among these were Richard Mitchell, Shubael Coffin, Paul Gardner, and Micajah Coffin and Sons. In June 1789 the latter firm shipped forty casks of right whale oil to Le Coutelux in Rouen. That Nantucket firms were selling oil to France which had been obtained from Dartmouth, Nova Scotia whalers may be found in a letter which Captain Micajah Coffin wrote to his brother Thomas, at Dartmouth, reporting that he had obtained for such sale a bill of exchange from Brook, Watson and Co., London for £74.

In May 1790 the Micajah Coffin firm bought the brig *Lydia* in Boston, for the purpose of shipping oil to Le Havre. Zenas Coffin, Micajah's son, was in Dunkirk, having arrived there on the Rotch ship *Greyhound* from a whaling voyage, in which he was serving as one of the mates. Captain Micajah Coffin took the *Lydia* to Le Havre, with whale oil consigned to Homberg and Homberg Freres. Returning from France on October 29, 1790, the *Lydia* reached Nantucket on December 13. Early in 1791 the brig sailed on her first whaling voyage from the Island under Captain Zenas Coffin, fresh from his experience on the Rotch whaler out of Dunkirk.

The commercial stability of the whaling world still revolved around London, both as the business center of the maritime world and as the marine insurance headquarters. Instead of receiving a complete payment for his oil at Le Havre, Micajah Coffin took a bill of exchange from Homberg and Homberg in the amount of £140, at thirty days' sight, which was promptly sent to Peter Thuellesen Company in London, to be placed in their account. In turn, the Coffin Company drew a bill of exchange for £139 on the London firm in favor of Captain Nathaniel Cary, "which bill we wish you to honor at sight."

William Rotch, Jr., now of New Bedford, was also shipping oil and bone to Le Havre on the brigantine *Hawk*, with a consignment from Seth Russell, a fellow townsman, as part of the cargo. Sperm oil was bringing from £28 to £33 per ton in France. An alien duty of six percent in livres was placed on all oil imports. In the summer of 1791, Captain Joseph Chase sailed from Nantucket with a cargo of oil for Le Havre, and a letter from Micajah Coffin to Homberg and Homberg recommended Captain Chase to them "as a person of considerable travel in the sea Captain occupation and otherwise in good repute with us." The Chase family was not unfamiliar with French ports. During the last years of the American Revolution, Captain Chase's brother, Lieut. Reuben Chase, had commanded a privateer out of L'Orient, having previously served with John Paul Jones on the *Bon Homme Richard*.

The London merchants were enjoying the success of their fleet, and deservedly so. Samuel Enderby, by his pioneering efforts, had advanced the cause significantly, and he continued to press the case for the British whalers by keeping Lord Hawkesbury and the Board of Trade fully informed. In one letter, (January 17, 1789) Samuel Enderby, Jr., stated:

I think it must give his Lordship Pleasure to see the Fishery he has patronized succeed so well under his direction. His Lordship first took the Fishery under his protection in 1785, the year prior to which sixteen sail of vessels had been employed in the Southern Whale Fishery; the value of the oil, etc., they brought have amounted to between 27 and 28,000 pounds, for which Govt. paid 18%, although the premiums were but £1500 per annum. The number of vessels which returned from that Fishery last year were 45 sail; the value of the oil amounted to £90,599, for which Govt. have and will pay £6,300, which is not 7% of the whole amount of the cargoes of oil, etc.[18]

The letter also advanced the basic premise of the London whalers—a policy suggested for the future—and one that their persistence in advocating was to prove successful by Hawkesbury's invaluable aid:

In my opinion, nothing is wanting to make this Fishery complete but an unlimited right of fishing in all seas. The British Adventurers would soon explore the most distant parts and the settlements of New Holland would be often visited as there are many whales in those seas.

While the whaleships were beginning their voyages around Cape Horn to cruise the South Pacific, there were voyagings of considerable importance taking place in the North Pacific which were to have an effect on the whaling industry.

It will be remembered that one of Captain Cook's objectives in his voyaging to the Northwest Coast of North America was to learn if a water-passage led through the continent from Atlantic to Pacific—thus becoming the legendary Northwest Passage. Approaching the northwest coastline, Cook had explored from Russian Alaska to the great island later to be named for Vancouver (one of his officers who later charted its length), and among Cook's own chartings was the famous harbor and area called Nootka Sound. Here, the crew had obtained the fine sea otter pelts which some months later were sold at fabulous prices at Kamchatka and Canton. Returning home with this information, but alas, without their remarkable leader, the expedition's experience led to plans by

18. Samuel Enderby to George Chalmers, January 17, 1789. *Board of Trade Records*, B.T. 6/95. p. 161.

other countries to follow up the opportunity to establish a fur trade route from the northwest coast of America to China.

One of Cook's men, George Dixon, and some of his companions had been temporarily engaged to lead an expedition out of Marseilles for a privately organized fur company, but this did not materialize. However, when the French in August 1785 sent out the expedition under La Perouse for the purpose of exploring the North Pacific and especially to investigate the northwest coast, the British were quick to respond to this challenge. A company was formed in London, called the King George Sound Company, with the fur trade as the motivating purpose.

This situation meant the establishment of trade routes directly from the northwest coast of America to China, and it was necessary to obtain the approval of the East India Company who had the rights to the China trade. The group appealed to Lord Sydney for aid, seeking a charter which would give them the exclusive privilege of developing this trade.

No newly organized company had a more elaborate set of plans—seeking the rights to the northwest coast fur trade; permission to trade with the islands of Japan as well as in Canton and the Siberian coast; allowance to bring back to England cargoes of whale oil and bone. Plans also included making use of the Sandwich Islands, as the Hawaiian Islands were called.

Captain Dixon was joined by Captain Portlock, who had also been to the area with the Cook expedition. The two men were selected as chief advisors for the London syndicate, headed by several members of the Etches family—Richard Cadman Etches and John Etches, of London, William Etches, of Northampton.

After much dickering the East India Company granted the syndicate rights to trade in furs but restricted other privileges in Canton and the eastern seas, fearing that, as with the whalers, this might lead to contraband shipments of goods.

Having gained the sanction of government and permits from the Honorable Company, the next step was to procure licenses from the South Sea Company. In the petition, it was stated:

The Petitioners therefore entreat the Company to grant them a license for Five Years to Trade to the said Coasts of America and likewise to grant them liberty to establish Factories on any parts of the said Coast or Islands within the Limits of the Companys' Charter, the Petitioners being willing to furnish the Company with a copy of the Journals of the whole proceedings.[19]

The petition was referred to a special committee, and after receiving government approval, it was granted.

Soon afterwards the Memorial which had been presented by the Southern Whale Fishery merchants was approved, and permission granted for them to proceed "through the Straights of Magellan, or around Cape Horn for the purpose of carrying on their Fishery."

It was Lord Hawkesbury's wish that the promoters of the Southern Whale Fishery be the ones to actually lead the way for the expansion of British trade in the distant seas, and that the North Pacific become another area for them to become pioneers, as had been the case in the South Pacific. However, it was the fur traders and sandalwood dealers who were to precede the whalers on the Northwest Coast of America.

When Captains Dixon and Portlock arrived in Nootka Sound they found they had been anticipated by an expedition from India—one led by Englishmen who had learned of the fur trade also and had organized companies in India. A month later another pair of vessels came in from Calcutta. Thus, three separate expeditions had sailed to Vancouver Island, all inspired by the account of the Cook voyages and determined to take full advantage of being first on the scene. Last to arrive at Nootka Sound were three American vessels—the brig *Eleanor*, ship *Columbia*, and sloop *Washington*, the vanguard of a score of ships later to be known as those of the "Boston men."

The territory in which Vancouver Island was located was but a part of the long extent of shoreline claimed by Spain, a coastal stretch extending as far north as 60 degrees of latitude. Alarmed by the invasion of her territory, the Spanish governor in Mexico sent two armed vessels to Nootka under Don Estaban José Martinez. The Spanish vessels sailed into Nootka Sound in May 1789 and after a preliminary period of sizing up the situation, Martinez notified the commanders of the several vessels that this was Spanish territory and they were poachers. One British craft flying Portuguese colors, and another also using a Portuguese flag, were

19. Minutes of the Board of Control, East India Co., August 11, 1785. Vol. 190, pp. 247–248.

allowed to depart. But two of these English vessels later returned and were captured and brought back to San Blas. The American craft, however, were granted the freedom of the port.

When the news of the capture of the two British traders reached London in January 1790 a wave of indignation swept the city. Some members of Pitt's cabinet advised preparation for war. Cooler heads prevailed. Pitt was cautious, and King George III did not want to resort to armed conflict. While considering the Spanish claims to Nootka Sound ridiculous, the British ministers were aware that the fur-traders had attempted to build a permanent settlement at Nootka and that this was a violation of Spain's announced ownership of the area. It was the abrupt seizure of the British vessels that aroused the ire of the ministry, and when the adventurer Captain John Meares arrived in London, bringing lurid details of the captures, there were many who predicted war.[20]

After some involved diplomatic exchanges, the Spanish government decided that France and Holland, her erstwhile allies, were not coming to her aid in case of war, and decided to reach an agreement. The British traders, who had established a settlement in Nootka, were thereby granted the right to "the buildings and tracts of land." The ships seized were returned and reparation promised for losses sustained (this latter was never fulfilled), and British and Spanish subjects were to be allowed to trade along the Northwest Coast without being molested.

The Southern Whale Fishery came in for a share of assistance with this Anglo-Spanish agreement. It was decided that the British whaleships would be allowed to cruise in South and North American waters adjacent to but not within ten leagues of places already occupied by Spanish settlements. Hawkesbury wanted free access to the waters of the Pacific whether the Spaniards had or did not have jurisdiction; Pitt was bent on the wider scheme of trade in all its phases. The Spanish claimed the coasts of two continents, all the way from Cape Horn to 60 degrees north latitude. This was based on ancient discovery and hardly in tune with conditions which were already breeding Revolution in the Spanish-American colonies.

When the terms of the agreement were being debated in Commons

20. Johansen and Gates, *Empire on The Columbia* (New York, 1957), p. 46.

18. Ship *George* of Nantucket, Captain Jethro Coffin
Leaving Le Havre, France, on a whaling voyage. Watercolor
by Montardier
[Photograph by Frederick G. S. Clow]
[Courtesy the Nantucket Historical Association]

none other than the famous Charles Fox asked why more advantageous concessions were not obtained for the British whalers. He quoted the government's own statistics, which showed that the produce of the Southern Whale Fishery had increased from £12,000 to £97,000 over a period of five years. On December 17, 1790 during a debate in the House of Commons over the Convention, one of the members, Mr. Poultney, stated the "whalefishery in the South Seas was carried on more for the purpose of smuggling than of catching whales."[21]

It was at this moment that Alderman William Curtis, a new member of Parliament, made his maiden speech. Having now become a whaling merchant out of London, Curtis was in the best possible position to talk on the subject. After congratulating the House and the commercial

21. *Gentlemen's Magazine*, London, Vol. 61, pp. 46–47.

interests of the country upon the security and prospect of a permanent peace, he then entered upon some account of the Southern Whale Fishery "stating it has increased much in the past five years, and he had reason to know, being a fisherman himself," and that the oil got from the Southern Fishery would fetch at market three times what the Greenland fishery brought in price.

There was an effort to refute Alderman Curtis' statements, and that gentleman quickly responded. Whereupon Sir William Young reminded the gentlemen that the chief object of the session was not trivial matters such as the price of whale oil. He recalled a statement by the Hon. Charles Fox that the Nootka Sound incident was not the object but the annihilation of all grounds for future dispute.

At this point Prime Minister Pitt arose to remark that the sentiments expressed by Alderman Curtis merited a considerable degree of attention. He remarked that it was probable that the members of Commons could thereby learn the opinion of a very large and opulent body of men. He pointed out that, through the Spanish Convention, "Britain had a full acknowledgment of our right the Southern Whale Fishery, which he considered as a matter of no very inferior moment."[22]

Thus, with William Pitt in the House of Commons, and Hawkesbury in the House of Lords, the London Whaling merchants were represented by two of the most powerful men in the Empire.

The Anglo-Spanish Convention was signed late in the year 1790.

Soon after this event (January 1791) the London merchants went into action again through a "Memorial" to the Board of Trade, asking for legislative permission to enlarge the scope of the voyages. The obvious intent was to throw off the necessity of obtaining further licenses from either the South Seas Company or the East India Company. Enderby, again one of the leaders in the memorial's presentation, had stated that the whaling merchants were intending to order their shipmasters to sail wherever they found the prospect of greatest advantage.

Enderby's ship *Rasper*, under Captain Thomas Gage, sailed early in 1791 and in July was at Amsterdam Island in the far reaches of the eastern Indian Ocean, where he was hoping to procure seals. The ship

22. Ibid.

Warren of Providence approached the Island and reported Captain Gage as "being an American but had sailed out of England ever since the peace. The ship being a Southern Whaler and had come to this island after seal, but greatly disappointed in his expectations, . . . Captain Gage informs me he means to load his ship with whale oil."

The American vessel gave another glimpse of the British whaler's activities at this time, the log book keeper reporting:

The Rasper *manned two boats in order to kill whales. They got two harpoons in one whale, the whale carried the Boat so far to the leeward they were obliged to cut away and leave him—they lost two harpoons and one lance.*[23]

Finding the weather bad and the whales wild and fractious, Captain Gage sailed for St. Paul's Island. Before leaving Amsterdam Island, he presented the American Captain with some fishing line and hooks, nails, knives, "one dozen onions, one cask salt, one bottle of porter . . . and some newspapers. Samuel Enderby, merchant of London, is the owner of the ship."

The 1790–1791 period was one of the most productive in the history of the Southern Whale Fishery. Out of the fleet then numbering fifty-nine vessels, London found thirty-three returning in 1791 with cargoes of oil and bone totalling in value £91,699, and with seal skins numbering 45,400, mostly from the Falklands and South Georgia, which brought £6,810 in addition.[24]

During this important period, additional numbers of Nantucket men were engaged to take out British whaleships and direct the attack and capture of whales. As a result, more than half of the London fleet of fifty-two vessels were commanded by Nantucketers, with officers and harpooners from that island also in the fleet.

Among the newly arrived Nantucket Quaker captains were Benjamin Coleman, to take out Alexander and Benjamin Champion's new ship *John and Susannah*; Jethro Daggett, to command the *New Hope*, Thomas Yorke's ship (brought over from Philadelphia) to sail from Falmouth;

23. Logbook of the ship *Warren*, Providence, Capt. J. Smith. Microfilm copy in G. W. Blunt White Library, Mystic, Connecticut.
24. *Board of Trade Records*, B.T. 6/230.

Frances Gardner, master of the ship *Nimble*, owned by John Jarrett, Jr., of London; Thomas Jones, in the *Nymph*, whose agent was George Russell; Samuel Marshall, as master of the *Elizabeth*, of London; William Bunker, to sail on the *Hope*; Benjamin Ramsdell, master of the *Harpooner*; Edward Starbuck, to take out the *Good Intent*; Stephen Skiff and Constant Mayhew who were from Martha's Vineyard, to command the ships *Barbara* and *Sally*, respectively; Abijah Lock, in the *Moss*; Elisha Pinkham, in the *Fonthill*; Mathew Swain, in the *Kitty*; Benjamin Swift, in the *Prince William Henry*; Joseph Russell, in the *Resolution*; Walter Clark in the *Sparrow*; William Folger, in the *Hero*; Stephen Coleman, in the *Hercules*; Thomas Clark, in the *Chaser*; Shadrach Keen, in the *Adventure*.[25]

From the port of Bristol, whaling merchant Sydenham Teast wrote to the Board of Trade, requesting information on "how to carry on the Trade and Fishery in the Southern Seas and Pacific Ocean, with benefit to himself, without infringing any part of the said Convention," meaning, of course the Anglo-Spanish Convention. The answer was quickly forthcoming.

That the Board of Trade had much more confidence in the Southern Whale Fishery than in the fur traders on the Northwest Coast was clearly shown by their next action. Hawkesbury, for the Board, put a question to the London merchants, asking if they had considered sending their ships north of the equator on voyages which might extend to the Northwest Coast of America.

The answer from Enderby and his associates was prompt—that the whaling ships should be granted permission to pursue such voyages north of the equator. This represented another challenge of the exclusive rights of the East India and South Seas Companies. But Hawkesbury was solidly behind providing the Southern whalers with the opportunity to expand their voyaging, and the Board of Trade, following his direction, voted to request of the monopoly companies the granting of further rights for the benefit of the whalers. In this Henry Dundas, head of the India Board, agreed with Hawkesbury and Lord Grenville also approved. Prime Minister Pitt was strongly in favor of the requests.

25. Hussey & Robinson, *Catalogue of Nantucket Whalers* (Nantucket, Mass, 1876), "From English Ports," pp. 53–54.

Recognizing the almost overwhelming political support which was enjoyed by the whaling merchants, the Directors of the East India Company were forced to concede some of their fundamental rights, although able to maintain their position as regards the China trade at Canton. In April 1791 a bill was drafted to the effect that the British whaleships and traders would be allowed to conduct voyages within the limits of the control of the East India Company and the South Sea Company. The stipulations, however, were too broad and Pitt did not push for its passage, being content for the moment in gaining smaller agreements from the two companies, which in themselves were important concessions. A few years later his patience was rewarded by the success of the bill which allowed the whalers to fish in Australian waters without restriction.[26]

The need of a naval vessel in the distant seas frequented by the whalers was in fact becoming increasingly apparent, and the London merchants petitioned for such an expedition vessel. Approval was given by the Pitt Ministry as early as April 1792, and it was decided to dispatch an armed ship to survey the coast of South America. The Lord Commissioner of the Admiralty nominated Captain James Colnett for the command. He was experienced in Pacific voyages, having been with the illustrious Captain Cook on his second voyage of circumnavigation. He had also been in command of the trading vessel *Argonaut*, when in 1789 he was one of the shipmasters seized by the Spaniards during the Nootka Sound confrontation, and was released only after suffering imprisonment at San Blas in Mexico.

The memorial of Enderby and his fellow merchants had requested the Admiralty's vessel be sent out "for the purpose of discovering such ports for the Southern Whale Fishers who voyage round Cape Horn as might afford them the necessary advantage of refreshments and security to refit."

The need for ports and places to procure fresh provisions and water was of paramount importance to the whaleships, whose voyages were now approaching the mark of two and a half years in duration. Among the diseases most dreaded on these extended voyages was scurvy, that

26. 32 Geo. III, Cap. 73, and 33 Geo. III, Cap. 90.

ailment which was caused by lack of fresh foods, notably vegetables and fruits, on board ships long at sea.

According to Colnett's account "there was no private commercial vessel available" for such an exploratory voyage, and so the Admiralty favored the loan of H.M.S. *Rattler*, then laying at His Majesty's yard at Woolwich, under repair. Naval regulations forbid the fitting out of the vessel as a whaleship and then, upon return restoring her to Naval Service, and it was therefore agreed to sell the *Rattler* to the Enderby firm. She was then removed to the Perry dock at Blackwell, in order to be repaired and fitted as a whaler.[27]

Brought around to Portsmouth, the *Rattler* was ready to sail early in November 1792. But the situation in Revolution-torn France delayed departure. Some dissatisfaction was expressed by the foremast hands, to whom the idea of the "lay" system in lieu of wages was not appealing.

Messrs. Enderby, Champion, Mather, and other London merchants were anxious for the voyage to begin so that their vessels in the Pacific would have some intelligence of domestic affairs in England and of the extent of the Revolution in France. Yet, delays continued another month and it was not until January 2, 1793, that Colnett sailed out of Portsmouth.

The voyage of the *Rattler* from the standpoint of the whaling industry was disappointing in light of Colnett's extensive prior experience in the Pacific. Throughout the months at sea the *Rattler* spoke few whaleships. Touching at Rio de Janeiro on the voyage out, Colnett rounded Cape Horn, sailed up the coast of South America to the Galapagos Islands, and went as far north as the California coast. In his narrative, he wrote:

There was now so large an extent of coast, in every part of which, I might meet with those British ships employed in spermaceti whaling to whom, I was instructed to communicate the circumstances and situation of Europe . . . I was indeed persuaded that the greatest body of fishermen were to the northward as they would find the best season there and might return with the Line.[28]

27. Captain James Colnett, *A Voyage To the South Atlantic and 'Round Cape Horn to The Pacific Ocean In The Ship Rattler*, (London, 1798), p. VIII.
28. Ibid., pp. 38–39.

In the first attempt to take a sperm whale one of the *Rattler*'s boats was stove to pieces, and they were unsuccessful on other occasions. Probably Colnett believed his real purpose was exploring for places for the whaleships to recruit, and by accomplishing this, he was doing all that could be expected. In any event, he was not a whaling master and this phase of the voyage reflected the basic fact.

Colnett made one outstanding contribution to whaling—his excellent chart of the Galapagos Islands and survey of these unusual outposts of the northwestern coast of South America, later made famous by Darwin in the *Beagle*. Names he gave to two of the islands are constant reminders of his visit. "I named one Chatham Isle, and the other Hood Island, after the Lords Chatham and Hood." He noted that Charles Island was "so-named by the buccaneers."

After twenty-two months of voyaging the *Rattler* returned to Portsmouth, actually first arriving at the Isle of Wight. Only one man was lost, and that due to an accident. In 1798, Colnett's account of the voyage was published in London, "printed for the author by W. Bennett."

12. Storm of Revolution: Dunkirk 1791-1792

WITH the advent of the year 1791 William Rotch and the Nantucket whaling colony at Dunkirk were deeply concerned by the grave problems that beset France. Looming darkly on the horizon was the great storm of Revolution. While the whaleships out of this fine old French port had embarked on voyages that found them meeting the challenge of storms at sea their whaling counterparts on the land were more fearful of the turmoil of the countryside. Rotch and his associates were aware of the forces that threatened a national upheaval. The Quaker merchant noted, in his letters home, his own forebodings, recognizing that what fate had in store for France was to determine the fortunes of the prosperous new whaling industry he had established at Dunkirk.

In one letter he commented on the action of the French Assembly in issuing paper money to the enormous total sum of thirty million pounds sterling—some twelve hundred million livres—calling it a "master stroke of policy," as it was funded on the total value of the estates of the clergy, which was reputed "to be one-third the wealth of the entire kingdom." He stated that many people originally opposed to the sale of the estates were now saddled with the paper currency and were "as sanguine as any that the estates would be sold." [1]

Rotch wrote that the news of King Louis' acceptance of the new French Constitution was for him "attended with pleasing sensations," but that he feared "an attack upon the Kingdom by the ex-Princes, nobles and clergy, allied openly or secretly by all the powers of Europe." He predicted that such attacks would come in the spring.

Thus, the impending storm promised a precarious future for the little whaling colony. This was particularly unfortunate as the Dunkirk venture had become a thriving success. In his letters this favorable market

1. Letter of William Rotch to Samuel Rodman, January 22, 1792. Rotch Collection.

Nantucket 3mo 19th 1815
Dear Uncle & Aunt
I received your acceptable
communication a few days after we heard
the Glorious news of Peace which supersedes
the necessity of my coming ~~here~~ to Hudson this
summer for there is now a bright prospect of
our getting a comfortable living here.
Barzillai and I work at coopering in the old shop
John is going master of the ship Thomas. Paul
has broken up his school. You can hardly
be sensible how lively every one was on the
happy return of Peace and what joy beamed
in every one's countenace all the public buildings
were beautifully illuminated the bell rang
several hours guns were fired and men and boys
giving 3 cheers and vociferating Peace, Peace
with as much joy as it is possible for you to
conceive and to crown all about 125 republicans
joined and had a genteel dinner at the
American Hall in Tristram Husseys store
and the Federalists acknowledged that it

19. Letter by Peter Macy
Describing Nantucket's celebration of the "Glorious news of Peace" ending the War of 1812
[Courtesy the Nantucket Historical Association]

was more compleat and carried in as good
order as any one ever was on the Island we
sat down (for I was one) at 3 O clock in the
afternoon. ~~and~~ rose a little befor 6 there were
a number of very applicable toasts drank and
~~and~~ a gun fired at every toast every one
was pleased and all retired in good order.
I have not time to write much so with love
to all I must conclude and subscribe
myself your affectionate nephew

Peter Macy 2d.

was evident but, always a realist, William Rotch, declaring his intent to "keep steadily on," reminded his sons to stay abreast of the exigencies of the times and to "keep our interest all insured in England, until an alteration in this Government," more encouraging, was forthcoming.

The Dunkirk whalers were sent out to three different parts of the world on their voyages. The larger ships were directed to round the Horn for the Pacific; another group sailed for the "Brazils" and the South Atlantic; the third cruised the coast of Africa, from Walfish Bay on the west coast; around Good Hope to Delagoa Bay on the western shores of the great continent.

The uncertainties of the French political situation made it necessary for the Rotch firm to keep their vessels, like the proverbial eggs, from being all in one basket. The *Penelope* was sent to Nantucket, to be registered as an American vessel; the *Maria* was directed to go around the Cape of Good Hope to the newly found "Madagascar Ground"; the *Barclay*, newly built, was used as a trading vessel, bring cargoes of whale oil and bone and sperm candles to Le Havre, consigned to Homberg and Freres; the *Diana* and *Mercury* sailed from Dunkirk for the South Atlantic; the *Ann* and *Fame* were dispatched to Walfish Bay; while the *Falkland* (newly coppered) with the *Hope* were sent into the Pacific.[2]

For one so strongly opposed to wars, William Rotch was sympathetic to the basic motives of the Revolutionists in France seeking ways to free the country from the oppressive dual yoke of the nobility and the hierarchy of the clergy. He wrote: "I feel a willingness to contribute my mite in any manner to Nations, kindred, tongues & people, the world over, for the promotion of the cause of Righteousness in the earth." What he feared for France did come to pass—"those flaming Patriots," who despised the ex-nobles (who had fled the country and were joining invading Austrian armies) and so refused any compromise with them. He went on:

I cannot better describe the present than to say they are much like our violent Patriots in America in the time of our Revolution; moderate men have no influence, therefore, they withdraw as much as they can. I never expected to be in this midst of a second Revolution, but am not sorry that I am here considering our own affairs requires it.[3]

2. Rotch to Rodman, July 29, 1792. Rotch Collection.
3. Ibid.

The manner in which William and Benjamin Rotch conducted themselves during these times is the best proof of the respect accorded them by the authorities of the Revolutionary government. But the greatest test came early in 1791, when they were called upon to appear before the National Assembly in Paris to present a "Petition to that body for some privileges and exemptions connected with our religious principles." The document was titled "The Respectful Petition of the Christian Society of Friends Called Quakers," and had been prepared not only for the sake of the Nantucket Quakers in Dunkirk but for other Friends who were living in several towns and villages in Languedoc.

On February 9, 1791, the two Nantucketers journeyed to Paris and joined their friend John Marsillac, a native Frenchman. Marsillac had aided them by drafting the Petition into French. By order of the Assembly the Petition was to be presented in person on February 10. Apparently, word that the two Quakers were to appear made the rounds, as when they arrived at the Entrance to the Assembly they had to face the considerable danger of passing through the "large Body of the lower classes about the Assembly." The danger arose from the fact that they would not wear the cockade which was the identifying mark of the Revolutionist. "We saw that our friends were full of fear for our safety," the elder Rotch recounted. But they continued on, "trusting to that Power that can turn the hearts of Men as a water course is turned."[4]

The National Assembly had moved from Versailles the previous November and was now meeting in what had been the Royal Riding School of Louis XV, or "Manege," as it was called. It was north of the Tuileries gardens. The Rotches, with their French companions, walked through the walled courtyard from the Rue Dauphin to reach it. A throng of the curious spectators crowded about them, but they passed through this great Concourse "without interruption."

The long, narrow room was filled to overcrowding. On six rows of benches, placed on each side of the hall, sat the deputies, and on the center of the south side sat the President of the Assembly, Comte de Mirabeau, the most influential figure in France at this time. On the side opposite was a small platform, and here William and Benjamin Rotch and their friends sat, prepared to address the President and the deputies.

4. Rotch, *Memorandum*, p. 55.

In the galleries, or place for the spectators, the usual conglomerate citizens of Paris had gathered.

"Nor did we wonder at their curiosity," observed Rotch, "considering the Object."

Almost immediately after they had been announced they were called upon to present their Petition. Because of the insecure command of the language, neither of the Rotches was prepared to read the document, but they stood erect beside John Marsillac as he presented it for them. Here was a spectacle so unusual as to cause an instant hush over the hall —the two Quakers, wearing their hats, standing straight and tall on either side of their French spokesman, the dignity of their bearing bringing a strange quietness of suppressed emotions to their observers.

Marsillac read the Petition slowly, with another Frenchman, Brissot de Warville, just behind him, giving certain guides for emphasis as he continued. The wording was simplicity itself, carefully observing that the French nation "had appointed you [the deputies] her Legislators, and your hearts having been disposed to enact wise laws, we salute the extension of your Justice to the Society of peaceable Christians [Quakers] to which we belong." Continuing, the Petition noted: "Many families which have come from America have settled at Dunkirk, under the auspices of the late Government, in consequence of the invitation given to the Inhabitants of Nantucket, for the purpose of extending the French Fisheries."[5]

William Rotch had carefully phrased one sentence in particular, which was carefully delivered: "These Islanders have proved themselves worthy of your kindness by their success, and the same motive will induce them to continue to deserve it." Another paragraph, which showed the sagacious philosophy of the elder Rotch, was delivered with an especial emphasis:

In an Age signal for the increase of knowledge, you have been struck with this truth, that Conscience, the immediate relation of Man with his Creator, cannot be subject to the power of man: and this principle of Justice hath induced you to decree a general liberty for all forms of worship. This is one of the noblest decrees of the French Legislature.[6]

5. Rotch, *Memorandum*, "Petition of the . . . Quakers," p. 71.
6. Ibid.

Noting that great persecutions had been inflicted on them, as Quakers, because of their principles, the Petition asked that they be allowed to conform to these principles to which Quakers had been "inviolably attached ever since their rise." In particular it mentioned that precept which forbade them to take up arms and kill men on any pretence—consistent with the holy scriptures. Some of the French influence then became evident in the wording, perhaps Marsillac's hand.

Generous Frenchmen, you are convinced of its truth; you have already begun to reduce it to practice; you have decreed never to defile your hands with blood in pursuit of conquest. This measure brings you—it brings the whole world—a step towards universal peace. You cannot therefore behold with an unfriendly eye men who accelerate it by their example . . . We submit to your laws, and only desire the privilege of being here, as in other countries, the Brethren of all men.[7]

Request was made for the continuance of their practice as Friends of their own manner of registering births, marriages, and deaths, and that they be exempted from taking oaths, "Christ having expressly forbidden them," as "it can give no additional force to the declaration of an honest man, and does not deter a perjurer." Following along with this thought, the petitioners trusted that "we shall not be suspected of a wish to evade the great purpose of the Civic Oath. We are earnest to declare in the place that we will continue true to the Constitution which you have formed."

With another of the practical touches so characteristic of William Rotch, the Petition noted:

We esteem employment a duty enjoined on all; and this persuasion renders us active and industrious. In this respect our Society may prove useful to France. By favoring us you encourage Industry and Industry now seeks these countries where the honest, industrious man will be under no apprehension of seeing the Produce of a Century of Labor snatched away . . . by the hand of Persecution.[8]

The reading of the Petition was often interrupted by applause from the gallery. On these occasions the officials, whose duty it was to keep

7. Ibid.
8. Ibid.

order, would utter a hushing sound to restore quiet. It was this sibilant sound which William Rotch misinterpreted as hissing, "from my ignorance of the language, and apprehended all was going wrong, until better informed."

At the conclusion of the reading, the Assembly's leader, the Comte de Mirabeau, arose to answer. This remarkable man, heavyset, almost grotesque because of his pock-marked face and his ungainly stance due to a twisted foot, directly contrasted with the tall figure of William Rotch, who stood erect, wearing his broad-brimmed hat, his handsome features slightly flushed, outwardly calm.

The Assembly's President began speaking quietly, and at once the Gallery became instantly still. Mirabeau's eloquence took command and his words became impassioned but well within bounds:

Quakers, who have fled from Persecutors and Tyrants, cannot but address with confidence those Legislators who have, for the first time in France, made the rights of mankind the basis of law . . . As a system of philanthropy, we admire your principles . . . The examination of your principles no longer concerns us: . . . we have decided on that point. There is a kind of property which no man would put into the common stock; the motions of his soul, the freedom of his thoughts.[9]

During the moments awaiting the translation of Mirabeau's words, William Rotch and Benjamin continued to stand erect, and the quiet dignity of the two men was in strange contrast to the noisy crowded galleries of Revolutionaries. At the pauses of the President of the Assembly, and after each subject, vigorous clapping broke out—all of which brought anxiety to the Nantucketers until the translator had provided the meaning of the remarks in a low voice.

Mirabeau's magnetism was not lost on William Rotch and, despite his inability to follow the French language, he had an instant admiration for the obvious mastery which this man and his words so clearly exercised over his listeners.

One of the delicate points in the petition had been that of taking oaths. Due to the variety of business transactions in Dunkirk, customs clear-

9. Rotch, *Memorandum*, The Comte de Mirabeau, "Answer of the President," p. 78

ances, and other matters, William Rotch was concerned as to whether the French authorities would insist on his subscribing such an oath to the documents involved. Mirabeau's next words enlightened him on this point.

Worthy Citizens, you have already taken that civic oath which every man deserving of freedom hath thought it a privilege rather than a duty. You have not taken God to witness, but you have appealed to your consciences. And is not a pure conscience a Heaven without a cloud? Is not that part of man a ray of Divinity?

The relief to the minds of the Nantucketers, father and son, may well be imagined. It is safe to say that, had the reverse been true, and they were faced with a demand that the oaths be taken, they would have quietly removed themselves from Paris and then from Dunkirk—thereby losing the fortune in the whaling settlement they had created—rather than act contrary to their beliefs.

Mirabeau continued, and with his next words there was a new timbre in his voice, a note of sternness:

You also say that one of your religious Tenets forbids you to take up arms, to kill on any pretence whatsoever. It is certainly a noble philosophical principle, which thus does a kind of homage to humanity. But, consider well, whether the defense of yourself and your equals, be not also a religious duty? You would otherwise be overpowered by Tyrants! Since we have procured Liberty for you, and for ourselves, why should you refuse to preserve it! The Assembly will, in its wisdom, consider all your requests. But whenever I meet a Quaker, I shall say: My Brother, if thou hast a right to be free, thou hast a right to prevent anyone from making thee a slave.[10]

Invited to remain for the Assembly's setting, the two Nantucketers were questioned by individual members and, after the closing of the session, returned to their lodgings. Remembering his experience in London, William Rotch did not abruptly leave Paris but remained for a few days so as to "visit the influential members of the Assembly, in their private Hotels, to impress them with the reasonableness of our requests."

10. Rotch, *Memorandum*, p. 80.

They found a cordial reception from all but two—one of these being the famous Tallyrand. After these visits they were invited to a number of other places where they met such men as Bishop Gregory, with whom they "had much conversation."

"The remembrance of these evenings," wrote William Rotch, "and of the feeling of divine influence which attended them, I believe will never pass away."[11]

The Assembly acted favorably upon the Petition, resulting in a guarantee of protection for the Quakers, of especial gratification to the Nantucketers in Dunkirk. It had other important effects. The bounties for the whaleships were continued, and in September 1791 the Assembly confirmed the original contracts which William Rotch had made with the government of Louis XVI. Of vital importance was the continuance of the bounties paid to those ships engaged in the whaling voyages out of Dunkirk.

Considering the activities and far-flung ranging of the whaleships very few were lost. Early in March 1791 the *Ville de Paris*, under the experienced Captain Shubael Gardner, was lost on the coast of Cornwall, England, while only a few hours sail from Dunkirk. Both vessel and cargo were lost, and three of the crew drowned, one being Reuben Gardner of Nantucket. "I wish the tidings may be judiciously conveyed to his nearest relatives," requested William Rotch. Francis Rotch's staunch ship *United States* was "cast away going out of the harbor [Dunkirk] all fitted for whaling . . . insurance paid; he bought 3 brigs." This information was contained in a Rotch letter home as of January 30, 1792.

David Starbuck, returning from the Pacific in the *Harmony*, brought a full cargo of sperm back to Dunkirk in February 1791 while the *Judith*, Captain Isaiah Hussey, was reported at Delagoa Bay with 120 tons of whale oil, and the *Ospray*, Captain Benjamin Paddack, was expected to return in April 1792. The *Penelope*, Captain John Worth, and the *Hope*, Captain George Whippey, returned from the coast of Peru in the spring of 1792, with the former having a narrow escape off the Azores, "losing all her rudder except 7 feet of the upper end," and some of the crew "being a little touched with scurvy, but capable of duty." The *Penelope*

11. Ibid. p. 59.

20. New Bedford—Successor to Nantucket as the Whaling Capital
From the Russell-Purrington Panorama [Courtesy the New Bedford Whaling Museum]

had left the South American coast on October 20, 1791, and arrived at Dunkirk the first week in March 1792. Worth reported the *Ann*, Captain Prince Coleman, had left the Pacific before him, and intended to put in at Rio de Janeiro on the way home. The *Favorite*, Captain David Folger, was reported at Walfish Bay. When the *Good Intent*, Captain Hawes, arrived late in March, he reported the *Bedford*, Captain Laban Coffin, had 700 barrels when he "left the coast."[12]

At this time William and Benjamin Rotch decided to invite two of their veteran shipmasters to "come ashore" and help in the firm's duties of outfitting and unloading the ships of the fleet. The two selected were Captain William Mooers and Captain Benjamin Hussey, both of whom

12. Rotch to Rodman, July 29, 1792. Rotch Collection.

were to lend invaluable aid not only at this period but later in the decade, when the Rotches had returned to America.

The prices for oil and bone remained high and this encouraged even more strongly the import of oil from Nantucket. But there was no escape from the growing turmoil of the internal affairs of France. The daily news from Paris was anxiously awaited, and the potential dangers increased as 1792 was ushered in by the threat of war with the Prussians as well as with the Austrians. In a letter to Nantucket, dated Feb. 18, 1792, William Rotch announced the safe arrival of the *Canton*, under Captain Coffin Whippey, from Nantucket, and of Captain David Starbrick, "from the Coast of Peru," in *Harmony*. In the same paragraph is described one of the tense situations in Dunkirk:

A great Riot took place here on the 14th inst. in which the front windows of 7, 9 or 11 houses, with the furniture and other property, destroyed . . . the Arm'd force was called out by the Magistrates, the Red flag displayed, & Martial law proclaimed . . . orders were given to fire on the Mob wherever they should be found . . . I never expected to be in the midst of another Revolution after that of America was completed, but here we are, and I am glad at present under every circumstance that we are here, though what sufferings may be is hid from our eyes; if things should continue in this agitated state, & the Government too weak to prevent lawless intrustion, I hope we shall have opportunity to shift our quarters to some place of tranquility.[13]

The Rotch firm and its French associates had reason for planning well ahead, with the turmoil in the nation threatening to engulf the whaling industry by the surging waves of the Revolution. Their business interests had broadened from Dunkirk to Le Havre. Added to the fleet from Dunkirk were the *Necker*, Captain G. Whippey; *Cyrus*, Captain Archaelus Hammond; *South Carolina*, Captain James Whippey; *Seine*, Captain Jonathan Worth; *Brutus*, Captain Reuben Macy; *Ganges*, Captain Charles Harrax; *Fox*, Captain Ebenezer Hussey; *Ardent*, Captain Peleg Bunker; *Thomas*, Captain Seth Folger; and *Hebe*, Captain Francis Macy. While all the masters and most of the mates, harpooners, and crews were

13. Rotch to Rodman, February 18, 1792, Rotch Collection.

native Nantucketers half of each crew, by agreement, was made up of French sailors. Out of Le Havre sailed the *Baleine*, *Charles*, *Harpooner*, and *Hero*; the *Sarah* sailed from Dieppe, while the *Victoire* had L'Orient as her home port.

In May of this year, 1792, William Rotch went to England, where he made a circuitous journey making visits to a number of Quaker meetings "in the north." Writing from London in July he requested his Nantucket branch of the firm draw on Thomas Dickason of London, or the De Barques in Dunkirk. Many of the whaleships of that period were sheathed in leather, but Rotch had become well aware of the value of coppering, especially with the prevalence of teredo worm in the warm Pacific, a worm which bored through a ship's hull with incredible speed.

Not only had the Rotch ventures in France proved successful but the opportunity for selling their oil and candles on the French market had provided a vital impetus to the build-up of the Nantucket whaling fleet. With the adoption of the Constitution by the United States, the conditions within the new nation improved rapidly and the growth of the seaports brought a new demand for whale oil and sperm candles. This improvement on the home front, plus the increasing ominous aspects of the French scene, induced William Rotch to plan for a return home to Nantucket, leaving Benjamin at Dunkirk, assisted by Captains Hussey and Mooers.

As of July 1792 these plans had so advanced that William Rotch could write to his son, William, Jr., and son-in-law, Samuel Rodman:

I hope soon to be able to see [hold] just our property here, and when I receive the inventory from thee, shall be better qualified to arrange our matters fully for a final settlement, so as for me to leave the business entirely. I think, at the return of the vessels, if they do return, will be the proper time; I shall then wish to know whether thou would choose to continue thy whole share here, or whether thou can do better with part of it in America.[14]

William Rotch, Sr., always looked well ahead. As events later demonstrated he recognized the natural advantages which the little Acushnet

14. Ibid.

River port of Bedford possessed. Soon after the elder Rotch arrived at Dunkirk the firm of William Rotch and Sons established a counting house branch of the business at Bedford, where the Rotch holdings had been retained after the death of the pioneer Joseph Rotch (at Nantucket) in 1784. It was Joseph Rotch, father of William, Sr., who nearly two decades before had joined forces with landowner Joseph Russell, to lay out the waterfront of the future town of Bedford. This property now became the nucleus of the Rotch firm's extensive holdings.

By 1791 the firm of William Rotch and Sons had established itself at Bedford, although the brick counting house at the foot of the Main Street square in Nantucket was still the headquarters. William Rotch, Jr., and Samuel Rodman, who handled the business as partners in the absence of the senior member, were also closely allied by marriage ties. William, Jr., had married Elizabet Rodman, sister of Samuel, and Samuel Rodman had married William's sister, Elizabeth Rotch. Both men became remarkably successful whaling merchants, guided carefully in the early years of their association by William, Sr., the veritable Nestor of the American Whaling industry.

In controlling his affairs, Rotch at this time was actually directing the whaling destinies of three ports—Nantucket, Dunkirk, and Bedford. He was finding his labors becoming too heavy, and his desire to get back to Nantucket—the true home port—was understandable. But before leaving Dunkirk and Benjamin, he was determined to see that the business was as well organized as his talents and the times would allow.

Being alert to the costs of materials in Europe, Rotch imported most of his bulk supplies from America. He had cedar boards for whaleboats, staves and heading for casks, and spars brought to Dunkirk, as well as barrels of salt beef and pork, beans, and molasses. Nor was any opportunity lost to ship freight on return voyages to Nantucket and Bedford on consignment. A firm advocate of insurance, William Rotch had all his vessels insured in London, and on one occasion chided his sons for having also insured the *Ospray* and *Canton* in America. Despite these costs, the Dunkirk venture was reaping a profit, with the bounties paid by the government adding to the Rotch treasury. The dangers of the sea were risk enough, but William Rotch insured his craft against man-made conditions which might affect them, due to the "present state of affairs in this country."

One of the amusing incidents (for even threatening danger has its lighter moments) occurred when some of the Dunkirk Revolutionary officers, appointed to inspect houses for extra food, forced their way into the Benjamin Rotch's house. His wife, Elizabeth, had anticipated the visit and removed all her preserves—fruits, jellies, etc.—and placed them in her feather bed, and then got into bed herself and under the blankets. The officers were ushered in, at her request, so that she might receive them in her bedroom. When the startled officers saw the Quaker lady ensconced in her bed she further disconcerted them by asking them to inspect the room's closets, apologizing for not arising because she was "in an interesting condition." After a hasty inspection, the officers quickly withdrew from the house.[15]

On another occasion, when it was ordered that the houses of the town must be illuminated to celebrate French victories over the Austrians (a customary way of public demonstration), the Rotches—as they had always decided—did not conform. Some time after the hour for the illumination had passed neighbors came to knock on the front door of William Rotch and congratulated him on his participation. It was only then that the family noticed a number of candle sconces, with the candles burned to the sockets, which had been carefully positioned on the front of the house. A number of their French friends had provided the array of lights, thus helping them to avoid a dangerous situation.

One of the few glimpses of the life of the Nantucket Quakers in the midst of the turmoil of that French town was provided by Eliza Rotch Farrar, the daughter of Benjamin and Elizabeth Rotch. In her "Recollections of Seventy Years," she wrote of what her parents had told her concerning various matters that had occurred during their residence. Benjamin Rotch never penned his own reminiscences of that stirring time. They would be invaluable today—the recollections of a man who had once matched wits with Robespierre, and who had been in Paris on the very day Marie Antoinette became the victim of the guillotine—that dreaded symbol of the Revolution that had gone beyond all the efforts of its creators to control.

15. Mrs. John Farrar (Eliza Rotch), "Recollections of Seventy Years," Bullard, *The Rotches*, pp. 314–315.

As shown by a letter to his son Thomas, written from London in June 1791, William Rotch had already prepared himself for the eventuality of being forced to leave Dunkirk. "If we are favored to return," he stated, "I believe we shall settle in Bedford."[16] The care of his planning was best seen by the way he shuttled his ships. After the new ship *Ann* had made her maiden voyage out of Dunkirk in 1791, she was sent to Bedford in June 1792 and made her initial voyage from that port under Captain Prince Coleman. When the *Ospray* arrived from Nantucket with supplies she was promptly fitted out and sent to Delagoa Bay on a whaling voyage. Registries of several vessels were changed from Dunkirk to Bedford, one of these for the new ship *Mary*. It was apparent that most of the Nantucket–Dunkirk vessels had double registries.

Across the English channel, the Board of Trade was keeping a close eye on the progress of the Dunkirk whalers. In February 1792 a dispatch from Paris to Lord Grenville reported that twenty-four ships had sailed on whaling voyages out of France during the year 1791. Oil was bringing some £60 per ton for sperm, and the French bounty amounted to 30 livres (42 shillings) per ton.[17]

While the British regretted the loss of the French market for their whale oil there was nothing in the "home market" picture of the whole Southern Whale Fishery to make them feel unduly concerned. The London fleet had increased from forty-four ships in 1790 to ninety-five ships in 1792—a remarkable growth. The total tonnage was in excess of 22,000 tons and over eighteen hundred men were on board these vessels. The value of Britain's whaling industry was reckoned at £189,000 in 1792 and she easily led the world both as a producer and a consumer in this industry.[18]

16. William Rotch to Thomas Rotch, London, June 18, 1791. Bullard, *The Rotches*, p. 248.
17. Lord Gower to Lord Grenville, February 10, 1792, "In Letters," *Board of Trade Records*, B.T. 1/3.
18. Southern Whale Fishery Listings, *Board of Trade Records*, B.T. 6/230.

13. New Worlds for Whaleships and Whalemen

NEW frontiers in the widening world of the Southern Whale Fishery were reached during the turbulent years that marked the last decade of the eighteenth century. While the first of the whalers to the Pacific were returning home, settlements had been established along the east coast of the new continent of Australia (known then as New Holland). The waters of the Indian and Pacific Oceans had opened a vast area for whaleships to explore. As has been noted, the first fleet of transports from England with convicts as cargoes had reached Port Jackson (Sydney) in 1788, to be followed two years later by a second fleet, with human cargoes of men and women packed like cattle in their holds. When a third fleet for the transportation of these "involuntary passengers" was assembled in 1791, the British whaleships were to become a part of this phase of the story.

Continuing its monopolistic control over the trade in this area of the Eastern world, the British East India Company had ordered all merchant craft used as transports to the Australian settlement to strictly obey certain definite instructions. Upon arrival at Port Jackson, as Sydney was first called, these merchantmen were advised that they would receive a legal discharge from government service, but must then prepare for a voyage to China or India for the Honorable Company, then return home. This policy had a strangling influence on the trade of the newest of England's colonies, as it prevented any free enterprise trade.

The instructions were explicit:

And whereas it is intended that several of the transport ships and vituallers which are to accompany you to New South Wales should be employed in bringing home cargoes of tea and other merchandise, from China, for the use of the East India Company, provided they can arrive at Canton in due time, whereby a very considerable saving would arise to the public in the freight of these vessels . . .[1]

Sturdy ships were needed for the transportation of the so-called convicts, and whaleships with their wide-beamed holds were well suited for such use. No doubt the experience of the first two whalers so used as transports had demonstrated their adaptability as carriers of human as well as oily cargoes. It was not surprising, therefore, that early in 1791 five more whaleships were so chartered. The convicts were taken aboard at Portsmouth and the ships joined the convoy bound for Port Jackson, in New South Wales. As the original place of settlement at Botany Bay had been found unsuitable, Governor Arthur Phillip (who had taken the first fleet there in 1788) had decided to locate at Port Jackson, the much better harbor to the north.

Reports by the East India Company vessels returning from India and China, of the large numbers of whales seen in the Indian Ocean, were a great inducement for the London whaling merchants to test the potential. There can be little doubt but that the opportunity for the expense of the ship's outward bound voyage to be underwritten by the government made taking whales on the homeward passage a possibility—and a chance to clear a handsome profit.

London whaleships in this (1791) third fleet of Australia-bound transport ships were the *Matilda*, Captain Matthew Weatherhead, owned by Calvert and Co.; the *Salamander*, Captain John Nicoll, owned by Joseph Mellish; the *Mary Ann*, Captain Mark Munroe, a Lucas ship; and the *William & Ann*, under the Nantucketer Captain Eber Bunker, belonging to John St. Barbe and Co. The long passage usually took five months, with stops at Rio de Janeiro and Cape Town if necessary, and often in India. First to arrive at Sydney (formerly Port Jackson) was the *Mary Ann*, with 141 female convicts and six children, as well as clothing and supplies, Captain Munroe reaching port on July 18, 1791. The ship was immediately reloaded with 133 male convicts and sent to nearby Norfolk Island, where another penal colony had been set up by the British.

1. Instructions from Lord Grenville to Governor Phillip, February 19, 1791. Historical Records of Australia, I, p. 215.

21. Nantucket Home of Captain Silas Jones (1791)
Master of the whaleship *Swan* of Dunkirk, France, 1787.
[Photograph by Cortlandt V. D. Hubbard]

Leaving Portsmouth late in March 1791 the *Matilda*, *William & Ann*, and *Salamander* arrived at Sydney in August. The *Matilda* arrived first, had 205 convicts and a detachment of soldiers; then the *William & Ann*, carrying 181 convicts and a company of soldiers; and the *Salamander*, with 156 convicts and some more soldiers. Two months after they came into port, a fifth whaleship sailed in—the *Britannia*, an Enderby ship, under Captain Thomas Melville—with 129 convicts on board.[2]

During the time the whaleships lay at anchor at Sydney a new vessel was then cruising in adjacent seas. Her name was the *Pandora*, Captain Edward Edwards. She was then on her famous voyage in search of the missing H.M.S. *Bounty*, which two years before (1789), under command of Fletcher Christian and the crew of mutineers, had sailed away to disappear into the South Pacific, destination unknown.

Little did either Captain Edwards of the *Pandora* or Captain Bunker of the *William & Ann*, realize what the next few months held in store

2. *Shipping Arrivals and Departures, Sydney, 1788–1825* (Canberra, Australia, 1963), p. 36. Compiled by J. S. Cumpston (hereafter cited as *Arrivals and Departures*).

for them. The latter was to sail south on a cruise for whales—a cruise that was to lead him into a series of adventures and to earn for him the title of "Father of Australian Whaling;" the *Pandora* was to resume her search by sailing north, eventually to be lost at the southeastern entrance of Torres Straits, and several of the mutineers, captured at Tahiti, were drowned in their deck prison, chained to ring bolts. It would remain for Captain Bunker's fellow Nantucketer, Captain Mayhew Folger, to one day solve the mystery of the *Bounty's* disappearance, when he visited Pitcairn Island in February 1808.

In this 1791 convoy to Australia, joined by the whaleships was the transport *Atlantic*, carrying Lieut. Governor King as well as 202 convicts and a company of soldiers. King delivered to Governor Phillip a set of instructions from Lord Grenville, which ordered the *Atlantic* converted as quickly as possible into a store ship, to be dispatched to Calcutta for supplies for the Colony. These supplies were desperately needed. The "involuntary passengers" were in most cases not geared for pioneer life, as a large proportion had been forced into such servitude for crimes that hardly deserved their being exiled.

The most important part of Lord Grenville's instructions, however, had to do with protection of the trading rights of the East India Company. Here was the glaring weakness of that vested interest, traditionally protected by government. Under his tight control only ships of the Honorable Company could trade here. As the "royal intention" ordered, no trade was permitted between "the intended settlement of Botany Bay or other places which might hereafter be established on the coast of New South Wales and its dependencies, and the settlements of our East India Company, as well as the Coast of China, and the Islands situated":

in that part of the world, to which any intercourse has been established by any European nation, should be prevented by every possible means . . . and that you do prevent any vessels which may at any time hereafter arrive at the said settlements from communication with any of the inhabitants residing within your government, without first receiving especial permission from you for that purpose.[3]

Bound by such regulations, whaleships of both American and British registry were not available in these waters as sources for provisions.

3. Ibid, p. 3.

There was a danger of starvation in the unusual settlements of New Holland. This situation became most apparent when one supply vessel, the *Guardian*, two years before, had been forced back after nearly being wrecked by an iceberg in the Southern Ocean. By arrangement of Lord Cornwallis, Governor of British India, provisions were available from Calcutta and Bengal. The *Atlantic* sailed in October 1791, but did not return from her voyage to India until June 1792, bringing rice, flour, peas, sheep, goats and cattle.[4]

That there would be clandestine trade was certain under such conditions. Also, as the whalers and the sealers were to be frequent visitors, they were often pressed into becoming suppliers legitimately, as well. Governor Arthur Phillip, a veteran of naval life, was well aware of the maritime dependency of New South Wales for survival.

It was Herman Melville who wrote: "The whaleship is the true mother of that mighty colony . . . in the infancy of the first Australian settlement, the emigrants were several times saved from starvation by the benevolent biscuit of the whaleship luckily dropped an anchor in their waters." Melville's whimsy was not without some body of fact. As history has recorded, the whaleships did play an important part in bringing supplies to the colony, acting in a temporary capacity as merchantmen. But it was their persistent activity as whalers and sealers which was to lead to the eventual throwing off of the shackles, forged by the East India Company, which controlled the trade in these seas—especially in this remote corner. Despite its grudging concessions, the Honorable Company still kept the whaleships' voyages restricted in these seas. Although this excluded them from participating in the lucrative China trade, the experience and information they obtained as to whaling grounds were to prove decisive in the final breaking of the shackles of the monopoly. And the Nantucket whaling masters in these London ships were to play important roles in this climax.

Now, as the whaleships and transports lay in Sydney harbor, two new vessels joined them in late September—H.M.S. *Discovery*, Captain George Vancouver, and H.M.S. *Chatham*, Lieut. W. R. Broughton—the exploring vessels dispatched from Great Britain for a voyage to the Northwest Coast of America. Three years before (1788) the two vessels of

4. Ibid, p. 5.

the French explorer La Perouse had also put in at Sydney—then sailed away on the resumption of his great voyage, from which they never returned.

However, it was the whaleships, which without sponsorship from government, were to be the most thorough of any vessels exploring these seas. They were to make voyages in regions where the official exploring expeditions had not ventured; they were to bring their owners riches from regions only known to them and their close associates, the sealers.

The whaling masters were well aware of the great potential for whaling in these seas, and word was dispatched to London. Captain Thomas Melville of the *Britannia*, reported to his owners, Samuel Enderby and Sons, in a letter:

We lost in all, on our passage from England, twenty-one convicts and one soldier. We had one birth in our passage from the Cape. After we doubled the Southwest Cape of Van Diemans Land [Tasmania] we saw a large sperm whale . . . Within three leagues of the shore [at Port Jackson] we saw Sperm Whales in great plenty. We sailed through different shoals [groups] of them from 12 o'clock in the day until Sunset all round the horizon, as far as I could see from the masthead. In fact, I saw great prospects in making our fishery upon this coast and establishing a fishery here. Our people was in the highest spirits at so great a sight, and I was determined as soon as I got in and got clear of my live lumber, to make all possible despatch of the Fishery on this Coast.[5]

When Captain Melville requested permission from Governor Phillip for a quick clearance of the *Britannia* so he could cruise for whales, the Governor was mindful of his governmental instructions. Melville told of the interview:

He took me into a Private Room . . . and told me he had read my letters. [Apparently those to the Enderbys] and that he would render me every service that lay in his power . . . which he did accordingly and did everything to dispatch us on our Fishery.[6]

5. Thomas Melville, Ship *Britannia*, to Samuel Enderby and Sons, Sydney, November 22, 1791. George Chalmers Collection, Mitchell Library, Sydney, Australia.
6. Ibid.

From the account given by Melville in this important letter it would appear that his vessel was the first whaleship to clear from Sydney for a whaling cruise. However, when he sailed on October 24, 1791, the *Britannia* was accompanied by the whaleship *William & Ann*, Captain Eber Bunker. Captain Melville wrote:

The secret of seeing the whales our sailors could not keep from the rest of the whalers here. The news put them all in a stir, but I have the pleasure to say we was the first ship ready for sea—not with standing they had been some of them a month arrived before us. We went out in company with the Will & Ann.[7]

Both Captain Melville and Captain Bunker were experienced whaling masters. During this historic joint pioneer voyage in Australian waters they lowered many times and killed several whales, but the weather was rough and they were forced to cut some adrift. On their very first lowering, the pattern of that initial voyage was established. The *Britannia* lowered two boats as did her companion ship, and in less than two hours seven whales were killed. But a rising gale forced them to leave their prey and return to their respective ships. "The *William & Ann*, saved one, and we took the other," recorded Captain Melville, "and rode by them all night." Riding a gale with whales alongside was a hazardous experience for whaleships, especially in the little known waters of the Tasman Sea.

It is an odd fact that history had failed to record the voyages of the three other whaleships in these waters—the *Matilda*, *Salamander*, and *Mary Ann*. After a cruise to Norfolk Island for Governor Phillip the *Salamander* made two whaling cruises in November and December, before sailing for home. The *Matilda* on her cruise south along the coast entered Oyster Bay and named it Matilda Bay, remaining there three days, with Captain Weatherhead remarking on the large number of kangaroos to be seen on the shore. The *Mary Ann* sailed on a cruise from Sydney on November 11, going as far as 45 degrees south latitude, but reported "no whales seen." In December the *Matilda* made another cruise along the coast as far as 36°, and then left New South Wales for the coast of Peru.[8] She was wrecked on Osnaburg Island in the Tua-

7. Ibid.
8. Cumpston, *Arrivals and Departures*, p. 27.

motu Islands, but her crew was saved, eventually reaching the safety of Otaheiti. She is said to have been the first whaleship lost in the Pacific Ocean.

The *William & Ann* and *Britannia* returned to Sydney early in November, and sailed again on November 22 for another combination of whaling and exploring in the waters off the southeastern coast of Australia. Captain Melville was first to return, reaching Sydney on December 16, when the *Britannia* was immediately chartered for trips to Norfolk Island by Governor Phillip—with permission to cruise for whales after completing his charter. Captain Melville returned to London in the *Britannia* in August 1793. The *Mary Ann* was also pressed into similar service to Norfolk Island, but upon clearing from that island sailed across the Pacific Ocean for the coast of Peru and the continuation of whaling before rounding Cape Horn and heading back to England.

The value of the services of these whaling vessels to the struggling settlements of New South Wales is all the more evident when it is remembered how remote New Holland and New Zealand were in those years. Governor Phillip was only too happy to grant them the whaling privileges they requested, despite the obvious possibilities that they might stray from the rigid regulations pronounced by the East India Company.

Following his return from his second cruise in company with the *Britannia*, Captain Eber Bunker sailed on a voyage of his own, taking the *William & Ann* across the Tasman Sea to the coast of New Zealand. Learning of Captain Bunker's intended course, Lieut. Governor King approached him with a proposal that he stop at Dusky Bay on South Island, New Zealand, and bring back two of the natives so that they could instruct the white colonies at Sydney in the proper manner of dressing and reparing flax so that it might be used for canvas, and so converted into clothing, or used for maritime purposes.[9] For this he was to receive £100.

Whether the *William & Ann* stopped at Dusky Bay or not remains a question. Captain Bunker never returned to Sydney on this cruise to

9. Governor Collins, *An Account of The English Colony in New South Wales*, Vol. I, (1798), p. 235.

produce the two natives and collect the reward. It is believed that he had made an exploratory coasting of New Zealand before heading his ship back for England, arriving in 1793. He was as enthusiastic about the prospects of whaling in the waters off Australia as Captain Melville, but he may have also recognized that the trend of the times was for the owners to send their ships into the Pacific by way of Cape Horn—that the effective control of Australian trade by the East India Company was still a factor with which to be reckoned. But his own interests were strongly in favor of the continuing perseverance of the whale fishery in these newly explored seas around New Holland and his confidence was to be substantiated.

The next British whaler to arrive at Sydney was another ship bearing the name *Britannia*. This has brought some misunderstanding to the account, as a number of historians have confused her with the Enderby ship of the same name. This *Britannia*, however, was owned by John St. Barbe, close associate of Enderby, and commanded by Captain William Raven, and sailed from Falmouth early in 1792, reaching Sydney on July 25, 1792.[10]

She had been granted a license from the East India Company for a three year voyage. According to a note in the Historical Records of New South Wales, Captain Raven is quoted: "My first plan, after discharging the cargo I brought to Port Jackson, was to have gone to Dusky Bay (New Zealand) to procure seals for the China Market." But this plan was promptly altered by an appeal to Captain Raven's humanitarian instincts. In command of the soldiers at Sydney was a Major Grose, who had become alarmed by the scarcity of provisions in the colony. Rations had become dangerously limited, and some of the men were without shoes. Calling together his officers, he found them equally concerned and as a result of the meeting it was decided to charter the *Britannia* and send her to Cape Town for supplies. Governor Phillip, after pointing out this would violate the rights of trade held by the East India Company, allowed matters to proceed without his official blessing.

Captain Raven agreed to perform the voyage but first he was to go to Dusky Bay and land a sealing gang. This was agreeable to the charter

10. Cumpston, *Arrivals and Departures*, p. 27.

22. Ship *Hero* of Nantucket, Captain James Russell (1819)
Captured by pirates at St. Mary's Island, Chile
[Photograph by Frederick G. S. Clow]
[Courtesy the Nantucket Historical Association]

parties, and the *Britannia* sailed from Sydney on October 23. Eleven days later, November 3, 1792, they had reached the shores of New Zealand but the weather was so bad that it was not until three days later that the ship was able to drop anchor in Dusky Bay. Going ashore they saw evidences of Vancouver's landing a year before, where trees had been felled, and after a survey Captain Raven decided to put his sealing gang ashore at a place called Luncheon Cove. In charge of the encampment was William Leith, Second Mate. A large house was constructed and provisions calculated to last the gang for a year were landed.[11]

On December 1, 1792, the *Britannia* sailed away, now on her voyage to Cape Town for the Sydney colony, taking a course which was to bring the men first south of New Zealand's Stewart Island. The day after leaving Dusky Bay they sighted the Snares, a group of islets which they named "Sunday Islands," not realizing Vancouver had discovered them a year before. Another of Vancouver's ships had come upon the islands

11. Logbook of the *Britannia*, Captain William Raven, Essex Institute, Salem, Massachusetts, U.S.A.

named Chatham. The first vessel to land and take seals here was the *Topaz* of Boston, in command of the Nantucketer Mayhew Folger. Later, on the same voyage, Folger stopped at Pitcairn Island and became the first to learn of the fate of the *Bounty* and her mutineers, and thus solve the famous mystery.

Captain Raven fulfilled his charter with the *Britannia*, reaching Cape Town and taking aboard livestock and provisions (not forgetting shoes, we trust), then made the return voyage to Sydney, where he arrived early in June 1793. Anxious to cross the Tasman Sea to New Zealand, as he was first thinking of his sealing gang, he was persuaded by Major Grose, now appointed Lieut. Governor, to take the *Britannia* to India. Once again, Captain Raven was allowed to first proceed to Dusky Bay, and this time Grose sent the newly-built colonial sloop *Francis* to be the ship's consort. Leaving Sydney on September 8, 1793, the two vessels crossed the Tasman to Dusky Bay, but the sloop met strong head winds and did not reach the destination until October 12, two weeks after the *Britannia* had already arrived.

Captain Raven was greatly relieved to find Leith and his men on shore all well, although they had only obtained forty-five hundred seal skins during the nearly one year isolation. During this time, however, the gang had not been idle, as they had busied themselves in constructing a small vessel some forty feet long and had nearly completed her. When the *Francis* finally appeared she was found in a badly weakened condition, and the *Britannia*'s crew went quickly to work repairing her—at the same time converting her into a schooner so that she was seaworthy enough to make the voyage back to Sydney safely.

The *Britannia* then continued on her voyage to India, stopping at Norfolk Island on the way, and returning to Sydney in June 1794 with more needed supplies. Captain Raven had been so successful in these voyages on speculation that he accepted another charter, this time sailing to Cape Town.[12]

During these busy years, when this whaleship was converted into a merchantman, other whalers had come upon the Australian–New Zealand scene. In November 1792 the *Chesterfield*, Captain Matthew

12. Ibid. Murray's *Journal* in this ship, ended 21 October, 1793.

Bowles Alt, came into Sydney to refit after a voyage to Desolation Island in the South Indian Ocean. She also was pressed into service as a provision ship, making a voyage to Norfolk Island and then Bengal in India, returning in April 1793.

Sydney was becoming a crossroads for adventurers, being visited by the Spanish exploring vessels *Discovery* and *Intrepid*, under Don Alexandro Malaspina, and the storeship *Daedalus*, of Vancouver's fleet, all the way from Nootka Sound on the Northwest Coast. Other whaling ships from London to reach the Australian grounds and put in at Sydney were the *William*, owned by James Mather, which under Captain William Folger, brought Rev. and Mrs. Samuel Marsden as passengers—two people who were to make their mark in the history of the neighboring New Zealand; the Enderby ship *Speedy*, with Captain Thomas Melville, who had been one of the first to seek whales off these coasts, in command; the *Resolution*, one of Tarratt and Clark's vessels, with the Nantucketer John Lock in command; and the veteran *Salamander*, Captain Wilham Irish, returning to her old haunts.[13]

The first American vessel to arrive was the brigantine *Philadelphia*, Captain Thomas Patrickson, on a voyage to Canton, reaching Sydney on November 1, 1792. A month later, December 24, the ship *Hope* of Providence, Captain Benjamin Page, became the second American ship to drop anchor here, also bound for China.

During these years the London fleet kept its steady growth. The 1791–1793 period saw Lucas and Spencer adding five vessels; Curtis and Co., six; Champion and Co., seven; and the Enderbys' four. These, in addition to those sent to sea by the new firms, such as Calvery and Co., Mather, Bennett, and Mellish, brought twenty-seven new vessels to the fleet—including two from Southampton, one each from Liverpool and Hull, and five from Bristol.[14]

The extent of the British Southern Whale Fishery was carefully tabulated by the enterprising Enderby firm, in a report sent to Lord Hawkesbury on January 24, 1793, by Samuel Enderby, which listed:

13. Cumpston, *Arrivals and Departures*, pp. 28–29.
14. *Board of Trade Records*, B.T. 6/230.

50	Vessels sailed in 1792
27	Vessels sailed in 1793
9	Vessels preparing the sail in 1793
7	Vessels at home
93	Total[15]

As regards the number of vessels actually returning in the year 1792, (a total of fifty-nine) this represented 22,314 tons of shipping, employing 1,805 seamen on board, and a total value in the catch of seal skins, oil, and bone of £189,135. This summary is of interest in showing how the products of whaling (including sealing) were proportioned.

2,096 tons of sperm oil	£75,456
3,389 tons of whale oil	67,780
5,485 tons	
2,473 tons of whale bone	£12,365
255,730 seal skins	31,966
2,620 ounces ambergris	1,572
Total	£189,133.5[16]

On May 15, 1792, Lord Sydney, Secretary of State for the Colonies, wrote to Governor Phillip of New South Wales:

It is proposed for the future to transport both the convicts and such articles for the settlement as shall be sent from hence by ships in the service of the East India Company . . . The quantity of spermaceti whales found on the coast may eventually become an object of great consequence to the settlement, and by a means of extending the communication betwixt it and this country (as well as others) much beyond that necessary degree thereof which attains at present.[17]

No one knew better than the Nantucket whaling master Eber Bunker the "object of great consequence" which whaling represented for the Australian settlements. His pioneer voyage in the *William & Ann*, in company with the *Britannia*, had provided him with an opportunity to

15. Ibid.
16. Ibid.
17. Historical Records of Australia, Series I, Vol. 1, p. 354.

gain this knowledge. Upon his return to London in 1793, Alexander and Benjamin Champion offered him command of their new vessel, *Pomona*, which he accepted, and sailed from London in May 1795.

Eber Bunker was uniquely qualified as a pioneer in this or any part of the ocean world. His father was James Bunker of Nantucket, and his mother Hannah Shurtleff of Plymouth, Massachusetts, who went with her husband in the Nantucket migration to Barrington, Nova Scotia, as has been described. Eber was born in Plymouth in 1761, the year of the migration to Barrington, Nova Scotia, probably on the way hence, as several Cape Cod and Plymouth families joined the Nantucketers in that migration. In 1768, when Eber was seven years old, his father was lost at sea, leaving a widow and four children. The following year his mother married again, and removed with her new husband from Barrington to Litchfield, Maine.[18]

Simeon Bunker, Eber's uncle, remained in Barrington, and the nephew first went to sea with him from that port. When the Dartmouth, Nova Scotia, colony of Nantucketers was active, three of his Bunker cousins were masters of whaleships. During the Revolution his cousin Isaac was captured while in the privateer *Black Princess*, out of Dunkirk, and put in the infamous Mill Prison at Plymouth, England, and upon being released returned to Nantucket.[19]

Eber Bunker probably made his way to London during that first influx of Nantucketers after the Revolution, and may have accompanied his cousin Captain Owen Bunker, who married an English girl in 1785—the same year he took command of the ship *Brothers* for Alexander and Benjamin Champion. Eber followed his cousin's example and in 1786 married Miss Margaret Thompson of Middlesex, only daughter of Captain Henry Thompson, of the Royal Navy, personal pilot for King George III. His wife's mother was the former Isabella Collingwood, a first cousin of the illustrious Lord Collingwood, who was to become Commander of the British fleet after Nelson's death at Trafalgar.

In June 1788 Captain Eber Bunker had his first command—the ship *Spencer*, owned by Lucas and Spencer—in which he made a voyage to the Brazil Banks and returned on August 18, 1789, with a valuable cargo

18. E. C. Moran, Jr., *The Bunker Genealogy* (Boston, 1955), p. 40.
19. Ibid, p. 39.

of sperm and whale oil. Captain Eber made his next voyage in the same ship *Spencer*, going around Cape Horn with the whaleships following in the wake of the *Emilia*, and filled the vessel. Upon his return, Alexander and Benjamin Champion offered him command of the ship *William & Ann*, chartered for its historic voyage to Botany Bay, the result of which episode has been already noted. Returning home in June 1793 he soon after assumed command of the new ship *Pomona*, also for the same company. Sailing May 6, 1795, he returned August 18, 1797, with a full ship.[20]

His growing family—three boys and two girls—no doubt made his stay between voyages a pleasant prospect, indeed. In his plans, however, he felt the further exploration of Australian whaling possibilities was a part of his future. When the Messrs. Champion offered him command of their new ship *Albion* he quickly consented to make the voyage from London to Sydney.

The *Albion* had been launched at Deptford, on the Thames, October 25, 1798; had been rigged, fitted out, and provisions taken aboard (900 tierces of salt pork); then set up quarters for another consignment of convicts. She sailed from England early in 1799, and arrived at Sydney on June 29.[21]

Following the time-proven sea course from Cape of Good Hope, far to the south, and below Australia and Tasmania, Captain Bunker took the *Albion* to Sydney in the record time of three months and fifteen days—a remarkable passage. Governor Hunter made special mention of this, the fastest voyage yet made between England and Australia, in a despatch. Collins gives the measurements of the *Albion* as follows: length of keel, 86 ft.; extreme breadth, 27½ ft.; depth in hold, 12 ft.; height between decks, 6 ft.; measured (old style), 362 tons.

Upon his clearance at Sydney, Captain Bunker sailed in September 1799, to resume his voyage whaling. He returned to Sydney in August 1800 with six hundred barrels of sperm oil, reporting having seen "immense numbers of whales." During his cruise he had put in at Broken Bay, on the Australian coast, for "wood and water."[22]

Bunker was in Port Jackson again in the *Albion* in 1801. On August 26

20. *Board of Trade Records*, B.T. 6/95, p. 249.
21. Cumpston, *Arrivals and Departures*, p. 35.
22. Ibid., p. 36.

he cleared for England with 155 tons of sperm oil on board. On his next voyage from England, Captain Eber took the *Albion* again to Australian–New Zealand waters. Leaving London early in 1802, he put in at Sydney on July 6, 1803, with sixty-five tons of sperm already on board, most of which had been taken on the New Zealand grounds. He also had on board sixteen rolls of tobacco, a box of hats, and an organ, all from England. It was on this cruise that he discovered a group of uncharted islands, reported as follows:

The Albion had been as far to the Southward [*actually it should read Northward*] *on this Coast as 22° S and discovered a group of Eight or Nine islands in 24° 5′ S which are from two to three Miles in length, may be seen 4 or 5 leagues off, are well wooded, and bound with Turtle, of which Captain Bunker brought some here. These islands lie about ten leagues N.W. from Break Sea Spit, and about 12 leagues East Northerly from Boston Bay* [*this should be Bustard Bay*]. *The westernmost of them was seen by Captain Flinders in his last voyage. Captain Bunker reports deep water and good anchorage all round them. Notwithstanding the length of time this ship has been at sea,* [*near 13 months*] *her crew are in perfect good health.*[23]

The islands are still known today as "Bunkers Group." But the inimitable Captain Bunker was destined for a more remarkable part in the history of this area. The *Albion* was chartered by Governor King to carry settlers (and convicts), stock and provisions to a new place of settlement on the island to be known as Tasmania, then called Vandieman's Land. They reached the Derwent River after a passage of twelve days, having anchored at Oyster Bay on the way, and killed three sperm whales during this period. Lieut. Bowen, who was to be the first Governor of the new settlement, was landed safely at Risdon Cove up the Derwent.

In a letter to Governor King under date of October 5, 1803 Captain Bunker wrote:

Mr. Bowne and myself thought it most prudent to run up the Risdon Cove, and by lashing two of my whaleboats side-by-side we got them ashore very well . . . I am happy to inform His Excellency that the Derwent River is one of the handsomest rivers I ever saw. There is not

23. *Sydney Gazette*, Sydney, Australia, July 10, 1803.

one rocky point all the way up, and every bay affords a fine run of excellent water; one that I saw runs with sufficient force to turn a mill. I am informed by Capt. Corteys (who, in the Lady Nelson, *reached the river first) that Ralph's Bay is shoal; my short Stay would not permit me to sound it.*
We lost one small cow on the passage, but I believe she was hurt coming on board. We had a good gale of wind on the 2nd night after departure from Port Jackson, so I have the ship to assist the stock. I hope Mrs. King and the little girl is well. Governor Bowen sends Mrs. King a pr. of black swans which he begs her acceptance.

E. Bunker

P.S. I think your excellency to let me have 150 ackers of land on the Hawkesbury River as I have only taken 5.0 at Vandaman's Land. I send your excellency a pair of Black Swans, which I beg his acceptance.[24]

As will be noted, Captain Bunker had been granted a most unusual privilege during the voyage from Sydney to the Derwent; he was permitted to launch boats and go after whales if the occasion arose, provided he saw to it that the convicts were properly guarded during such an event. One may imagine lowering boats to chase whales with a hundred wretched prisoners left under closed hatches until the crews' return!

It was the grant of 150 acres on the Hawkesbury River, to which was added another 150 acres, that gave Captain Bunker his first holdings in Australia. They were to become a farm and a future home. He was probably thinking of this as he sailed from Sydney on August 21, 1804, bound for London, with 1,400 bbls. of oil and 1,300 seal skins under hatches. Upon arrival at his London home, his account of the prospects in this new land must have convinced his wife, as when he again sailed—early in 1806, this time taking command of the Campbell and Co. ship *Elizabeth*—Mrs. Bunker and the five children were on board, bound for their new home. On August 5, 1806, the *Elizabeth* arrived at Sydney and the Bunkers took up residence ashore. In honor of his wife's now famous cousin, Captain Bunker named his property "Collingwood."

Although he now signed himself "E. Bunker, Planter" he did not

24. Captain Eber Bunker to Gov. P. G. King. Copy made by Captain Harry O'May, Hobart Town, Tasmania, for E. A. Stackpole, February 9, 1957.

23. Whaleship *Syren* of London, England, Captain Frederick Coffin Shared with the Nantucket whaleship *Maro* the honor of taking the the first whales on the "Japan Grounds." Painting by Walters [Courtesy the Nantucket Historical Association]

leave the sea for a number of years. He took the *Elizabeth* whaling to the southward stopping again at the Derwent, returning to Sydney in April, 1807, and again in September. He dispatched the vessel to London in November 1807 under Captain Alexander Bodie.

The property owned by Bunker in Sydney was situated in the southern end of the city, in later years to be known as Princes Street. It was an excellent plot, overlooking the harbor at both sides of the hill from Cumberland Street to George Street, and on the early layouts of Sydney the name "Bunker's Hill" was given to the locality.

During the term of office of Governor William Bligh, former master of the mutiny ship *Bounty*, Captain Bunker finally joined the ranks of those who deposed Bligh on January 1808. The British authorities continued to regard Eber Bunker with considerable respect. In 1808 he was intrusted with a mission—the pursuit of the new brig *Harrington*, piratically seized by some forty convicts who had sailed toward New Zealand. For this mission Captain Bunker was placed aboard the brig *Pegasus*, and sailed from Sydney in May 1808 for the Bay of Islands. The *Harrington* was not found. In fact, it was not until March of the following year that one of His Majesty's ships forced her ashore at Manila, where she became a wreck.[25]

In this year (1808), while he was at sea, Mrs. Bunker died. Governor Macquarie engaged Captain Bunker's services in various duties, one of which was a voyage to Tahiti; another, in command of the *Venus*, a voyage to India for supplies. While on the voyage with the *Venus* he anchored in Adventure Bay and found buried at the foot of a tree a bottle containing letters from La Perouse, dated one month after that unfortunate explorer had left Sydney. He did not lose his whaling connection, taking the ship *Frederick* on a voyage in November 1810. The following year found him in command of the *Governor Macquarie* bringing spars from New Zealand. One of his most interesting assignments was bringing the whaleship *Seringapatum* from Sydney to London late in the year 1814. This ship had been captured by Porter in the *Essex*; retaken by the British whalers (who were prisoners) at the Marquesas Islands; sailed to Sydney by these whalemen; then placed in charge of Captain Bunker for the voyage back to England. As a cargo he placed aboard 8,594 seal skins, 92 casks of oil, 276 bundles of bark, 11 casks arrowroot and bullocks' hides.

Upon returning to Sydney he sailed the ship *Enterprise* on a sealing voyage for her American owners. The following year (1818) found him commanding the *Dragon* on a voyage to Bengal. In 1821 he was promised a grant of six hundred acres on the Hunter River and given a permit to take one hundred cattle there. Next, we find him back in England, where he purchased the ship *Wellington*. Always seeking new whaling grounds he continued his restless search, taking out the ship *Alfred* in 1824 for a voyage to the Santa Cruz Islands.

Captain Bunker was married three times, his second wife being Margaret McFarlane, widow of an officer of the East India Company, and his third wife being Ann, widow of an army officer, Captain William Minchen. At his pleasant home "Collingwood" he died on September 27, 1836, at the age of seventy-five, and was buried in the old Church of England cemetery at Liverpool, New South Wales. By his first wife, he had six children, five of whom lived to maturity, three being sons—Henry was lost at sea. James became master of a whaleship, Charles

25. Cumpston, p. 63.

married but had no children. The elder of the girls, Isabella, married Captain Thomas Laycock, and the younger, Mary Ann, married Captain Arnold Fisk, an American from Rhode Island.[26]

Like so many active mariners he kept no diary, or journal, and he would have liked Governor Macquarie's description of him: "A very able and expert seaman, and of a most respectable character." As the acknowledged "Father of Australian Whaling" Eber Bunker deserves much more recognition than he has received. The childhood of the man prepared him to not only follow in the footsteps of his Nantucket ancestors but to join his Nantucket cousins in the long pursuit of their calling.

There is a sequel to the Captain's story. During World War II Colonel Laurence Eliot Bunker, also a descendant of the Bunkers of Nantucket, was in Sydney on the staff of General MacArthur, and did some personal research on Eber's life. Writing to his mother in Wellesley, Massachusetts, Col. Bunker reported some of his findings, and of finally locating the ancient mariner's grave:

On our way from Canberra to Sydney, I persuaded the General to stop long enough in Liverpool to see if I could find Captain Eber's grave. St. Luke's is right on the main Highway and having been built in 1819 is one of the local antiquities. At first I thought I was going to be quite unsuccessful as there was no sign of a grave near the church. But I found it about a block down the road and there in the front row was the big flat stone which was made much later than the time of his death, as it also commemorates two of his grand-daughters.[27]

As the leader in whaling in these seas Captain Eber had seen the whaleships of his adopted country (many under command of the Nantucketers) take a major share in bringing about the Parliamentary Act of 1802, which finally broke the monopoly of the East India Company in these seas. The Act in question (42 Geo. III, Cap. 77) was entitled: "An Act to permit British Built ships to carry on the Fisheries in the Pacific

26. Captain Eber Bunker, genealogical family record; copy by A. P. King, Pukekohe, New Zealand. Vol. XI, *Royal Australian Historical Society Journal and Proceedings* (1925), Part I, p. 15.
27. The Grave of Captain Eber Bunker, description in letter from Col. Lawrence E. Bunker, February 6, 1945, to his mother, Mrs. Clarence Bunker.

Ocean without License from the East India Company or the South Sea Company."

Britain had unknowingly reached its peak as the leader of the Southern Whale Fishery. With the new century the Nantucketers began again their second surge to ascendancy in the industry they had originally established as its pioneers.

14. Quaker Whalers' Haven at Milford

In Pembrokeshire, Wales, the magnificent harbor of Milford Haven stretches far into the pleasant countryside. Tales of its early history involved Welsh and Norman castles and Cromwell's armies, but the most unusual chapter in its story had nothing to do with warriors or feats of arms. On the contrary, it was concerned with the building of what is today the present town of Milford—a venture more extraordinary, as the promoter planned the creation of a whaling port through the invitation to a group of American whalers—the original Nantucket Colony that had settled at Dartmouth, Nova Scotia. The whalers were invited to bring families and ships; to establish themselves at a quay-side area at Milford, where houses would be erected, warehouses and shops built, and their bluff-bowed vessels safely moored.

No more remarkable an incident had yet occurred in this Old World setting. All the variegated colors of an ancient tapestry were woven into the story. The hues were of light and shade, highlighted by the dreams of a determined young Englishman; the financial support of his uncle, a titled Lord; the economic opposition of London whaling merchants, supported by the head of the powerful Board of Trade; and the courage of the families of the whalemen, twice removed from their native land. Now but a footnote in the larger history of whales and whaling, the story of Milford's beginning is the more intriguing for what might have been its sequel—success. Fiction would probably have treated it differently but could not have improved its facts.

In the sequence of events the story contained all the aspects of an eighteenth-century drama. No playwright could have better selected those concerned in the action, a cast guaranteeing its unusual development. In these roles were handsome Charles Greville of London, a member of Parliament, one-time sponsor of the beautiful Emma Hart who was to become Lady Hamilton; Greville's uncle, Sir William

Hamilton, owner of the land at Milford where the new town was to be built; the whaling merchants Samuel Starbuck and Timothy Folger, Quaker leaders of the Dartmouth Colony, and their Nantucket whaling family neighbors; Jean Louis Barrallier, a French engineer who had escaped the Revolution, and who was the architect of the new town; Charles Stokes, mysterious agent, who had crossed the ocean to sound out the Nantucketers as to the removal; certain London whaling merchants, who feared the rivalry a successful port at Milford might bring; and Lord Hawkesbury, dominant, powerful in his headquarters of the Board of Trade at Whitehall.

Even those who came briefly onto the stage played notable parts: Emma, Lady Hamilton, whose love affair with England's great naval hero was the gossip of the hour; Lord Nelson, who returned her love, well shown in their visit to Milford; Abiel Coleman Folger, wife of Timothy, whose almost cryptic diary presents the contemporary scene; Benjamin Rotch, who came from Dunkirk to Milford and was a momentary saviour of the economic scene; and Elizabeth Rotch, his wife, whose account of their life in the Welsh town is one of the rare records illuminating the stage.

There was a true prophecy in one of Shakespeare's plays, *Cymbeline*, when Imogen remarked:

How far is it,
To this same blessed Milford? and by the way,
Tell me how Wales was made so happy as
To inherit such a haven.

Could the playwright have obtained his accurate word-picture of this place from some Elizabethan mariner? It would appear that only a seaman's enthusiasm must strike such an inspiration.

Sir William Hamilton had acquired the manor of Hubberston in Pembrokeshire through his marriage to the widow Catherine Bardou. The future town of Milford Haven was then a stretch of almost vacant land, lying between two deep water indentations—Priory Pill, to the west, and Castle Pill to the east—a steeply sloping expanse of green fields dropping down to the shores of the harbor. Sir William had some nebulous plans of building wharves or quays, but with acceptance of an appointment as Envoy to Naples, Italy, he put this aside. When his wife died in 1782, he was left as sole owner of the manor. Being now quite used to an indolent life, it is doubtful if he would have promoted any development here had it not been for his nephew, Charles Francis Greville. Then in his late thirties, Greville was made Sir William's agent for Hubberston Manor, and as Hamilton was childless he looked upon Greville as his heir.[1]

A complex man, Charles Francis Greville might have been recorded by historians as the young aristocrat who befriended a young, beautiful English girl named Amy Lyon, better known as Emma Hart, who became his mistress. Although she had little or no schooling, the sixteen-year-old girl had a naive charm and natural presence. It was Greville who introduced Emma to his uncle, Sir William, and five years later (1791) the mistress of the nephew became the wife of the uncle.

Rumors had it that this transfer of position and fortune for Lady Hamilton was an arrangement based on the payment of the nephew's many debts by Sir William. In any event, Greville was something more than a London man-about-town. He had been a member of Parliament for ten years; he numbered among his close friends men of culture; he was an art collector, mineralogist and amateur astronomer. But his vision of a whaling port in Pembrokeshire became his historical monument.

Sir William Hamilton's support of nephew Greville's plans, to build a town at Milford and to invite the Nantucket whalers, must have been considerably encouraged by Lady Hamilton's once close association with Charles. The importance of that former relationship cannot be wholly discounted. About this time, 1789 (two years before he married his twenty-five-year-old-mistress), Lord Hamilton discussed with his nephew the latter's visionary prospect. There were other reasons why the idea had a practical aspect. A growing trade with Ireland (a packet plied regularly to and from Waterford), and some evidence of American trade, prompted Sir William to seek legislation to enable him to actually build a town. By Act of Parliament in 1790, he was empowered to "provide quays, docks, piers and other erections, and to establish a market with proper roads."[2] Among key structures planned were an inn for

1. Flora Thomas, *The Builders of Milford* (Haverfordwest, Wales, 1951), p. 25.
2. Parliamentary Acts, "30 Geo. III, Cap. 55."

travellers, a custom house, and wharfside warehouses. But it was a town planned rather than a town begun, until Greville provided the basic idea which actually brought this "proprietary town" into being.

The drama of Milford Haven had a prologue. As was noted in a previous chapter, when William Rotch made his pilgrimage through Britain in the summer of 1785, seeking a port where his proposed Nantucket Colony might settle, he wrote of Milford Haven:

I suppose it is the best harbor in England or Wales. It is surrounded with a fine fertile country . . . abounds with provisions and coals, and has more wood in its neighborhood than any other harbor that I am acquainted with. Ship building is also cheaper here . . . It is very near Ireland, which I look upon as some advantage. Upon the whole I believe this port the best adapted for a number of people, but for my part I prefer one of the ports on the English Channel.[3]

Whether Greville met William Rotch in London, or whether he learned then of the Nantucketer's proposal to government is not clear. But it is significant that Greville's original idea of creating a whaling port at Milford had its inception at this time—the idea that Milford should become a haven for the wandering Quaker whalemen from Nantucket.

Greville presented his case in a letter written to the very London whaling merchants who, later, were to oppose his plans—Messrs. Enderby and St. Barbe, and their associates. He described the "object of my establishment at Milford," noting he had undertaken it as he "considered it as a National object"—furthering the whaling industry of Britain by inducing Nantucketers to settle there. With pointed malice, he noted that the "influence of your committee in 1785 [when they sought to block the proposal of William Rotch] with Lord Hawkesbury tended to limit the liberality and justice of this country to the Nantucketers . . . deprived me of a chance of settling Mr. Rotch and his family at Milford."

This statement is important in establishing the fact that Greville might have gained the Rotch colony of Nantucketers that had become so

3. William Rotch, Sr., to Samuel Rodman, London, November 2, 1785, Bullard, *The Rotches*, p. 211.

successful in Dunkirk. He further emphasized this point by observing that the frustration of the Rotch proposal had "Confirmed his prediction to the Privy Council . . . that France would gain a whale fishery from our folly, and we should lose that market and establish a competitor formidable in the European market."[4]

Greville, in his disappointment at losing first opportunity with Rotch, decided to go directly to the source, the Island itself, and invite the whalemen; also to approach the group in Dartmouth, Nova Scotia, with similar invitations—since Parr had been forbidden to allow any more emigrants from Nantucket. Greville wrote Lord Grenville, Home and Colonial Secretary, outlining his plans for Milford, suggesting that an agent named Charles Stokes be sent to Nantucket and Dartmouth to investigate the attitude of the whalemen. As expected, the letter was referred to the Board of Trade.[5]

Hawkesbury, at this time well aware of a possible error in shunting aside William Rotch, decided to follow up Greville's proposals. Stokes was interviewed and then promptly commissioned to go to America. "This judicious person," as Greville described him, sailed during the summer of 1790 and his dispatches, to both the Board of Trade and Greville, reported the Nantucketers more inclined to follow Rotch to France—but that the Dartmouth whalers, favoring the Milford invitation, had received the approval of Governor Parr for such a removal. Stokes returned early in 1791, filled with enthusiasm for this development. He urged that the British Government encourage the removal of the Nantucketers as "these valuable people will be in due course of a few years entirely lost to every country."[6]

No one was more cognizant of the power of the London whaling merchants than Governor Parr of Nova Scotia. After their appeal to the Board of Trade had been advanced by Lord Hawkesbury, and Parr had received the peremptory order that further immigration from Nantucket to Dartmouth must stop, he saw that the future of this or any other Colonial Fishery was doomed. His letter to Lord Sydney is a fine example of controlled hurt. Under date of November 21, 1790, he wrote:

4. Charles Francis Greville, letter to "The Committee of South [Sea] Whalers," London, July 7, 1792. *Liverpool Papers*, 38,288.
5. *Board of Trade Records*, B.T. 5/6, p. 261.
6. Ibid.

24. Captain Frederick Coffin
[Miniature in the possession of the Nantucket Historical Association]

I have to lament exceedingly that Government did not inform me of their wishes five or six years ago, or empowered me to encourage our imigration from Nantucket to England. If they had I am persuaded that few of the people called Quakers would have remained there and that none would have gone to Dunkirk, or any other place in France.[7]

Greville's proposal to the Dartmouth whalers, that they remove to Milford, followed by Stokes' visit. This brought about a formal petition by the Dartmouth group—with Parr's sanction—to Sir William Hamilton asking what terms might be offered for such a removal. Having learned by their own experience the necessity of approval directly from the British government, their proposal was carefully worded, and submitted by Greville through Lord Grenville's office, noting that the aid of government was solicited in the removal of their property to Milford Haven, and "that the same families have sent proposals to Sir William Hamilton by Mr. Stokes." Dated February 21, 1791, the communication carefully noted:

That it was the wish of Sir William Hamilton to accommodate them on terms of mutual advantage; that they think the situation on his Lands between Hubberstone and Castle Pill will answer their purposes, and desire him to transmit the conditions as to Rents, lengths of leases, etc., with which he will accommodate 25 families, on an average of seven persons, and 13 ships whose compliment is 182 men, the average tonnage of the 13 ships is 75 tons which, fitted for a whale voyage, they estimate at £11, (£11 per ton) which makes the Capital they propose to have afloat this year for Milford amount to £12,305. As the nature of their adventure admits of no delay, they desire that the port at which their vessels are to enter, may be ascertained, and the address of Sir William Hamilton's agent to whom on their arrival they may apply; as on the receipt of a favorable answer there would probably not be time to hear again before their arrival.[8]

Greville had no doubt drawn up the proposals of Sir William, which offered the settlers leases of plots of ground equal to their plots at Dart-

7. Gov. Parr to Lord Sydney, Privy Council Records, P.R.O., P.C. 2/136, p. 92.
8. Petition of the Colony of Whalemen at Dartmouth, Nova Scotia, May 18, 1791. Privy Council Records, P.R.O., P.C. 2/136, pp. 93–113.

mouth; that ship owners "shall enjoy all the accommodation for shipping;" and a guarantee of land for the site of a Meeting House for the Society of Friends.

But there was another clause in the Greville "communication" that read as follows:

A plan to enable the Nantucketers who abandon their property in Nantucket to remove their floating property and carry on their Fisheries from a British port.[9]

Most carefully, it was stipulated that those vessels coming directly from Nantucket might bring a cargo of oil, and "on its first outfit from Great Britain shall be entitled to Registry and all the privileges of British ships." Only forty ships would be permitted such entry. Greville pointed out that the French were anxious to "turn the distress of the Nantucketers to their own advantage if the liberality of Great Britian does not open its Ports to them."

In his shrewd appraisal of the situation then prevailing on both sides of the Atlantic, Charles Greville advanced his arguments so as to play upon the wish of the London merchants to eliminate their pesky Colonial rival in Nova Scotia, and Lord Hawkesbury's desire to obtain the services of whalemen directly from Nantucket rather than have them join the already successful Rotch venture in France. With these facts on his side, Greville appealed for support from both the astute Lord Hawkesbury and the committee of London merchants.

Greville's letter contained a listing of the families at Dartmouth that desired to be removed to Milford, the number of their vessels and the quantity of tons, and signed by twenty-three of the former Nantucketers who represented a total of 161 persons "to be removed to Great Britain," and as was noted: "These Families are Nantucket Families except Benjamin Robinson, who settled in Dartmouth under advantages of Nantucketers."

Ship masters listed were Silas Paddock, Jonathan Barnard, Libni Barnard, and Edmund Macy; Obadiah Worth was listed as a sail maker; Nathaniel Macy as ship carpenter; Benjamin Robinson and John Foster, coopers.

9. Ibid.

	Vessels	Tons
Samuel Starbuck and Family		
Daniel Starbuck and Family	3	280
Samuel Starbuck and Family		
Timothy Folger Esq. and Family		
Barnabas Swain and Family		
Peter Macy and Family		
David Grieve and Family	7	585
Timothy Folger Jun. and Family		
Benjamin F. Folger and Family		
David Coleman and Family	1	70
Seth Coleman and Family	1	70
Stephen Waterman and Family		
Jonathan Coffin and Family		
William Ray and Family	1	70
John Chadwick and Family		
	Vessels 13	*Tons* 1,075

Each vessel 14 *hands*
13 *vessels*

The vessels employ about 182 *hands*[10]

It was further carefully noted that, as the Dartmouth ships were constructed entirely for the sperm fishery, "they could derive no benefit from the bounties and premiums now given to the Greenland and Southern Whale Fishery," and it was humbly proposed that a bounty of "30/per ton, for every ton the vessel is registered," would be in order.

Coupled with the petition of the Dartmouth group was a separate but similar proposal from Benjamin M. Holmes, of Halifax across the harbor from Dartmouth, who wished to join in the venture. He presented the following statement of his property:

10. Ibid.

	tons		men	
Vessels—Ship–Romulus	*160*	. . .	*15*	*Now at*
Brigs–Joseph	*140*	. . .	*15*	*Sea on*
Argo	*130*	. . .	*15*	*Whaling*
Industry	*140*	. . .	*15*	*Voyages*
Schooner–Resource	*75*	. . .	*14*	
A new Ship to be launched about the 1st of May next	*480*	. . .	*15*	
	825		*89*	

The Captains, Officers and most of the crew of the within mentioned Vessels are from Nantucket.[11]

This made the potential Dartmouth migration to Milford Haven comprise nineteen ships and 271 seamen, and twenty-three families—"and entire class of Nantucketers of those who are now entitled to all the privileges of British subjects in their persons and ships—exclusive of the said 55 to transport each family."

Greville submitted the total package as follows:

Value of the property in Lands and houses certified by Governor Parr	£*6,000*
Compensation to M. Holmes for loss in transferring his property	*500*
Freight of 25 Families or average of 7 persons total 175 Persons	*1,375*
Do to Mr. Holmes for 5 Families of 7 persons each 35 .	*275*
	£*8,150*
Deuduct of the value of property certified as above by Governor Parr which 25 Families offer to convey to government	£*6,000*
Balance against Government	£*2,150*[12]

11. Ibid.
12. Charles F. Greville to Lord Hawkesbury, London, March 26, 1791. *Liverpool Papers*, Addit. Mss. 38,226, f. 98.

Greville's carefully prepared presentation contained a most important point. With the removal of the Dartmouth ships Britain would gain a control of the valuable spermaceti oil and head matter, as the Nantucketers in Nova Scotia specialized in the sperm whale fishery. During the year 1790, some 1,383 tons of sperm oil and head matter were exported to Great Britain by the Dartmouth fleet of twenty-two vessels. If the vessels of this fleet were granted a bounty, he contended, it would induce other Nantucket ships to join them at Milford.

The migration proposal concluded with an itemized account of the estimated values of the property owned by each of the principal families at Dartmouth, amounting to £5,985.

Not losing any time, Greville wrote to Lord Hawkesbury in March 1791, enclosing a sketch of the site of the proposed town at Milford. He also passed along the information that one of the London correspondents of William Rotch had learned of the French National Assembly's favorable action, on the Nantucket man's request for renewal of the bounties at Dunkirk, and the continuance of the prohibition of foreign oils.

"This," Greville asserted, "will enable him to bring the remainder of the Nantucketers to France unless Great Britain counteracts the French plan, as the prohibition in France extends to American as well as British oil. The whole of my reasoning therefore applies in the fullest . . . they must emigrate and if your Lordship does not open [Milford] to them [the Nantucket Islanders] they must go to France."[13]

As on a previous occasion when faced with a vital decision, Lord Hawkesbury directed his able secretary George Chalmers to appraise the situation. Chalmers had been highly critical of William Rotch's proposal, and remembered, no doubt, that his recommendation to reject it had given the Dunkirk venture an opportunity to succeed. This time he favored the Greville proposal.

It would bring the Southern Whale Fishers away from Nova Scotia, where they are a nuisance to England . . . nothing could be more important than the settlement of such a people to Milford.[14]

13. Ibid.
14. George Chalmers to Lord Hawkesbury, London, March 14, 1791. *Liverpool Papers*, Addit. Mss. 38,229, ff. 13–19.

The Board of Trade followed suit by recommending financial support. In April 1791 the Treasury granted the proposed migrating families £50 each if five or more removed, payment to be made on arrival at Milford; £2,000 to be paid as recompense for losses incurred by the removal, and a just cargo of oil to be admitted duty-free.[15]

The request for £5,985 in payment for their property at Dartmouth was refused, the Board of Trade stipulating the settlers must sell the property as best they could. Pensions of £150 per annum were allotted Timothy Folger and Samuel Starbuck, the two leaders of the group, for the duration of their lives and the lives of their wives because of their Loyalist reputations.[16]

During his negotiations with Hawkesbury and the Board of Trade, Greville had arranged for a representative of the Dartmouth group to visit Milford Haven and inspect the prospect. In the summer of 1791, Samuel Starbuck, Jr., the twenty-eight-year-old younger son of Samuel, arrived in London and found a letter awaiting him from Greville:

I will wait for you in Wales, and will show you on the spot what I can do and what I will do . . . I alone solicited the provision for your Father and Mother and for Mr. Folger and his wife, and have conducted the whole of the interests of your Friends. Your uncle Rotch can tell you that I have been equally sincere in my good wishes to him and in my desire to settle them at Milford . . . I am persuaded of your discretions and good sense from your being entrusted by your Friends of Dartmouth to report on their Interests . . . The mail coach comes through to Milford Haven, and Haverford West is 8 miles before you arrive at the Haven. I should therefore meet you there and I will take you to the situation which Sir Wm. H. proposed to allott. You will probably not be sorry to rest at Havenfordwest, as it is a long trip.[17]

Samuel Starbuck, Jr. found Milford Haven as Greville hoped he would, an opportunity to build a whaling town in the finest harbor in England. There was another factor entering the picture. Samuel had gone to

15. Minutes of the Board of Trade, B.T. 5/7, p. 131.
16. Ibid.
17. Thomas, *The Builders of Milford*, pp. 22–25.

Dunkirk to visit his cousin Benjamin Rotch before returning to Dartmouth, and he found the Revolution had already brought about definite plans on the part of the Nantucketers there in case the French Revolution forced them to leave. As Greville mentioned in his letter, he saw clearly the opportunity of gaining the Dunkirk Rotch group for Milford Haven if they were forced out of France.

Not wishing to bypass any opportunity to bolster his grand scheme, Greville wrote directly to Prime Minister Pitt, proposing that he (Pitt) sponsor a special measure, whereby the Dunkirk group would be granted the privileges of British citizens upon arrival at Milford, and that ships, not exceeding twenty in number should be then given British registries. This gave Greville two strings to to his bow—the Nantucketers at Dartmouth and the Nantucketers at Dunkirk—a combination that would pose a strong competition for the London whalers.[18]

As was to be expected Pitt immediately consulted Hawkesbury who, in turn, pointed out that London merchants must be consulted, as it was "their spirited endeavours" which had given Britain its present ascendancy in the Southern Whale Fishery. As Parliament would not again meet until late in the year, no action could be expected until that time, as the Board of Trade could not support any measure that would injure the business of these London merchants by "introducing foreigners into the Kingdom as Rivals to them."[19]

That Greville was now fully aware of his convictions is evident by the letter he then wrote to the "Committee of the South [Sea] Whalers from London," under date of July 7, 1792. This letter was in every way a remarkable one, as it presented in a clear and forceful way not only "the object of my Establishment at Milford," but also the actual results of such a removal to that port by the Nantucketers who had agreed to settle there. Both the sincerity of the promoter and the shrewd analysis of the venture itself constitute a document of perception on the whole subject of whaling out of Britain. One point Greville stressed was this:

18. Charles F. Greville to Prime Minister Pitt, *Liverpool Papers*, Addit. Mss. 38,228, f. 21.
19. Lord Hawkesbury to Charles Greville, *Liverpool Papers*, Addit. Mss. 38,228, ff. 34–35.

It must be obvious to the Committee that the whole Produce of the Fishery so to be removed, in fact does not increase the Quantity of Oil in the British market, for all the Nova Scotia oil is already disposed of as British oil. And your information will probably suggest that not only the possibility of smuggling Nantucket oil will be defeated by the removal but it may be possible for you on this Removal taking place from Nova Scotia to prevent a Revival of the Whale Fishery in the British Colonies.[20]

Greville also declared he had not previously consulted the Londoners because: "I neither wanted your weight of interest nor your Capital, and I prefer creeping on with Industry and small Capitals." To this sharp statement he added that the Nantucket system of shares or "lays" for the crews of ships "will be departed from by Great Capitals who engage from the River with fresh crews every voyage, and I know, if the system of Shares is not kept up somewhere within its original Purity, there will be soon a want of skillful Masters, Harpooners and Fishers"—consequently that Milford would be a veritable nursery for such men "as he intended to support the 'lay' system to the utmost."

There could be no doubt as to the basic purposes of Greville's plan, which are best stated in his own words:

1. To cut up both French and Colonial competition in the Whale Fishery. 2nd To separate the detached Nantucketers from the Nantucketers remaining in America. The 3rd Object is personal to me, vizt.—To give Permanency to my Establishment by a further addition of original Whalers, either by a general Invitation or by a partial one.[21]

The reply from the London merchants, through John St. Barbe, came within a month's time, and was a carefully worded rejection of Greville's desire for their cooperation. One paragraph summed up the Londoners' sentiments:

We cannot join with you on any account. The Fishery being now perfectly established in Great Britain, we are under no apprehension

20. Greville to "Committee of South [Sea] Whalers," London, *Liverpool Papers*, Addit. Mss. 38,228, f. 9.
21. Ibid.

of its being injured by the Trade carried on out of America . . . And as to the Fishery at Dunkirk we have not the least doubt but the moment the French fail in their encouragement (the period of which we think at no great distance) the fishermen settled at that place will readily embrace any invitation this country may hold out to them.[22]

Although admitting the fishery established out of Milford Haven "must be attended with many advantages," and that, if encouraged, a number of the Americans then in England might be induced to settle there, the London merchants disagreed with Greville "as to the invitation to Nantucket whalemen." They declared the "Expense and pains to establish the Fishery out of this Kingdom, and at a very great risk whether we should succeed, and now to solicit foreigners to participate who cannot exist much longer in their own country in that branch of Commerce, would not only show our weakness but must in a short time be the ruin of the fishery."[23]

Greville was not deterred by this rejection. His next appeal was directly to Hawkesbury, and once more the careful President of the Board of Trade requested his trusted Secretary George Chalmers to examine the state of the Southern Whale Fishery and report. With typical thoroughness, Chalmers reviewed the matter and, because he was not to lose his prejudice, he again failed to gauge the potential of the Nantucketers as a colony rather than as individual whalers. As Hawkesbury's right hand man, he favored the London merchants. He pointed out that the government had met its obligation to Greville's commitment at Milford by providing £2,000 from the Treasury, £1,342 of which had been paid to six Dartmouth whalers already arrived at Milford, and that it was not the Pitt's ministry's intent to go beyond a limited project as a means of providing financial aid to Nantucketers from Nantucket or from Dunkirk.[24]

In a later report, Chalmers noted: "The pointed answer to Mr. Greville is: What you propose is not a good to be desired, but an evil to

22. Committee of London Whaling Merchants to Greville, August. 2, 1792. *Liverpool Papers*, Addit. Mss. 38,228, f. 28.
23. Greville to "Committee of South [Sea] Whalers," f. 28.
24. Chalmers to Hawkesbury, ff. 13–19.

25. Table Bay and Cape Town—Cape of Good Hope
Pencil copy of a print found in the pages of a whaler's logbook
[Courtesy the Nantucket Historical Association]

be avoided. The Government do not want to have more ships at present in this Fishery, but more Markets for the produce of it." [25]

There was sound reasoning for Chalmer's argument. Under the sponsorship of Lord Hawkesbury the British Southern Whale Fishery had become a strong branch of maritime industry, with headquarters in London, and "ought not to be distressed or disturbed to promote individual projects," and, of special significance, was that phrase mentioning "an implied trust between Government and Adventurers" when the Fishery was first established. The London merchants had every confidence that Hawkesbury would continue to support them.[26]

As Greville had pointed out in his letter to these same merchants, however, it was their influence with Hawkesbury which "tended to limit the liberality and Justice of this country to the Nantucketers, and in fact deprive me of the chance of settling Mr. Rotch in Milford in 1785." This same influence was even stronger in 1792.[27]

During this period (1792) some fifty-nine whaleships had returned to British ports, the great majority to London, bringing home oil, whale

25. Ibid., f. 79.
26. Ibid.
27. Greville to Committee of South (Sea) Whalers, July 7, 1792.

bone, and seal pelts to a total value of £189,000. Information received by the Board of Trade indicated that the Dunkirk–Nantucket connection was their great rival, and that at this time another American port was looming as a potential competitor: Bedford, in Massachusetts. The man who had made it possible for Nantucket to rebuild its whaling fleet through the importation of American oils by way of Dunkirk's "back door" process, had also begun developing Bedford.

Hawkesbury felt the London fleet's success had given Britain a clear superiority in the Southern Whale Fishery, and that the French Revolution was sooner or later to drive Rotch from France; therefore he need not support the Milford Haven venture beyond the limited steps already taken.

Time was on the side of the able President of the Board of Trade. But the sands in the hour glass were running in one direction and, once the bottom was filled, the glass itself must be turned.

Perhaps this seemed too far ahead for even that astute man to feel worried.

In August 1792 Greville hastened to return to Milford to greet the first settlers from Dartmouth. He wrote to Hawkesbury, requesting his opinion and that he speak to Mr. Pitt. The reply was patterned after Chalmers' opinion—it must await the consideration of the London merchants investments. Another letter from Greville, under date of September 11, 1792, prodded Hawkesbury again. Some two weeks later, the reply brought to Greville the "honest and deliberate sentiments" of Hawkesbury on the subject, which were basically unchanged from his previous ones. It was the duty of government not to encourage rivals, "who may so overstock the Market as to bring Ruin on the present Adventurers."

The London merchants clearly held control of the chess game.

Early in 1793 Greville wrote a letter to Hawkesbury which reviewed the Milford Haven project thoroughly. He started out boldly:

I have desired Mr. Pitt to authorize me, as far as he can with propriety, to treat with a limited number of Foreign Fishermen. I have seen Mr. Rotch twice this year, and I am convinced that if the confusion

in France forces them to remove they will look to other situation than Great Britain from their considering the door shut to them.
I have stated the question fairly to the South Sea Whalers in London, and the answer from Mr. St. Barbe shows that they will oppose a general invitation.[28]

Reflecting his deep resentment, Greville followed up his comment with this statement: "Nothing but the mismanagement of the London merchants keep up the American adventure."

Another request was that Milford Haven be allowed to import all American produce such as lumber from Virginia for whale boats (cedar), rice, tobacco, and other goods. "It is hard," he wrote, "to see the tobacco ships of Liverpool and Bristol waiting orders in Milford and not a merchant here able to import a cargo . . . The principality of Wales has not a port licensed to carry on [American] trade, though best adapted by its port's situation."

Such practical ideas were bound to meet with opposition by ports which saw in Milford a potential rival, but it was the Committee of London Whaling merchants who continued to be Greville's chief opponent.

Samuel Starbuck, Sr., and Timothy Folger had gathered about then in Dartmouth some twenty families, who agreed to accompany them to Milford. As Samuel Starbuck, Jr., wrote to Greville:

It is natural for my Father, Samuel Starbuck and Father-in-law, Timothy Folger, and persons of advanced age to prefer Great Britain, where the stability of its Government gives security and permanency. But to young men, who have their fortunes to make, the French bounties of 50 livres per ton for ships on entry from every whaling voyage, without any limitation or restriction; the prices of oils, which will net when realized for sperm and head matter about £40 per ton, the black oil £20 per ton, are tempting advantages, and these from a free port not likely to suffer the changes of Government having prospered under the old and new Constitutions.[29]

28. Greville to Hawkesbury, Addit. Mss.
29. Samuel Starbuck to Charles F. Greville, August 22, 1792. *Liverpool Papers*, Addit. Mss. 38,228, f. 24.

As the senior Starbuck and Timothy Folger had been granted yearly pensions of £150 each by the Crown, it was certain they would prefer Milford and bring many of their close associates with them. However, the appraisal of the two ports by the sons of the older settlers at Dartmouth show how clearly Dunkirk appealed.

In regard to the Mother Island of Nantucket, the invitation to go to Milford Haven found no response. There, the crisis had already begun to ease. It was Richard Uniacke, of the Nova Scotian Assembly, who, in a report to Governor Parr in 1791, saw the situation now more clearly than even the Dartmouth or Milford promoters. He wrote:

Instead of the circumstances of these people [the Nantucketers] being distressed," *the report pointed out,* "*it is rather the reverse—they are at the present Day building many new Ships and extending their Fishery, so that the Terms which some years ago they would with joy have accepted for their removal, I am of the opinion, although the same should be offered with treble advantage, they would now refuse. Therefore, to make an estimate of the expenses that it would require to remove the whole bulk of the Inhabitants of Nantucket to Great Britain, would be to* calculate on a Sum which no nation could afford to give.[30]

And then the world of maritime affairs suddenly became changed again by war. At a time when the first settlers from Dartmouth were arriving in Milford Haven the French Revolution burst out in renewed fury and once again the Nantucketers in England, France, and America would feel the tides of war catching at the keels of their far-flung whaling ships.

30. Richard Uniacke to Governor John Parr, August 5, 1791. Public Archives of Nova Scotia, Vol. 302, Doc. 20.

15. Once More the Tides of War

1. MILFORD HAVEN

On one occasion Lord Stanhope, the brother-in-law of Prime Minister Pitt, wrote that the period between 1784 and 1793 was "the most prosperous and happy, perhaps, that England had ever known." This prosperity was in large measure due to Pitt's careful handling of commercial affairs. His adroit control was directly a result of the reestablishment of the Board of Trade, with Lord Hawkesbury his right-hand man. Other members of his cabinet were also close personal friends—Lord Carmarthen, Foreign Minister; Lord Sydney, Secretary of State, Henry Dundas, able Treasurer, and Lord Grenville, Foreign Secretary, his cousin and staunch supporter.

Trade had succeeded war; business had taken the place of the usual intrigue involving Whigs opposing the "King's Friends." Pitt's handling of the Spanish crisis in 1790, and his success in secret diplomacy with the French Revolutionists, added much to his prestige. The 1790 election gave him a strong majority in Parliament.

The brilliant Pitt, however, tried to ignore the French Revolution. At a time when England's military aid for the Austrian, Prussian, and Netherlands troops battling the Revolutionists might have defeated the French, the English Prime Minister remained aloof. His apparent indifference to the fate of the French clergy and nobility matched his complete unwillingness to face some of the wrongs in his own country—the press gangs, the transportation of people guilty of petty crimes, the debtors' prisons, religious intolerance. It was no wonder that the plight of the Nantucket whaling groups, who were willing to migrate to Milford Haven upon Charles Greville's direction, was but a minor issue to the austere and powerful Pitt—perhaps the most solitary of England's Prime Ministers.

But Greville had managed to reach Pitt, as was evident from Greville's letters to Hawkesbury. Also, quite obvious—after the initial success of Greville's bill in Parliament—was Hawkesbury's decision to lend no further aid to the Milford Haven venture after Greville had won Parliamentary approval giving him authority to invite the Dartmouth Colony to Wales.[1]

The first of the Nantucket–Dartmouth vessels to reach Milford was the *Sierra Leone*, owned by Folger, Starbuck and Co., which under Captain Elisha Clark, arrived in June 1792. On board were Samuel Starbuck, Jr., his wife Lucretia Folger Starbuck, and children.[2] As there were no habitations then at Milford, the family took up residence at Haverfordwest until the little hamlet on the shores of the magnificent harbor, between Hubberston Pill and Castle Pill, began to take form. The ship almost immediately sailed on her first whaling voyage from Milford.

Timothy Folger and Samuel Starbuck, the two leaders of the removal—this marking the second of such projects—sailed from Dartmouth on August 2, 1792, on the ship *Aurora*, under Captain Andrew Myrick. Accompanying the families of Folger and Starbuck were several other Nantucket families who had joined the original move from Nantucket. The time of their arrival at Milford was mentioned in a letter written in October 1792 sent from Dunkirk by William Rotch, Sr., to his son-in-law Samuel Rodman, in Nantucket: "I received last evening a letter from Brother Starbuck. . . . they arrived at Milford the 23rd inst . . . after a passage of 24 days—all well, except the women being much fatigued with the voyage." The voyage across the Atlantic had taken only three weeks and three days. In that time they had gone from one world to another.

No one worked harder than Greville. His was a single-handed task, as his uncle, Lord Hamilton, aside from granting him the use of the land, ended any financial aid after the initial costs of constructing a wharf (or quay, as the English call it) and a Custom House. Greville also was able

1. Correspondence between Greville and Hawkesbury, July 1792, *Liverpool Papers*, Addit. Mss. 38,228, f. 7.
2. Letter of Samuel Starbuck, Jr., to Charles Greville, *Liverpool Papers*, Addit. Mss. 38,228, f. 175.

to have the embryo "New Milford" declared a post town, an official clearing place for letters between Ireland other places and the country.

But two formidable barriers to Milford's success had already been erected—Lord Hawkesbury's policy of ignoring the settlement, and the open hostility of the London whaling merchants. Greville knew of the odds against him but with a stubborn courage the promoter persisted, feeling the merits of his plan would ultimately win.

In the late autumn of 1792, Timothy Folger and his wife Abiel Coleman Folger, and Samuel Starbuck, Sr., and his wife Abigail Starbuck, with other families from Dartmouth had become well settled near Milford. The Folgers were living at a place called Robeston Hall. A visitor to Milford at this time noted that some seven families had taken up residence at Milford.[3] The Folger and Starbuck firm had five whaleships and it is certain that the "Artificers" they brought with them consisted of coopers, sail-makers, whaleboat builders, and candle-makers. The two leaders were fully knowledgeable as to oil refineries.[4]

That most of the whaleships at Dartmouth, together with many families, did not follow the two leaders to Milford was due to two major developments. One of these was the resentment and suspicion of the Dartmouth Colony for the British home government. Speaker Uniacke, of the Nova Scotian Assembly, expressed it best in his letter to Richard Cumberland, the Province's Agent in London:

There was cause to suspect the intention of Government to remove the Whale Fishery from Dartmouth to Great Britain [Milford Haven], notwithstanding the assurances to the contrary . . . This is a fatal blow, aimed against this Province and if pursued will be universally considered as an Act of the highest injustice.[5]

Uniacke reminded Cumberland that the whalers from Nantucket had been established by the Province of Nova Scotia at a considerable expense, and that "property of at least £1000,000 depends on its success." It was this same astute Uniacke who had prophesied it would be impossible to lure other Nantucketers from that island to Britain at this

3. J. F. Rees, *The Story of Milford* (University of Wales Press, Cardiff, 1954), p. 24, quoting Mary Dudley.
4. Ibid.
5. Uniacke to Cumberland, August 16, 1791. *P.A.N.S.*, Vol. 302, Doc. 20.

time (1791) as the whale fishery had taken on a remarkable revival there.[6]

Despite his own bitterness at the connivance of the home government, the Nova Scotian Governor, John Parr, still provided advice, in his letter of reply to Lord Sydney:

I would strongly recommend to your Excellency . . . to procure the active services of these two men [Starbuck and Folger of the Dartmouth colony] . . . their influence and connection with the Colony settled in France, as well as with the Inhabitants of Nantucket, will enable them to . . . by degrees assembled from Nantucket at Dartmouth a sufficient Colony which may be made serviceable to Government. . . . The Island of Nantucket being within the Dominion of a foreign power, it would be attended with hazard for any set of Agents to go there for the avowed purpose of executing a project that would tend to remove the whole of the Inhabitants and Fishery to Great Britain; therefore if anything is done for their removal it must be by those Gentlemen gradually removing such as are willing to engage in the enterprise. . . . I see no other prospect but that the Southern Fishery, for want of being pursued in a proper manner, will be lost altogether to British subjects, as this [Dartmouth] is the only part of His Majesty's Dominions wherein since the peace it has been carried on to any advantage . . . I speak not from my own knowledge but from the knowledge of Men who have spent long lives in gaining experience in that Branch of Business alone, and from attentive observance of eight or nine years I am strongly of the same opinion.[7]

Then followed the granting by the Privy Council in London of pensions of £150 for life to both Timothy Folger and Samuel Starbuck and their families, and an allowance of £50 to each family removing from Dartmouth to Milford, together with a common fund of £2,000 indemnification for loss of value in property left behind. Duties on first cargoes brought into Milford were remitted.

6. Ibid.
7. Gov. Parr to Lord Sydney, Privy Council Records, May 18, 1791. P.R.O., P.C. 2/136, p. 92.

That only a few of the Dartmouth colony of Nantucketers followed Folger and Starbuck is the best evidence of their lack of confidence in both the venture and its leaders. Those who were to remain (at least temporarily at Dartmouth) did so because, as Quakers, they believed Samuel Starbuck, Sr., and Timothy Folger had entered into private agreements for the removal to Milford after their fellows at Dartmouth had instructed them "with sending a public one."[8]

Most of this bitter feeling was directed against Samuel Starbuck, Sr. Word of this cleavage in the ranks of the Quakers reached Nantucket, and as the Nantucket Monthly Meeting was the parent organization of the Dartmouth Friends, this group wrote to Dartmouth in the summer of 1793 and the elders inquired as to the trouble. The reply of the Dartmouth meeting was interesting. It recounts that upon the appearance of Agent Stokes (Greville's man) from England, the subject of removal was discussed, and that "after some days said inhabitants of Dartmouth convened and agreed that each man's property should be estimated and proposals fixed on which said, Inhabitants would be willing to move . . ."

Samuel Starbuck proposed that, if the above "proposals were acceded to, said Inhabitants should one and all go, but if not to their satisfaction all should stay, which he requested each to express their approbation." Continuing, the report reads: "People here find fault with Samuel Starbuck for being accessory in sending some private proposals while he was intrusted by the Inhabitants with sending a public one."

Some time later, it was learned that "a private Memorial had likewise gone forward," and that Starbuck and Folger had both received "partial annuities," although they had "undertaken to negotiate the public business." Upon receipt of this information, the Nantucket Meeting declared it could not grant a certificate to Samuel Starbuck until the embarrassment caused by the accusation was cleared. As a result, certificates as members of the Nantucket Monthly Meeting of the Society were "granted all except Starbuck at this time."[9]

Writing from Milford in June 1794 Samuel Starbuck expressed his "grevious hurt" by the circumstances which were so "injurious to my reputation." He declared that upon learning of the opportunity of

8. Records of the Dartmouth Meeting of The Society of Friends, Library of The Peter Foulger Museum, Nantucket Historical Association, Nantucket, Mass.
9. Ibid.

Captn M Gardner St Helena Sep 6th 1815

Respected Sir

I am now laying at the Island of St Helena Arrived yesterday from the Coast of Africa with 70 tons of black Oil theres ben Eight Ships on the Coast and All Done but little we anchored Along the Coast from Walwich Bay to Tiger Island I shall lay here Eight days to refresh my Ships Company and then off for the Coast Brzile —

Sally when I left England was well and her little Boy and as fine a little fellow as you Ever saw, the News from Timor Most of the Ships Done small the Hannah & Eliza Is one Among the low Numbers you Can Inform Eastons friend he was well the last Accounts from them March last I think most likely I shall return to America After this voyage for the whale fishery Is Nearly Done out of England the Gass lights has upset us All If you Dont get that 44 dollars from W. B Coffin I thank you to Acquaint me or Sally my Mother Brothers and Sisters I wish you to Inform them I am well yours

Fredk Coffin

26. Letter from Captain Frederick Coffin, September 6, 1815
[Courtesy the Nantucket Historical Association]

removal to Milford, he and his sons had decided to "act disconnected from any other person," although they had a "natural affection with other settlers." Timothy Folger, however, had advised that all should act together. A committee had thereupon acted for the colony, Starbuck declared, "to which I was not named nor consulted." After an agreement had been reached, Starbuck and Folger petitioned the Crown for an annuity "for ourselves and wives in consequence of what we had suffered in the American War." To this compensation they were as justly entitled as any other prominent Loyalist.[10]

However, as Seth Coleman, of Dartmouth, wrote: "Was this doing as he would be done by?" There will always be the question: As a principal promoter in the migration from Nantucket to Halifax, had Samuel Starbuck, Sr., abused his privilege by "furthering his own ends" at the expense of the Dartmouth colony? It must be remembered that, as this was their first migration, the whalemen "could not obtain the least favor from Gov. Parr without the sanction of Samuel Starbuck or Timothy Folger." Naturally, they still considered Starbuck a leader when the time came for removal to Milford. Starbuck himself had declared in Meeting, ". . . it was the families of the whalemen that the British government wanted, and not their land at Dartmouth."[11]

Several of the Nantucket families remained in Dartmouth, the most prominent that of Seth Coleman. Here was a man who left a strong imprint on the community, and his diary existed up until a few years ago when it was destroyed by fire. His descendants at Dartmouth lived until recent years. One of the most interesting members of the family, born in Dartmouth, was Captain John B. Coleman, one-time master of the whaleship *Charles & Henry* of Nantucket. Early in November 1842 while the vessel lay at Eimeo, in the South Seas, Captain Coleman took on board a young man who was "on the beach," and signed him on as a boatsteerer. His name was Herman Melville.[12]

Around the little cove at Dartmouth, in the winter of 1793, a strange silence fell over the deserted wharves and warehouses, the empty, simply

10. Ibid.
11. Ibid.
12. United States Consular Records, National Archives, Washington, D.C., in "Lahaina Records, Protests, Letters, etc., 1842–1855, No. 7625.

built homes, and the Quaker meeting house. Across the sweep of the great ocean the little fleet of whaleships finally sailed into the magnificent harbor of Milford Haven. From their decks, the Quaker whalemen and their families looked up at the craggy hills of Wales. This was a new and splendid opportunity: the land was rich, the harbor unexcelled. And, although in their hearts they would have preferred the sandy shores and rolling heath of their real homeland of Nantucket, they knew their destiny lay in the voyages of their ships in pursuit of the whale.

At Milford Haven history now came full cycle. A century and a half before, the English forefathers of these Quaker settlers had left Britain for life in the wilderness of the New World; now their descendants—although in a small colony—were returning to the Old World to seek the very thing the original Colonists had also braved three thousand miles of Ocean to find—freedom of opportunity.

It is of interest to note the names of the individual family groups who made the journey: Samuel Starbuck and his wife Abigail Barney Starbuck; their sons, Daniel Starbuck and family, Samuel Starbuck, Jr., and family; Timothy Folger and his wife Abiel Coleman Folger, and Timothy Folger, Jr., and family. Also, Barnabas Swain, Peter Macy, David Grieve, Benjamin F. Folger, David Coleman, Jonathan Paddock, Nathaniel Macy, Uriel Bunker, Frederick Coffin, and James Gwinn—all with families.[13]

The arrival of the Nantucketers actually marked the beginning of the town of "New Milford," as it was then called. Greville wrote that a mason and his men had been engaged, and that "he is a good mason . . . and is to begin work immediately . . . in the meantime I have referred him to you for advice and to no other person."[14] The elder Starbucks were then living at Robeston Hall, some two miles from the location of Milford. A quarry was opened for stone and a brick kiln built for the construction of the new homes. The building of the town now proceeded under the direction of the Nantucketers, with Greville's approval.

Lumber and stone and "artificers" were available locally, and houses went up rapidly. A single wharf or quay was constructed and the first few houses built. But it was not until the arrival of the Frenchman,

13. List of Nantucket Whaling Masters in Foreign Ships, Hussey and Robinson, *The Inquirer and Mirror*, July 7, 1876.
14. Thomas, *The Builders of Milford*, p. 30.

architect Barrallier, that the little town gradually took shape. Parallel to the shore three main roads were laid out, called Hamilton Terrace, for Sir William; Charles Street, for Greville, and Robert Street, for the latter's brother and nephew—all with intersecting lanes. It was a fine prospect. Wooded cliffs and deep, calm harbor blended with St. Ann's Head, the promontory to the west, like a guardian to the entrance to the most spacious harbor in Britain. Although it had been the scene of much history serving as the only Norman stronghold in Wales, the landing place for Henry II and Richard II, as well as Cromwell, its great harbor had never sheltered such a fleet as the whaleships from America, and the Quakers were the first to occupy a town on this site.

The migration of the Nantucket whalers to Milford Haven was as odd a twist to American history as one might find. However, none other than Thomas Jefferson had already pointed out why such a development could logically take place. Noting the success of William Rotch and his Nantucket whaling colony at Dunkirk, Jefferson had written to John Jay:

> . . . *The way to encourage purchasers is to multiply their means of payment. Whale-oil might be an important one. In one side are the interest of millions, who are lighted, shod and clothed by the help of it, and the thousands of laborers and manufacturers who would be employed in producing the articles which might be given in exchange for it, if received from America. In the other scale, are the interests of the adventurers in the whale-fishery, each of whom, indeed, politically considered, may be of more importance to the State than a single laborer or manufacturer, but to make the estimate with the accuracy it merits, we should multiply their numbers in each side into their individual importance and see which predominates.*[15]

One may only surmise what William Rotch would have created at Milford Haven had he been granted the right by Lord Hawkesbury and his Board of Trade to bring his Nantucketers to Britain.

15. Thomas Jefferson, *Correspondence and Miscellanies From The Papers of Thomas Jefferson*, Edited by Thomas Jefferson Randoplph (2nd Edition, Boston and New York), *Memoirs*, Vol. 2 (Boston, 1830), Letters CLXXI, Jefferson to John Jay, pp. 358–359.

With their acceptance of their new domicile, the Dartmouth-to-Milford Quaker whalers nonetheless continued to retain their ties with Nantucket, Dunkirk and London. The removal from Dartmouth marked the second departure from their established home in less than a decade. To the whalemen the sea was their home as well as the land, but to the women the separation was touched with sadness. As Abiel Coleman Folger, wife of Timothy Folger, wrote in her diary some years later, of the anniversary of that arrival, ". . . this day Sixteen years I step'd my feet on Welch land and a grevious day it was to me."[16] At the age of fifty-five, leaving her grown children in Nantucket, she was never to return to her homeland, and her sad appraisal of her lot is understandable.

2. DUNKIRK

Then war came again—like a plague to the world of the whalemen. On February 1, 1793, the French Republic declared war on England. If any conflict could have been anticipated it was this one. In Dunkirk, William Rotch had seen the inevitable disruption and planned accordingly. As early as June 1792 he had written his son Thomas from London that he wanted to return to America and if favored, he would prefer to settle in Bedford.[17] The elder Rotch had many friends in England, and had shrewdly gauged the potential of war with France by re-registering a number of his ships out of London, thus anticipating capture by British cruisers. His familiarity with the mercantile situation was aided by his many connections in Britain through his travels and visits to various Friends Meetings.

A visitor to Dunkirk gives evidence of this relationship among the Quakers. Job Scott, a travelling Friend, left Boston on the Rotch whaleship *Mercury*, Capt. Benjamin Glover, "a ship bound on a whaling voyage from Dunkirk." Approaching this French port on the first day of the year 1793, Scott recorded:

16. Diary of Abiel Coleman Folger, property of Mrs. George Jones, Nantucket, Massachusetts.
17. William Rotch, Sr., to Thomas Rotch, Bullard, *The Rotches*, p. 248.

Got within four miles of Dunkirk. Being desirous to be with a few friends at Dunkirk next day at Meeting I went aboard the pilot boat . . . and got safe ashore, and soon found the house of my dear friend William Rotch and family, of Nantucket, now resident here . . . where I met a very cordial reception.

Jan 6: I sat with the few friends at this place in their meeting and rejoiced in the Devine presence.

Jan. 18: Crossed to Dover from Calais with William Rotch and Robert Grubb attending Friends Meeting with them.[18]

William Rotch left Dunkirk for London "in order to save our vessels if captured by the English," as he wrote in his *Memorandum*. Two days after he left France the unfortunate Louis XVI was beheaded, and two weeks after he arrived in England, the French Republic declared war on England. His timing was remarkable. Before another month passed two of the Rotch ships, the *Ospray* and the *Mary*, were captured in the English Channel with full cargoes. Although bound for Dunkirk, the papers which William Rotch handed Customs when the vessels reached London enabled him to recover his ships, producing their American registries.

There was a special touch to this incident. Being human, even the usually reticent William Rotch could not resist noting how he had out-foxed his old enemy. He mentioned it, however, in his typical manner of understatement:

My going to France to pursue the Whale-Fishery so disappointed Lord Hawkesbury that he undertook to be revenged on me for his own folly, and I have no doubt he gave directions to the cruisers to take any of our vessels they met with going to France. When the Ospray *was taken by a King's ship, the Officer who was sent on board to examine her papers, called to the Captain, and said: 'You'll take this vessel in, Sir. She belongs to Mr. Rotch!*[19]

Coupled with the danger to which the whaleships were now exposed, the British navy began stopping all American ships bound to and from

18. *Journal of The Life & Travels of Job Scott* (New York, 1797), pp. 300–307.
19. William Rotch, *Memorandum*, p. 67.

France. A number of these ships were detained over long periods, an irritating practice. The French also put unnecessary restrictions on American ships. The *New Bedford Medley*, for Nov. 1, 1793, reported:

France has a decree that no vessel shall carry out a cargo that has not brought one in, which leaves 30 sail of American ships in Bordeaux without business.

Benjamin Rotch had remained in Dunkirk while his father went to London. His mother and sisters went aboard a vessel at Dunkirk secretly, and reached England safely, where they rejoined the anxious head of the Rotch household and went on to London.

During the next few weeks Benjamin carefully arranged for the business of the Rotch firm in Dunkirk. The faithful French house of the De Bauque Brothers was a strong ally, while Captains Benjamin Hussey and William Mooers, together with Jeremiah Winslow, were to remain in France as representatives of the Rotch–Nantucket ships, as well as to plan for ships which might safely operate from Dunkirk when the war tensions eased.

At last the time came for Benjamin to try his "removal." Still under the close watch of the Revolutionary officers, he found it necessary to make this departure as unobtrusive as possible. In later years his daughter, Eliza Rotch Farrar, told of the experience:

Not until the fall of Robespierre was my father permitted to leave France; then he embarked on one of his own vessels with his family and all his valuables on board. He was aware that a number of persons had secreted themselves on board in order to escape from France, and as two custom house officers accompanied the vessel down harbor my father was afraid they would discover the fugitives.[20]

By providing the French officials with a hearty lunch and plying them with much wine, Benjamin diverted their attention until the ship was well down the harbor and it was necessary for them to board the pilot boat for a return—not having time then (or the inclination) to actually search the ship thoroughly. Many years later Captain John Hawes, who

20. Eliza Rotch Farrar, "Recollections of Seventy Years," Bullard, *The Rotches*, p. 320.

commanded the ship in which Benjamin Rotch left Dunkirk, told his daughter that a large amount of specie was hidden in the ceiling of the ship's cabin, none but Mr. Rotch and Captain Hawes knowing of its whereabouts.[21] The ship landed Benjamin and family in England.

The outbreak of the war between France and England found many of the Rotch–Nantucket fleet at sea, and it was feared many would be captured before warned of the danger in attempting to return to Dunkirk. As an example, the *Three Friends*, Captain Abel Rawson, while off Walfish Bay on the Western coast of South Africa, sighted a Britsh cruiser, was chased, but managed to escape. Fortunately, Captain Rawson had learned of the war from Captain Amaziah Gardner, who had lost the Rotch ship *Hebe* to an armed British whaler. Capt. Gardner had gone ashore with a boat's crew and escaped, later boarding the first whaler coming into the Bay—Captain Rawson's command. The *Three Friends*, not wishing to risk the return voyage for Dunkirk, put away for America, arriving at Bedford in mid-September 1793. Another American shipmaster returning on this whaleship was Captain Isaiah Cahoon.[22]

At this time, the armed British whalers became actually letters-of-marque, or legalized British privateers. One of these was the *Liverpool*, of London, well armed with twelve guns, which sailed into Walfish Bay in July 1793 and captured the Dunkirk whalers *Phebe*, Captain Edward Coffin, and *Judith*, Captain Paul Ray. Five other whaleships were in the Bay at the time—all with Nantucket registries, however, as well as as French. Three were also of the Dunkirk–Nantucket fleet—the *Harmony*, *Hero*, and *Harlequin*—but were not molested. During that night, Captain Ray cut his cable and escaped, the *Judith* arriving at Bedford a month later with the news.[23]

Another Dunkirk ship, the *Favorite*, under Captain David Folger, had reached the Bay of Biscay when a ship hailed her to break the news of the war, and Captain Folger immediately changed course for America, reaching Nantucket safely. The *Favorite* had been on a twelve-month

21. Rebecca W. Hawes, *John Hawes*, in Old Dartmouth Historical Sketches, No. 22.
22. *New Bedford Medley*, September 20, 1793.
23. *New Bedford Medley*, August 23, 1793.

voyage to the Brazils, and her crew were in sad shape with the scurvy, when she arrived.[24] Equally fortunate was the whaleship *Eliza*, also a Rotch ship, which left Dunkirk in July 1793 and managed to elude the British cruisers, reaching Nantucket in September. Captain Benjamin Coleman reported "great rejoicing at Dunkirk," celebrating a victory of the French army over a combined Austrian-Prussian force.[25]

On the Southeast coast of the African continent, at Delagoa Bay, another group of Dunkirk whaleships was running the gauntlet of war. Early on the morning of June 30, 1793, the Nantucket whaler *Asia* sailed into Delagoa after having been as far distant as the west coast of Australia, and to Desolation Island. At anchor in the Bay were eleven whaleships—all under command of Nantucket men—among them the *William Penn*, Captain Obed Fitch; *Leveret*, Captain Isaiah Bunker; *Greyhound*, Capt. Obed Bunker; *Negar*, Captain John Hawes; *Edward*, Captain Micajah Gardner; *Benjamin*, Captain Isaiah Hussey; *Dauphin*, Captain Stephen Gardner—all from Dunkirk. News of the French declaration of war against England had not as yet reached the east coast of Africa.[26]

When the news did arrive it was through the very backwash of the war itself. On September 2, 1793, a Dutch privateer from Cape Town entered Delagoa Bay and proceeded to capture the *William Penn* and *Greyhound* as French vessels. When the Dutch boarding party went on board the *Benjamin*, however, the registry produced gave her as a Nantucket whaleship, and Captain Hussey was allowed to sail. During the night Captain Gardner got the *Dauphin* to sea and escaped, reaching Dunkirk safely, while the *Benjamin* crossed the Atlantic to Bedford.[27]

The *Dauphin* then went through another adventure which somehow was typical of the varied careers of these Dunkirk-based whalers. Francis Rotch, brother of William Rotch Sr., was still in France and, being more hardened by the changing fortunes of his life than his Quaker relatives, chartered the vessel to the French government for a voyage to Haiti with

24. *New Bedford Medley*, September 25, 1793.
25. *New Bedford Medley*, September 20, 1793.
26. Logbook of the ship *Asia*, entry of June 30, 1793.
27. *New Bedford Medley*, February 17, 1794.

troops, munitions, and supplies for the French army. After completing this assignment the *Dauphin* returned to Bedford to outfit for another whaling voyage—her ultimate oil cargo to go to Dunkirk once more. At Bedford, William Rotch, Jr., wrote a letter to Uncle Francis which is remarkable for its polite yet strong criticism:

I shall take the liberty to give thee hint upon what I have heard was the destination of the Dauphin, *vizt., to carry troops to Hispaniola. However the insurrection in that quarter may be viewed in France, with the thinking people of New England whose minds are unclouded with the dark deeds of slavery, it is looked upon that the struggle on the part of the Negroes and Molattoes is as just as was the American struggle for liberty. How much then must thy feelings be hurt by taking a step for the sake of gain that will be considered by thy countrymen as unworthy an American . . . I have pledged myself a faithful friend to the abolition of slave trade and slaveholding, and am almost daily concerned with protecting the injured Africans and promising them liberation . . .*[28]

The exigencies of war may have caught one member of the Rotch family—but it was attributed by William, Jr., to his uncle's "submission to the choice of those connected with thee than to thy own wish." As for William Rotch, Sr., he had long been an advocate of the black man's freedom, and on his ships they were given the same "lay" or share as the whites.

Among the numerous adventures in escapes, captures, and near-captures the Dunkirk fleet experienced, one of the most bizarre had to do with the *Edward*, under young Captain Micajah Gardner. On November 20, 1793, she arrived at Bedford from St. Helena, reporting that the governor of that British-owned island tried to decoy Gardner into the anchorage. The Captain had sent his mate and a boat's crew ashore, unaware of the outbreak of war, having just come from the Indian ocean. Instead of his boat returning another came out with a message from Governor Brook, announcing that "France was at war with the world,"

28. William Rotch, Jr., to Francis Rotch, July 10, 1792, Bullard, *The Rotches*, p. 263.

27. Friends Meeting House, Milford Haven, Wales
Erected by the Quakers from Nantucket who came to Wales from Nova Scotia [Photograph by S. Robinshaw] [Courtesy the Nantucket Historical Association]

that the American ambassador at Paris had been beheaded, and offering generous terms if he brought his vessel in and had it taken over by His Majesty's authorities. Captain Gardner remained off the island, hoping to recover his boat's crew, but when all he received was another note from the governor, threatening him with "severity of treatment" unless he surrendered, he sailed away—shaping his course for Bedford.[29]

As the long war dragged on the Dunkirk fleet became scattered, finally gaining safety for the moment in the several homes in which the Nantucketers had been able to find refuge. Of those captured by the British, William Rotch in London was able to reclaim six; eight became victims of French, British, and Dutch privateers; ten found a new home in Nantucket and six in Bedford.

29. *New Bedford Medley*, November 23, 1794.

Despite the gloomy prospects of the conflict, William Rotch still kept three of the Dunkirk whalers based in that port and had an equal number serving as cargo vessels between Bedford and Dunkirk. His son Benjamin remained in France until November 1793 when, as related, he crossed the Channel to Portsmouth, bringing his wife and three children —two of them, Eliza and Benjamin, born in Dunkirk during the three years of residence. Francis Rotch, Captains Hussey and Mooers, and Jeremiah Winslow remained in France to conduct the affairs of the Rotch firm.

Once more the tides of war had carried the fortunes of this remarkable whaling family toward the dangers of a lee shore. But while much of their hopes and plans were washed away, these tides could not destroy them. In London, William Rotch completed his business and, when Benjamin reached his side, he turned the firm's affairs over to his son. A new ship, named the *Barclay* for Robert Barclay, one of the close English friends of Rotch, had been constructed in Bedford by George Claghorn and launched in October 1793. She was sent to France with oil and an assorted cargo early in 1794, with orders to then proceed to London and bring William Rotch, his wife, and two daughters back to America. The elder Rotch noted:

We embarked on 24th of 7th month, 1794; had a long passage of 61 days, and arrived in Boston 23rd of 9th month. The night before our arrival an awful circumstance took place during a squall, Calvin Swain, brother of our Captain David Swain, fell from the main topsail yard into the long boat and was instantly killed . . . We proceeded to Bedford, and after spending a few days there, returned to our home at Nantucket, finding all our children and grand-children well that we left more than four years before, and six added in Samuel and William's families.[30]

William Rotch, Sr., remained only a year in his old home before moving to Bedford, there to join his son William and son-in-law Samuel Rodman in transferring (to that virtual colony of Nantucket) the firm of William Rotch & Sons. With this removal—the last to be made by this

30. Rotch, *Memorandum*, p. 68.

great champion of international whaling—the port of Bedford was to acquire a stimulus to the economic surge that was to eventually create the "Whaling City."

The Nantucket–Dunkirk–Bedford fleet had its triangular trade still in operation. But a fourth port was soon to be added—Milford Haven. Benjamin Rotch was not to return to Dunkirk, leaving his agents there to carry on the Rotch firm's activities while he and his family continued to live in England. During that year (1794) he remained in London attempting to resolve a suit in chancery involving one of the Rotch fleet, and when this was completed he decided to settle in Milford Haven where his uncle and aunt, Samuel Starbuck, Sr., and Abigail Starbuck, and his cousins were already living.

It was the coming of Benjamin Rotch to Milford which gave the hamlet a new lease on hope, as with the enterprising young man there also came a number of the Rotch ships.

Before leaving London, however, Benjamin and his family enjoyed many opportunities to be with English Quaker families. Among their friends was the American artist Benjamin West, who was President of the Royal Academy, and saw to it that Nantucket Quakers met many Americans who were living as exiled Loyalists in London. On one occasion, the Rotch family was admitted to an exhibition at Somerset House, and through West's intervention, attended a preliminary session at which the King and Queen, Princesses, and young Edward, Duke of Kent, were to appear.

Little Eliza Rotch was disappointed because King George wore no flowing robes or diamond-studded crown, but a brown bob wig. And as for the King, he too had a bit of disappointment, and little Eliza wrote about it many years later:

My father and mother, two other Americans who were their friends, and myself, formed the favored few. All wore the Quaker dress but myself, and the King said in his nervous way, 'Are these your Quaker friends, Mr. West?' and repeated it twice before Mr. West could reply. 'Is that little girl a Quaker? She has not got a Quaker bonnet on: if she had the Quaker bonnet, I would have spoken to her. Tell her so, tell her so, Mr. West.' Meanwhile the Princess Elizabeth admired the bonnet of my mother, and said she wished she could have one like it. The Royal

Party stopped a few minutes in the hall, on purpose to give the strangers opportunity to look at them, which was a kind attention of Mr. West.[31]

The friendship of Benjamin Rotch and Benjamin West was a lasting one, and in West's famous painting of "Christ Healing the Sick," the head of the demoniac's father is a likeness of Benjamin Rotch.

Another of the memorable socials found the Rotch family being invited to the well-appointed home of Charles Greville. Here they met Sir William and Lady Hamilton for the first time. The account by Benjamin's daughter Eliza is most revealing:

Lady Hamilton was still very handsome and sang sweetly. My mother refused to be introduced to her, who was the lion of the evening . . . and though she [Lady Hamilton] saw that she was being avoided by the beautiful Quakeress, she was not to be deterred from speaking to her, and . . . said she hoped to see her in Wales that summer, as Sir William and she hoped to visit Milford.[32]

3. MILFORD HAVEN—WHALING PORT

Charles Greville was delighted with the prospect of Benjamin Rotch removing to Milford. In November 1794 he had written to Lord Hawkesbury, seeking his help in obtaining further concessions from the Government.[33] "The settlement of Milford has been severely depressed by the lack of your Lordship's support," he wrote. War had now given Milford a new opportunity to be recognized as a naval base, especially should the French attempt to go to Ireland and establish a headquarters in that country. After a year of appeals, Greville finally persuaded the Navy Board to use Milford to build ships for the Navy.

Without question the coming of Benjamin Rotch to Milford Haven gave a new impetus to the revival of the proposed town.

31. Alfred Rodman Hussey, *An Exile From Home*, a paper delivered at the annual meeting of Nantucket Historical Association, July 18, 1947.
32. Eliza Rotch Farrar, "A Daughter's Recollections," Bullard, *The Rotches*, p. 340.
33. Charles Greville to Lord Hawkesbury, November 2, 1794. *Liverpool Papers*, Addit. Mss. 38,229, ff. 101–102.

At the same period a young French engineer, Jean-Louis Barrallier, came on the scene to assist in the construction of the naval vessels and a dock yard. Forced to flee Toulon when the Revolutionists took over that port, Barrallier was appointed an assistant to the Inspector-General for Naval Construction at Milford, a reward for assisting the Royal Navy invading and occupying Toulon. With his arrival at Milford he was immediately pressed into service by Greville to lay out the streets of the new town, in a plan agreed upon by Sir William Hamilton. The plan was "that of three parallel streets, well terraced, with right-angle intersecting side streets—often described as the gridiron lay-out—and it still gives Milford its distinctive character, despite much haphazard building in later years." [34] The progress made in the next six years was such that the Inn (now called the Lord Nelson Hotel), the Hamilton House, and other larger structures, were constructed. This shows how ill-founded is the belief that the Nantucket Quakers actually built the town.

There is, however, evidence that Greville expected the new town to be primarily for the development of the colony of Nantucket whalemen. His letter of June 1797 shows this clearly. Writing to Samuel Starbuck, Sr., he stated:

I have explained to Mr. Barrallier my intention to give every preference to your accommodation in my power. He has the plans of Milford and he is limited by me to certain points which are essential to you as well as all future settlers; vix., to keep the quay line clear for warehouses throw the embanking opposite to the warehouses on the persons who lease those spots. [35]

What the whaling families from Nantucket did do was to bring to the new town a tradition. The fact that Milford was planned and designed originally for them, and first occupied by them as the original settlers from Dartmouth, is a unique demonstration of this whaling tradition's value. The industry they brought to this corner of Wales could have become a signal success in Britain but for the antagonism of the London merchants and the failure of Hawkesbury and the Pitt Ministry to further support Greville in his projected planning.

34. Rees, *The Story of Milford*, p. 26.
35. Thomas, *The Builders of Milford*, p. 37.

Having lost their advantage in Dunkirk, the Rotch–Nantucket combination might have given Milford Haven the necessary stimulus. As it was, Benjamin Rotch brought several of the Rotch fleet to Milford Haven, and the connection with America was maintained through Bedford.

Benjamin's daughter Eliza, then twelve years old, described the activities of the Rotch family after reaching Milford, to take up residence after nearly three years of pleasant living at Islington near London:

My father began immediately to build stores and a dwelling house, and ships began to arrive from America, [*Bedford and Nantucket*] *full freighted with sperm oil. The business attracted the artisans necessary for carrying it on, and houses sprang up on every side, and Milford became a scene of activity unknown before. The author of so much prosperity was deservedly popular, and his prompt pay secured him plenty of workmen. The oil imported from the United States was landed, the casks coopered, and then re-shipped in small coasting vessels to London. This, with the outfitting of his ships for the South Seas, made a thriving business for a variety of trades, and introduced some new shops into the town.*[36]

Exercising the same careful attention to the details of getting his vessels documented under the British registry, Benjamin Rotch was thus able to bring in several cargoes of oil which were actually obtained when the ships were under Dunkirk, Nantucket, or Bedford registry. While this was circumventing the heavy British import duty, it was necessary to acquire the funds necessary to launch the whaling industry on a more substantial basis than the first settlers from Dartmouth were able to do.

The *Ann*, under Captain Prince Coleman, sailed from Bedford on December 20, 1799, loaded with 681 casks of sperm oil and 426 bundles of whalebone, consigned and delivered to Benjamin Rotch at Milford, the cargo valued at $36,465. A month later another of the former Dunkirk whaleships, the *Wareham*, left Bedford for Milford Haven under Captain James Gwinn, with sperm oil valued at $27,317.75. She

36. Eliza Rotch Farrar, "Recollections of Mother," Bullard, *The Rotches*, p. 341.

also carried staves valued at $808, and casks of salt beef, pork, molasses, coffee and rum.[37]

Supplies for whaling voyages were far cheaper in the United States than in Wales. It is of interest to note what the *Ann* and *Wareham* brought with them to Milford Haven, aside from their own outfits, other supplies, including such cargo as, per ship *Wareham:*

30 bbls prime beef, at $9	$270.00
21 bbls prime pork, at $15	315.00
9 casks molasses, 321 gals	165.00
1 bag coffee	31.00
2 Casks N.E. Rum	100.80
	$877.30
Shooks of staves, valued at	$808.37

In the *Ann's* cargo there were casks of bread (hard tack), raisins, chocolate, cranberries, and two kinds of soap.

Early in March 1800 the two vessels completed outfitting at Milford and sailed for the South Seas, with Captain Gwinn in command of the *Ann*.

The manifest of the *Hannah & Eliza*, Captain Micajah Gardner, sailing also to Milford, showed 526 casks of sperm oil, valued at $29,650, plus a quantity of goods "necessary for whaling outfits," including white oak staves, 8 casks sugar, 3 casks coffee, 40 barrels beef, 20 bbls. pork, 10 bbls. flour, 9 casks molasses, 1 cask rum (125 gals.), 8 barrels tar, 10 whale lines (cork), 1 bbl. spruce beer, 1 cask dried apples, and a small quantity of "tobacco for the crew." Delivered personally to the home of the ship's owner, Benjamin Rotch, at Milford, were 1 barrel nuts, 3 barrels brick wheat flour, 1 small bag dried apples. Leaving Bedford in December 1800 the *Hannah & Eliza* reached Milford Haven without incident and soon after cleared for the United States again, Captain Gardner making another round trip during the next few months.

Late in April 1804 the *Hannah & Eliza*, with Captain Gardner still in command, sailed from New Bedford for the new Australia–New Zealand whaling grounds. Five days out she was chased and forced to heave

37. Rotch Account Books, New Bedford Public Library, New Bedford, Mass.

28. A. Milford Haven, Wales (1794)
From an engraving by J. Sewell in the *European Magazine*
[Courtesy the Nantucket Historical Association]
B. Milford Haven, Wales (1812)
The town built for the Quaker whalers from Nantucket
[Courtesy the National Museum of Wales]

C. The Hon. Charles Francis Greville
From a painting by Romney
[Courtesy the Nantucket Historical Association]
D. Lady Hamilton
From a painting by Romney
[Courtesy the National Maritime Museum, Greenwich, England]

E. "Castle Hall," Milford Haven
Home of Benjamin Rotch, no longer standing
[Courtesy the Nantucket Historical Association]

to by the British frigate *Leander*, and ten of the crew were taken off by a pressgang. Forced to return to port, Captain Gardner sailed again on May 23, making a passage around the Cape of Good Hope and reaching the island of Tasmania on October 21, 1804. Continuing on to the New Zealand grounds, the ship eventually put in at Norfolk Island for provisions, having previously met the Milford whaler *Ann*, Captain James Gwinn, a fellow-Nantucketer, and accompanying him on a cruise.

In May 1805 a year from home, the *Hannah & Eliza* put in at Broken Bay on the Australian coast near Sydney, where the ship was "hove down," her hull cleaned, and her "bends" tarred down or "blackened." After another cruise along the coast, the ship returned to Broken Bay to procure provisions from Sydney, and sailed for home in company of the *Ann*, arriving at Milford Haven late in September 1806. It is probable that the *Hannah & Eliza* had both an American and British registry, as the impressment of her crew indicates the latter. The *Ann* was owned by Benjamin Rotch, and had formerly sailed out of Dunkirk.[38]

Also returning to Milford was the *Grand Sachem*, Captain Whippey, (August 1808) having been gone over two years on a voyage to the New Zealand grounds. Another of the original Halifax fleet, the *Jefferson*, Captain Andrew Brock, made a similar voyage, while the *Aurora*, Captain Andrew Myrick, came back a year later after reaching the same grounds. The *Grand Sachem*'s crew of twenty-two were all British subjects except two. This was no doubt the proportion in the other Milford ships.

It is significant to note that the first whaleships to explore the New Zealand grounds were from four different ports—London, Nantucket, Milford, and Bedford—but all were under Nantucket masters. Included were the *Vulture*, Captain Thomas Folger, and the *Elizabeth*, Captain Bunker, both of London, but formerly Rotch ships out of Dunkirk which had been captured by the British in the French war and sold to London merchants.

One may only imagine the thoughts of these veteran whaling masters when they brought their bluff-bowed ships into the spacious harbor of

38. Collection of Ship's Papers, E. A. Stackpole, Nantucket. Lloyd's List, 10, 6, 1806; Mediterranean Passes, P.R.O.

Milford Haven and finally moored alongside the solitary quay. The beauty of this location must have struck an immediate appeal. Yet, they were whalemen and their lives were more in the open sea, and under their command the Milford fleet proudly maintained their Nantucket record.

Because whaling materials and supplies could be obtained more cheaply and readily in America, the Starbuck and Folger interests at Milford followed the pattern of the Rotches at Dunkirk, and a pattern of trade was established in which ships intended for whaling also served as merchantmen. The whaling voyages also followed the courses to grounds frequented by the Nantucket ships. The *Romulus*, under Captain Matthew Pinkham and then under Captain William Slade, went to Delagoa Bay, Africa; Captain Joseph Clasby took the *Resource* to the Pacific; and the *Hibernia*, under Solomon Coffin, was among the first in the Galapagos Islands' whaling region.

During the next decade the Milford fleet made an enviable number of voyages; Captain Andrew Myrick took the *Aurora* around Cape Horn and into the Pacific, returning to Milford in January 1806; former Dunkirk ship *Ann*, under Prince Coleman, made her next voyage as an armed vessel due to resumption of the war with France; another Nantucketer, Captain Laban Russell, made two excellent voyages in the *Charles*, and then went into the employ of the London fleet; the *Duke of Kent* was sent to the South Seas under Captain Ammiel Hussey of Nantucket, who was succeeded by Captain Edward Clark. One of the original Dartmouth whalers, the *Maitland*, commanded by Captain Uriel Bunker, also made some good voyages between 1796 and 1806, mostly to the Pacific.[39]

The decision of Benjamin Rotch to cast his lot with the Nantucketers at Milford was to Greville a happy anticlimax to his dream. Benjamin had made considerable money at Dunkirk and, through shrewd investments in London banks, was able to settle in Milford with his wealth intact. His arrival stimulated the settlement's economy.

A contemporary Welsh historian presented an excellent word-picture of this revival in the fortunes of Milford Haven. In 1811 he wrote:

39. Felix Farley's *Bristol Journal*, August 27, 1808, November 14, 1807.

The Quakers from the Island of Nantucket, who accepted Mr. Greville's invitation to settle there, were a valuable accession to his new colony, and everything like commerce and enterprize that has discovered itself at Milford may be dated from their arrival. They are a most industrious, well disposed people, with the dignified simplicity of manners and strong understanding that their sect is generally distinguished for.

Mr. Rotch, the principal of the new settlement from America, is a gentleman of great commercial knowledge, connextions and property, and first fixed himself and his family in the town of Milford . . . on the terrace a house on a very large scale to suit his handsome establishment, but before he had nearly finished it an opportunity offered of purchasing a beautiful villa within a short distance of the town, just on the other side of the eastern Pill, called Castle Hall, built and inhabited for some years by the late Governor Holwell, one of the survivors of the diabolical imprisonment (the Black Hole) at Calcutta . . . afterwards purchased by a wine merchant of Havorfordwest, who sold it to Mr. Rotch, where he now resides, having enlarged and beautified the house, grounds and gardens.[40]

Castle Hall became a true show place, with an ingenious hot house where Benjamin raised exotic plants and fruit trees. Despite his Quaker upbringing he was fond of society, and his beautiful wife, dressed in her sober but handsome gowns, made the Rotch home a place of dignity and culture, the fame of which spread throughout Wales.

The first bank in Milford Haven was founded by Francis Rotch (eldest son of Benjamin), Samuel Starbuck, Jr., and S. L. Phillips of Haverfordwest, and was to be known as the Milford and Haverfordwest Bank. St. Katherine's Church was consecrated in 1808, with Benjamin Rotch among the laity present. Nelson's trophy, from his victory on the Nile, the truck of the French Flagship *L'Orient*, stands near the west door, a mute sentinel of a glorious era. Few realize that this was brought to Milford by Lady Hamilton long after Nelson's death.

But only meager accounts remain of the American colony in Wales and these are tantalizing in what they could present as a fascinating

40. Richard Fenton, *A Historical Tour Through Pembrokeshire* (London, 1811).

story. Included in these vignettes are bits of the mysterious trade between Bedford and Milford; the activities of Samuel Starbuck, Jr.; the miller and baker, introducing American corn bread to Wales; the range of the whaling fleet in the South Seas and the ironic twist which found several captured during the War of 1812 by Commodore Porter during the *Essex*'s cruise in the Pacific; the later years of the Rotch family at Castle Hall; and the visit of Lord Nelson and Sir William and Lady Hamilton.

In the latter instance, Nelson and the Hamiltons were guests of honor at a banquet held at Milford Inn in August 1802. On this occasion Nelson presented a portrait of himself to the Inn and inscribed his name on a pane of one of the windows. Ever since that time the Inn has been called the Lord Nelson Hotel. While in Milford Haven, Nelson, with his seaman's eye, noted the superb harbor and its advantages, and reported this to the Admiralty. Greville, who had early seen the opportunity of eventually establishing a naval base, continued to press the subject with Hawkesbury and Grenville.

Lord Nelson, during the banquet, was aided by Lady Hamilton in cutting his meat. Sitting opposite at the long table, Mrs. Benjamin Rotch noted (with disapproving eye) the unconcealed devotion of the beauteous Emma. The scene must have presented a study in contrasts—the Quakeress and the Mistress, each living according to the dictates of their strong wills.

When it appeared that Milford might prosper two economic changes brought reversals in her fortunes. The first was a more gradual lessening of her trade, chiefly the result of the failure of government to aid her whaling merchants—the inevitable influence of the London merchants. Next was the business failure of Benjamin Rotch. As a crowning blow the War of 1812 caught them as in a vice; the loss of three ships was like the loss of a fleet for any other port.

Benjamin's business collapse was ironic. During the last years of the war between England and France he had accumulated a large quantity of sperm oil, and had left orders with his London agent to sell when the price reached a high level of £120 a ton. But the agent, dreaming of an even more astounding price, held off. Then, with all its dramatic suddenness, Napleon was defeated, France fell, and the London whaling fleet came back with full cargoes. This, with imports from America, brought a sharp fall in the price of sperm. Forced to sell, Benjamin found no market, and when other investments failed to materialize he found himself bankrupt. To pay his creditors he had to sell Castle Hall and all but one ship, and soon after these misfortunes he moved to Bath with his family.

After years of holding meetings in various dwellings the Nantucket Quakers at Milford built their Friends Meeting House in 1801, which still stands and is still used. A short time later occurred the first death of those in the original colony—Abigail Starbuck, wife of Samuel Starbuck, Sr.,—to be followed, one by one, by her contemporaries. In the little Quaker burial ground at the rear of the Meeting House, are the graves of the original settlers of Milford, their resting places marked by low, grey stones, bearing simply their initials and dates. Here are interred Timothy Folger, his wife Abiel Coleman Folger, his daughter Lucretia, wife of Samuel Starbuck, Jr., and other Nantucketers.

The diary of Abiel Folger, already mentioned, was found only recently in a Nantucket attic by one of her Coleman relative's great-great granddaughters. Its almost cryptic entries constitute a precious contemporary account, albeit fragmentary, of the life of the Quaker settlers—exiles from their homeland.

Barrallier, the French engineer, whose skill had been employed to lay out Milford as well as to build the dock yard, was a neighbor of the Folgers. In her diary, Mrs. Folger mentions the launching of one of the British frigates, named the *Milford*. At five o'clock in the morning, catching the high tide, the *Milford* slid down into the Harbor.

I went down with nibor Barrelear," wrote Abiel Folger, "... She [the ship] looked like a wraith on the moors, and did not make the least noise but glided along ... it was a butiful sight."[41]

Three weeks later, April 22, 1809, Charles Greville died. In St Katherine's, the church he proudly saw consecrated, is a memorial, and the inscription serves the man's epitaph: "He sacrificed his fortune in his endeavours to promote the prosperity and develope the resources of this place."

41. *Diary of Abiel Coleman Folger*, "Historic Nantucket," Nantucket Historical Association, (April, 1955), p. 24.

With Charles Greville's passing the whalemen lost their staunch advocate in London. Although Robert Fulke Greville, who succeeded his brother to the estate, tried his best to help, he lacked the persistence of Charles. Yet, it was Robert who gave a telling climax. Writing to the Starbucks, he stated:

The situation at Milford . . . was as encouraging to perservering speculations as any seaport could be, and the disappointments which have happened to your Friends must be attributed to the overpowering influence of the capital which, becoming jealous of their success, at length crippled cruelly that enterprise which the situation of Milford was rendering prosperous to fair speculation.[42]

Milford Haven's early nineteenth century's potential will probably always remain a matter for conjecture. Certainly, the coming of the railroad era in England would have been a powerful influence on its development. At the moment it needed government aid, however, that necessary strength was lacking. Had Parliament's initial support been continued it is possible that the most perfect harbor in Britain might have obtained that impetus to enable it to gain the strategic American cotton trade—a trade which actually developed its rival, Liverpool.

As an ironic sequel to Milford's story, today the harbor is a rendezvous for other oil-carrying ships—the great tankers—which bring vast quantities to the refineries opposite the town. Twentieth-century oil has succeeded eighteenth-century whale-oil; and huge motor vessels have replaced the square-rigged whaleships. But, as for the whalemen from Nantucket—they cannot be replaced, neither at Milford nor at the home port of Nantucket. The breed has vanished; the story should not.

42. Thomas, *The Builders of Milford*, p. 45.

16. The Home-Comers and a New Home Port

UPON his return to Nantucket in September 1794 William Rotch, Sr., found himself in a community still divided in its opinion regarding his efforts to salvage the whaling industry. Many of his former associates had left Nantucket, men such as Samuel Starbuck, Timothy Folger, and Alexander Coffin; and a new generation had come into active business life during his decade of absence. Despite his quiet strength and indomitable will, Rotch found the criticism of his fellow Islanders difficult to accept. Facing the politicians of England and France; competing with rival merchants in London, Boston and Le Havre; matching wits with naval forces—all were challenges he was willing to risk. But, at home, the undercurrent of criticism (despite the fact that the critics were in the minority) was of grave concern to him. He was weary and the drain on his energies was depressing.[1]

While this situation at home made him unhappy, his decision to again remove from the Island was based strictly on the advantage which he saw with the development of nearby New Bedford. As mentioned earlier, it was to this village on the Acushnet River that the firm of William Rotch and Sons had already established a branch (three years before) with the removal from Nantucket of William Rotch, Jr., and Samuel Rodman, the son-in-law of William, Sr. This change of location was natural enough in view of the pioneering efforts of Joseph Rotch, the father of William, Sr., who in the decade before the Revolution was, with Joseph Russell, a co-founder of Bedford, or New Bedford as it was to be known.

Well might Virgil have described the course of the Rotch firm's fortunes, which were now to find a true haven in this place on the mainland:

1. Rotch, *Memorandum*, p. 68.

29. "The Priory," Milford Haven
Built by Samuel Starbuck, 1793. Photograph by S. Robinshaw
[Courtesy the Nantucket Historical Association]

If from your course astray, or tempest driven—
Such haps as often the sailor undergoes
On the deep sea—you now are come within
Our river-banks and at our landing lie.

From its beginnings New Bedford was a town built for whaling. In its setting on the gently sloping bank, with the Acushnet River's wide mouth before and the forested hillsides behind, it was a natural site for a seaport town. Nonetheless, it was the last of the areas in Dartmouth township of Bristol County to be settled. When the original Plymouth Colony was at length divided into counties, Bristol was one of three carved from that Colony, and Dartmouth township came into being in 1664. A century was to elapse before Bedford Village had its first building near the water's edge.

The growing importance of Colonial whaling created Bedford. The original settlers were first generation Americans of English descent, and the rich farm land was their first interest. Then came the call of the sea.

As one of the two men responsible for its initial success in this establishment of Bedford, Joseph Russell can be called the founding father. The Russells were pioneer settlers of "Russell's Mills," at Apponeganset in Dartmouth township, and Joseph had established his homestead west of the original family holdings, in land partly inherited and partly purchased. By 1755 he had developed a large farm, establishing his home site in a section now at the head of William Street in New Bedford.[2]

To the north, on another farm, was his neighbor, Ephraim Kempton. The Russell and Kempton farms sloped down to the Acushnet and the future town was to occupy their sites. Within the farm limits, and through others to the north and south, ran the main County road, from Clark's Cove to the head of the Acushnet River's rather short length; then proceeding over the headwaters to join the junction of the main road just beyond the head of the river and thence to Plymouth on the one hand and Newport on the other. One wonders if Joseph Russell actually realized then the potential of the Acushnet as a natural harbor for shipping in competition with Fairhaven directly opposite. In any event Russell was able to persuade neighbor Kempton to join him in setting out house lots. Then, like a perfect partner, came Joseph Rotch, the senior member of the well-established Nantucket whaling firm.

It was to this embryonic village that Joseph Rotch came in 1765 to transform it into a port. He saw the potential, especially the easy access to the sea (in contrast to the sand bars which often blocked the entrance into Nantucket harbor) as a determining factor. Opportunity was grasped immediately, and with the aid of the Rotch money and business relationships with other ports, there were erected a wharf, warehouses, shipsmith shop, and oil refinery.

Joseph Rotch had established his whaling firm in Bedford somewhere around 1765. Accompanying him from Nantucket came his sons Joseph, Jr., and Francis—the career of the former was cut short by death while traveling in England; that of the latter has been already mentioned, both in London and Dunkirk. Of the early activities in Bedford of Joseph Rotch, an historian has stated:

2. William A. Wing, *John Russell*, Old Dartmouth Historical Sketches, New Bedford, No. 69, p. 12.

Under the mighty impetus given by this energetic business man, with his abundant means and skillful methods, the wheels of industry began to move. Houses and shops multiplied, highways opened, wharves were built, the population increased and the waterfront became the center of active business . . . Under a grove of buttonwood trees that stood by the river bank . . . the keel of the first ship was laid.[3]

The ship was the *Dartmouth*, built in 1767, and one of the three to participate in the Boston Tea Party six years later.

It was not until November 1769 that Joseph Rotch described himself as "of Dartmouth." All of his extensive purchases of land in the future Bedford were by "Joseph Rotch of Sherborn." It was this Nantucket merchant who suggested the name of Bedford for the hamlet—the family name of the Duke of Bedford in England being "Russell," and Rotch felt this Russell land should be so named.

With a swiftness which surpassed their planning, the little port blossomed, and by 1775 a fleet of sloops from "Dartmouth" made Bedford their headquarters. Then came the Revolution—and the ending of the infant industry by the pressures of the war.

Prevented from carrying on the business of whaling, and confronted with the swift gains of privateering, it was natural enough for the men of Dartmouth township to accept this new means of livelihood. The adventuresome characteristics of whaling lent themselves admirably to privateering. Bedford's location was a geographical asset, and privateers from other ports now made their way into the Acushnet. With Newport occupied by the British, and Boston blockaded by the Royal Navy, Bedford became a rendezvous for the Continental craft.

The little town bristled with activity, as the coming and going of the armed vessels, the shore-parties between sailings, the transfer of goods to warehouses, all brought prosperous times. Encouraged by the trends of the times, local men joined in the enterprises. Both Russell and his co-founder of Bedford, Joseph Rotch, as members of the Society of Friends were opposed to the war.[4]

3. W. M. Emery, *The Epic of New Bedford*, a lecture delivered before the Old Dartmouth Historical Society, January 19, 1934.
4. Ibid.

Here, indeed, was a paradox. Nantucket—the leader of the world's deep-sea fleet—through its Quaker leadership managed to remain neutral throughout the Revolution, despite its exposed location, and the loss of a whaling fleet. In a far better position, Bedford disdained neutrality and became a headquarters for privateering—also, despite the fact that the community's two founders were Quakers, their example as non-combatants had little following in the village.

The Acushnet now became a haven for privateers and their prizes. Soon the presence of these vessels and their crews, and the sale of prize goods, made Bedford a headquarters. Encouraged by the prosperity, more Dartmouth men joined in the enterprise. One of the new privateers fitted out in 1776 was the whaling brig *No Duty on Tea*, formerly owned, ironically, by the Quaker Joseph Russell. Other whaling craft were converted for this new and lucrative practice and the whalemen, unable to follow their profession, joined the coastal mariners and adopted the role of privateersmen.

In September 1778 the power of the Royal Navy made itself brutally felt. A fleet, under Lord Grey, coming up the coast from New London with transports to bring off cattle, sheep, and supplies swept down upon hapless Bedford. Troops were landed which came into the town, set fire to its waterfront, and destroyed warehouses filled with contraband and marine supplies, burning every vessel in the wharves, and some of these with cargoes still unloaded. In reporting the raid to Viscount Howe in New York, Captain Fanshaw of the frigate *Carysford*, flagship of the fleet, wrote: "I cannot particularize the Damage done, but by the Appearance of Shipping before dark and the conflagration, I suppose it must be great." [5]

Disheartened by this catastrophe, Joseph Rotch had returned to Nantucket, not returning to Bedford until 1782. He died in 1784. But his example had not been lost and his eldest son took up where he had stopped. While still in Dunkirk, William Rotch, Sr., as already noted, directed the removal of the Rotch firm from Nantucket to Bedford in 1791. With his own removal from Nantucket in 1795 the firm took on a new impetus.

5. Dispatch of Captain Fanshaw to Viscount Howe, September 6, 1778, Admiralty Records, P.R.O., London. Adm. 1/488, p. 244.

Very early in the two-way trade (Nantucket–Dunkirk–Bedford) the Rotches were able to more firmly establish their business in Nantucket and later in Bedford. Writing from France, William, Sr., directed his son and son-in-law to have a ship built, commissioning George Claghorn to construct the vessel—the same man who was to build the famous frigate *Constitution* years later. The new Rotch ship was named the *Barclay*, after the English Quaker who had befriended the Americans. The *Barclay* was built from live oak, procured for the Rotches by Zachariah Hillman, who went to Georgia with five ship carpenters especially to procure it.[6]

Her timbers were allowed to season for a year before building. The vessel was launched in October 1793 and after fitting out was dispatched to Le Havre under Captain David Swain, with a cargo of 238 casks of whale oil, 50 casks of sperm, and 99 bundles of whale bone weighing 3,313 pounds. The consignees were Homberg and Homberg, Freres, they paying freight of 80 shillings per ton sterling and 5% primage.

The *Barclay* continued in the merchant (New Bedford to France) service until August 1797 when, under Griffin Barney, she was sent on her first whaling voyage, making a highly successful voyage to the Pacific and returning on June 26, 1799, with 1,200 bbls. of oil and 21,000 seal skins. The *Barclay* was destined to become one of the most famous whaleships of her time, continuing her active career until 1857.

The firm purchased a new brigantine, the *Mary*, in 1793 and under command of Capt. William Taber she was sent to London with seventeen tons of whale oil and eighty-one casks of sperm.

As was customary, careful attention was paid to the manufacturing of sperm candles as well the supply of sperm oil and head matter. William Rotch, Sr., had written from Europe:

We must not lose the Spermaceti Co., as I hope, if we are favored to return, to be able to make more of it than heretofore. You may remember I hinted . . . I wanted your advice; it was on this subject. The Secret of —being only in one person at present, who I was in treaty with for it, but the enormous sum he demanded would not be complied with. Since

6. William M. Emery, *Col. George Claghorn*, Old Dartmouth Historical Sketches, No. 56, New Bedford, January 1931, quoting *New Bedford Medley*, October 19, 1793.

which [time] we have been in search after it, through other means, & believe have got so much of the principles, as to make some advantage, though by no means equal to him.[7]

Thus, upon arrival in Bedford, one of William, Jr.'s first acts was to see to it that a candle manufactury was reestablished, and the West India market pursued. He did not neglect his original business links with Europe, and in fact he enlarged upon this initial connection. In London, Champion and Dickason continued as agent for the Rotch oil, and the De Bauque Brothers in Dunkirk remained the French representative, despite the Revolution and its upheavals.

Some idea of the range of the Rotch firm's activities at Bedford in 1794 may be gleaned from the shipments to other ports. The schooner *Tabitha*, Capt. Allen, took 1,504 pounds of candles in fifty boxes to Boston in June, consigned to Joseph Hussey. Capt. Barstow went to Baltimore in July with 1,190 pounds for John McKim. In August, the sloop *Mayflower*, Josiah Kempton, took forty-seven casks of oil and twenty-one bundles of bone for Joseph Hurd and ninety-five boxes of candles and four mahogany logs for Joseph Hussey. The same sloop, a month later brought a cargo of 109 tons of sperm and whale oil across the Western Ocean to Le Havre. Other firms and ports supplied with oil and candles were Philip Wanton in Alexandria, Virginia, William Burroughs in Norfolk, Virginia and John Habersham in Savannah. The schooner *Eliza* was sent to Corunna, Spain, with 221 casks of whale oil, consigned to Laogonerer and Co.[8]

The need for new ships was paramount. When the Rotch fleet returned to America from Dunkirk, the ships were divided between Nantucket and Bedford. To the island port at Nantucket went the *Favorite*, *Warren*, and *Hector* which were owned mostly in Sherborn but had been able to use Dunkirk for the selling of their oil under the Rotch agency. To New Bedford went the *Dauphin*, Capt. Stephen Gardner; *Ann*, Captain Prince Coleman; *Delaware*, Captain Tuckerman; *Janus*, Captain Obed Folger; *Lydia*, Captain Ray; *Susan*, Captain Barzillai Hussey; *Union*, Captain James Barney, and *Diana*, Capt. Jared Gardner.

7. William Rotch, Sr., to Samuel Rodman, Bullard, June 7, 1786, *The Rotches*, p. 223.
8. Rotch Account Book, New Bedford Public Library, New Bedford, Mass.

The full significance of the great impetus this fleet gave to the port of New Bedford may be determined by the fact that while in 1794 only one ship (the *Rebecca*, not a Rotch vessel) sailed from the Acushnet—and none of these to the Pacific—in 1795 eight full rigged ships cleared from the wharves of New Bedford and six of these were of the original Rotch fleet out of Dunkirk.

The manner in which William Rotch protected the registry of his ships was demonstrated through the voyages of the first of his ships named *Ann*. Shortly after this ship returned to Dunkirk from the Pacific he wrote (under date of May 5, 1792) to Samuel Rodman in New Bedford:

The Ann *is preparing for Bedford & the fishery, she will be completely fitted for whaling . . . William Moores is taking her over but she will need a master for whaling . . . Take out a new register for the* Ann *in thine and my name of Nantucket, & make no mention of my being resident of Dunkirk, only a visitor, and make no mention of Benjamin's (his son with him in Dunkirk) name, in the case.*[9]

As has been noted, the *Ann* was later transferred to Milford Haven registry for Benjamin Rotch, and made voyages out of that port to the Australian–New Zealand grounds. She made six voyages, always successful, and was succeeded by a second *Ann.*

During this year, a new vessel had been built for William Rotch at Le Havre, which had a less fortunate history. In one of his letters from Dunkirk, the elder Rotch mentioned "boaring the holes for the trunnels before or after the planks are on will contribute much to the seasoning."

It was in this same year (1792) that the firm lost the *Ville de Paris*. In the last month of 1792 William, Sr., wrote to William, Jr.: "Silas Jones at Delagoa Bay lost his second mate, a son of George Bunker; I think his name was Cromwell, killed by a whale. He left a widow."[10]

The turn of the century found the little port of Bedford actively engaged in the lucrative but dangerous European trade, running the

9. William Rotch, Sr., to Samuel Rodman, May 5, 1792. Rotch Collection, Old Dartmouth Historical Society.
10. Ibid.

blockade of British frigates and French privateers. Typical of the adventures of some of these whaling masters, becoming merchant captains for the time being, was that of Captain Andrew Pinkham, a native of Nantucket. Taking the ship *President* from her regular vocation as a whaleship, the New Bedford owners chartered her to the New York merchants Minturn and Barker, which firm sent Captain Pinkham to Gibraltar with 3,000 bbls. of flour, 80 casks of butter, and 113 firkins of butter.

On March 29, 1801, when within sight of her destination, the *President* was captured by French and Spanish privateers and taken into Algeciras, Spain, where the cargo was confiscated and the ship seized. For seventy-seven days Captain Pinkham was held a virtual prisoner by, of all people, the American Vice-Consul, on a charge that the cargo was illegally entered. It was only after several months, through letters to President Jefferson, David Humphreys, the American minister in Madrid, and petitions circulated by American residents at Algeciras, that Captain Pinkham was released. Writing to Minturn and Barker, the ex-whaling master noted:

The ship President *is at present in possession of the Spaniards . . . in Consequence of being bound to Gibralter, which they say has been in a State of blockade by the Spaniards since the 15th of February, 1800; you have no Cause to regret the ship's not Clared for any other port in the terraqueous globe as that Could not have Saved her as will appear by the following lines . . . on the 20th of March, about 70 leagues West of Capt St. Vincent I was boarded by an English cutter who Examn'd my papers and let me proceed but no doubt would have detained me had I been bound to any other port or at best a French port. On the evening of the 29th, I entered the Straits . . . and becalmed . . . surrounded by six gunboats and privateers one of which was French the others were Spanish (the Spanish took the* President's *papers first) . . . demanded the papers, which he took with great Eagerness, looking over his shoulder at the same time at the Frenchman who, by this time had boarded us from the other Side & were tumbling on board with drawn Swords, but Seemed rather dejected when they saw the papers in possession of the Spanish captain, who held them fast & refused to let the French Officer even look at them; the Same as a little boy who has Caught a bird of*

At the Court at St. James's.
the 18th of May 1791.

Present

The King's most Excellent Majesty

Earl of Denbigh — Viscount Sydney
Earl of Chatham — Lord Grenville
Earl of Leicester — Lord Amherst.

Southern Whale Fishery

Order approving a Report of the Committee of Council for Trade upon certain proposals made by persons concerned in said Fishery carried on from Nova Scotia to remove themselves & their Families to Great Britain, and transmitting the same to the Treasury.

Whereas there was this Day read at the Board, a Report from the Right Honorable the Lords of the Committee of Council appointed for the consideration of all Matters relating to Trade and Foreign Plantations dated in the words following – Vizt –

"The Lords of the Committee appointed for all Matters "relating to Trade and Foreign Plantations have taken into "consideration the Letter of Lord Grenville, one of Your Majesty's "principal Secretaries of State, inclosing two Letters from "John Parr Esquire Lieutenant Governor of the Province of "Nova Scotia, and proposals made by certain persons "concerned in the Southern Whale Fishery, who are settled "in the said Province, to remove from thence to Milford Haven "under certain Conditions therein mentioned; which Letters "and proposals were referred to the Consideration of this "Committee by Your Majesty's Order in Council of the 2nd "of last month, directing Their Lordships to Report Their "Opinion thereupon for Your Majesty's information."

"Governor Parr in one of the before mentioned Letters, dated "31st December 1790, says, "He has to lament exceedingly, that "Government did not inform him of their Wishes five or Six "Years ago, or empower him to encourage an Emigration from "Nantucket to England: If they had, he was persuaded, that "few of the People called Quakers, would have remained there, "and that none would have gone to Dunkirk, or any other "place in France"."

"On this point the Committee think it right to observe, "that

30. Order Approving a Report of the Council for Trade Upon proposal made by persons concerned in the Southern Whale Fishery to remove themselves to Great Britain [Courtesy Public Record Office, London, England]

which he is afraid, and when he is afraid his play-fellows are about to take from him . . . I think no Clearance would have clear'd her unless it had pleased our merciful Creator to have favored us with a brisk breeze.[11]

During the two months Pinkham was a captive, four other American ships bound for Gibraltar were taken and brought in to Algeciras, one of these ". . . the ship *Molly* of Philadelphia, with 14 guns and 40 men, who fought five privateers for two hours, but they were too heavy for him."

After his shabby treatment by the consular representative of his own country, Pinkham was suddenly and unexpectedly released and promptly sailed the *President* out of Algeciras on August 11, 1801. A few hours later he *spoke* the U.S. frigate *Philadelphia*, bound on her ill-fated voyage to Algiers, and was advised to put in at Gibraltar and wait a favorable wind. The trials of Captain Andrew Pinkham were not over, however, for in attempting to reach Gibraltar he was again attacked by three privateers, "and as soon as they came within range they commenced firing and continued for two hours. My situation now became peculiarly unpleasant, my crew all went below to avoid the shot; they (the privateers) could have rowed to the ship . . . I kept on until a ship called a Kelpie, armed to oppose these gunboats, rowed out from Gibraltar and drove them back . . . a breeze soon springing up enabled me to run into Gibraltar." A few days later the *President* left that port in a convoy guarded by English ships of war. Pinkham wrote:

I came across the Atlantic in the month of September, with the ship half ballasted and that with sand, and no shifting boards and five 32-pound shot holes between wind and water, with only a piece of inch-board nailed over them.[12]

On September 28, 1801, he reached New Bedford. His action in thus saving his ship was typical of the whaling masters who lived with their vessels.

I engaged to take charge of the President *as soon as she was launched . . . in October, 1796, . . . and I sailed in her first whaling voyage [to*

11. Letter of Captain Andrew Pinkham to Minturn and Barker, New York, April 12, 1801.
12. Logbook of Ship *President*, in possession of John A. Diehl, Cincinnati, Ohio.

the Pacific] and have performed a number of voyages in her Since that time . . . the thought of leaving the Ship here [Spain] almost overwhelmed me for She seemed more like a home to me than any place . . . I should part with her with far more reluctance than ever I left my home, for then I left with a view of providing for my family, but now I am in prospect leaving . . . perhaps the only means of providing for them." [13]

After ten years at his chosen home, William Rotch, Sr., saw the years of his efforts now coming to fruition. Bedford had become *New* Bedford and by 1805 was second only to Nantucket as a whaling port, and the advantages of her geographical location—which were one day to provide her with the opportunity to supplant the Island as the great whaling port of the world—were now becoming more and more apparent.

For over two decades, the elder Rotch was semi-retired. But his business judgments were always to be called upon, even by his able successors to the Rotch firm, William Rotch, Jr., and Samuel Rodman. To the Rotch Counting House, under Johnny-Cake Hill, came a number of young men, and as each proved his worth he was invited into the firm. Among these clerks were men who eventually formed their own firms and, with the great expansion of the industry in the 1820–1830 period, several became outstanding merchants. In this group were such men as Charles W. Morgan, James Arnold, Andrew Robeson, and Thomas Hazard, Jr.—all who were to launch their own whaling fortunes as the heads of their own business houses.

Firms which were to soon rank with the Rotch company now became strengthened by the growth of New Bedford. Isaac Howland and Co., Seth Russell and Sons, John Avery Parker, Gideon Allen, and T. S. and N. Hathaway became stalwart builders of the town's fortunes.

From the cupola of his home on William Street, William Rotch, Sr., watched the town grow, a keen observer of the increasing number of New Bedford's fleet. Fully aware of the rapid growth and potential of New Bedford he must have often felt that pleasure derived from his own part in the development, particularly his firm's ventures that had become so greatly successful. As he recalled the past, there was certainly one man he remembered more vividly than others—the lean-visaged Lord Hawkesbury, that careful, able manager of Britain's commercial

13. Ibid.

fortunes, whose eyes were like a candle's flame, always flickering, with brief moments of warmth, but for the most part rarely illuminating his mysterious inner person. Now, the Nantucket merchant could feel his cherished dreams coming true.

His faith in the power of economic forces over political had been sadly weakened during his decade of effort in Europe. First had been the fierce competitive struggle with the London merchants and their natural support by the Pitt ministry; then had come the great success in Dunkirk and as swift a crumbling of the fortunes by the political fortunes of France. Returning to his native land, he found that the growing pains of the Republic had at last given way to easier times, and with the adoption of the Constitution the prospects were bright. Now, the commercial ascendancy of the young nation was having a stabilizing effect on the political direction.

The true genius of William Rotch lay in his strong faith in the application of Nantucket methods to whaling. In the relationship of workmen ashore in the cooper shops, sail lofts, rope walks, and chandleries, and the men on the whaleship the Nantucketers had established not only an industry but a method of maintaining it. True, the Quaker philosophy played its part but it was the combination of the grim realities of whaling and the application of honest business associations in a community of closely related families which created Nantucket as a virtual whaling kingdom.

William Rotch was now to feel a mantle of contentment come over him. In the serene years of his retirement he would see his home Island of Nantucket and his adopted home of New Bedford lead the New England whaling industry to new triumphs. Of particular significance, he would see the Nantucket and New Bedford men finally wrest from England the control of the Southern Whale Fishery and, by so doing, regain for America that branch of whaling she had originally created.

17. The South Seas: Ocean Home of Whaleships

THE year 1795 marked a turning point in the political affairs of Europe. Following the terror-marked months of the Revolution, the French government had become controlled by better leaders, while the success of French armies in the field had brought about treaties with Prussia, Spain, and Holland. The situation for England, however, was sharply altered. Her former allies were now neutralized, and the fleets of Holland and Spain were now to be allied with France, and Britain's sea power was to be seriously challenged.

The captains of the first whaleships to the Pacific Ocean, in their cruisings up the western coast of South America, soon found the unfriendly ports of Chile and Peru were a definite handicap in provisioning their ships and crews. Coupled with the distrust of British trade was the realization by Spain that the rising trend of Revolutionaries in these colonies could be abetted by the English. The ports of Valparaiso, Callao (Lima), and Paita were then closed to British vessels, and because the Americans spoke the same language they were also warned off.[1]

Fully aware that the Pacific Ocean was to be the greatest potential for whaling both Britain and the New England ports of America were determined to send their ships to this ocean. Forced to seek places where refreshments could be found the whalers were at first hard put, and the dread scurvy (caused by lack of fresh foods) took its toll on many a ship. But as the merchants in the ports of Chile and Peru soon found, trading with the whaleships was profitable, and clandestine operations resulted. However, the commandants at the few ports available, where ships could anchor safely, continuously warned the newcomers away and prevented provisions from reaching their decks.

During the last decade of the eighteenth and the first decade of the

1. Stackpole, *The Sea-Hunters*, p. 153.

nineteenth centuries Britain's navy became the greatest marine force in history. Forced by the sheer necessity of maintaining her "wooden walls," the British were able to accomplish two vital feats—the building of her naval forces and the protection of her maritime trade. From her merchant fleet she was able to recruit, willingly or unwillingly, sailors for her navy. One of the arguments used by the London whaling merchants for government support of the Southern Whale Fishery was that this branch of commerce provided an extensive nursery of seamen for the Royal Navy.

It may be truly said that the efficiency of the press gangs in British ports was a major means of keeping the fleets manned. The fleet operations were extensive and discipline was necessarily harsh. A point was reached where tyranny by the officers became the rule, eventually leading to the fierce mutinies in 1797 at the fleets in the Nore and at Portsmouth. Certain reforms which followed helped to brighten the dark picture considerably.

Of lasting benefit to the British cause was the heralded success of the blockading fleets in keeping the French in check during 1794. Although her naval fortunes were not distinguished in 1795, Britain at least maintained her commanding position. The failure of the French government to provide proper dock-yard accommodations for their fleets, coupled with the inefficiency of her officers and low morale of her sailors, reduced the effective services of the French navy. Her fleets remaining almost constantly in port was a vital mistake for France, as the necessary exercises of sea duty were not utilized. On the other hand Britain's fleets were constantly ranging the European coasts in both the Atlantic and Mediterranean, their crews always on the alert.

In the South Atlantic, the Island of St. Helena was a veritable crossroads of the sea, and here many whaleships were met by frigates of the Royal Navy providing protection as they joined convoys sailing for the English Channel. Adoption of the French naval policy, that decided them to abandon fleet engagements and concentrate on commerce-destroying sorties, had much to do with the necessity of Britain adopting a practice of frigates convoying merchantmen to ports. Blockading French ports was the chief function of the main British fleets after Admiral Jervis had defeated the Spanish fleet off Cape St. Vincent.

But it was the vigilant patroling of the sea roads of the world of commerce that kept Britain's supremacy constantly in the international scene. With the growing heat of the Napoleonic Wars neutral trade—especially that of the United States—began to suffer, as both French and British captured trading vessels bound to either country from a neutral country. This use of force became the source of friction that was soon to develop a sense of outrage in America, as well as frustration in the free trade markets of the world.

One of the signal victories of the British fleet was attended with no loss of ships or men—the capture of Cape Town in 1795. This provided a strategic port of the highest importance for merchantment and whalemen outward bound to India and Australia and homeward bound from eastern seas. When in August 1796 the Dutch fleet, under Admiral Lucas, was surrendered to the English Admiral Elphinstone in Saldanba Bay, Cape of Good Hope, another threat was removed. This occasion marked the first time in naval annals that an entire squadron had capitulated without firing a shot. After a stormy passage from Holland, the Dutch fleet was so badly mauled by the sea, and the sailors so weakened by the lack of food that Lucas decided it useless to bring about an engagement. He sailed into the Bay to give up and seek help.

After the victory at St. Vincent in February 1797 the British controlled the seas. But this did not prevent the swift French and Spanish privateers from frequent sallies, and whaleships leaving or entering the English channel were vulnerable to attack. Convoys from the West India Islands and St. Helena were well protected, and American whaleships often joined their British brethren to sail with Royal Navy frigates escorting. One such convoy sailed from St. Helena in September 1796 and the British whaleships *Emilia*, *Trelawney*, and *Ann and Sally*, were accompanied by the Nantucket whalers *Olive Branch*, *Harlequin*, *Union*, *Commerce*, *Hero*, and *Minerva*.[2]

But the Pacific was still remote and outside the protective fringes of the Royal Navy. Something of the risks which the whaleships experienced in sailing 'round Cape Horn may be found in the voyage of the ship *William*, under Captain Mott. Departing from London on the 12th of April, 1796, the ship joined the outward bound fleet protected by several men-of-war. A week later the *William* parted with the West Indies

2. *The London Times*, Marine Column, November 9, 1796.

section of the fleet, as well as the Mediterranean group, and continued on with the East India vessels being conveyed by H.M.S. *Jubilee* and H.M.S. *Goliath*. At the Cape de Verdes, the *William*, together with the *Beaver* and *Moss*, other whaleships, left the convoy and headed across the South Atlantic for Cape Horn and the Pacific.[3]

The logbook of the *William* is very valuable for the details it provides. One illustration shows her on "a passage to the Pacific," as a square-sterned ship with five windows, her hull painted red above and black below the waterline, with a yellow strake the length of the ship below the rails, her sails including a main royal, foretopgallant, and mizzen topsail. Another drawing depicts three of her boats "going after the whale," one being painted similarly as the ship, another showing a red strake and the third black and yellow. Below the water line they were painted white. The oarsmen all wore pigtails similar to the man-of-war's sailors.

Rounding the Horn in early August 1796 the *William* sighted Massafuero Island off South America's west coast on August 26. After cruising on the "on-Shore Grounds," the ship reached the Galapagos Islands on November 6. The log reported:

At 4 p.m. came to anchor at Chatham Island, Stephens Bay, in 15 fathoms, bottom soft sand with small shells, the large rock bearing west and the eastward end of the Bay bearing W. by N. by compass. We may run in here without danger, only keep nearest to the rock, as there are some small rocks . . . This island has a few terrapin on it. You can get a great many fish about the rock & alongside the ship. You will have a breeze of wind from the N.W. & in the afternoon from the south'd that is off shore.[4]

A number of whaleships were spoken, and the log included illustrations of the *Betsy*, *Lydia*, and *Atlantic* of London. The latter two had hulls painted in a pattern similar to that of the *William*—the upper portion of the hull red, the lower section to the water line black, with a yellow strake the length of the ship along the rail. The drawing of the *Atlantic* had a small star-shaped gilt ornamentation on her after quarter.

3. Logbook of the whaleship *William*, Capt. W. Mott, National Maritime Museum, Greenwich, England. Voyage, April 12, 1796, to June, 1799.
4. Ibid., entry of November 6, 1796.

One boat hung over the stern, and two on the stern quarters. The spars were painted white, with doublings black, top-gallant and topmasts showing clearly.

On January 8, 1797, the *William* and *Atlantic* in company, a school of whales was sighted and both ships lowered all their boats taking sixteen whales—each ship taking eight. A week later, eight whales were taken, and on January 19, the two ships killed thirteen, saving eleven. Six days later five more were taken. An illustration showing a "spermaceti" indicates the whales taken were of this species.

February 1st found the *William* making her way toward James Island. "Most of our people down with scurvy," noted the logbook. Three days later the ship had reached this island in the Galapagos Group and had found "turpin plentiful," going ashore to "make a tent for sick." One of the remedies for combatting the scurvy is mentioned in the days' entry for February 18: "buried the sick in hot mould."

The *William* cruised off the Galapagos Islands during the next three months in company with the London whaler *Greenwich*, also owned by the Enderby firm. Fortunately, the newly arrived whaleship had some potatoes, pumpkins, and onions which she shared with the *William*. Late in June 1797 the latter ship had reached the Tres Marias Islands—the scurvy continuing to take its toll—reporting: "Several of our people so bad that their gumbs have grown over their teeth."

Evidence of the animosity shown by the Spanish settlements along the coast of Mexico is revealed in the entry for July 8, when the ship was at Cape Corrientes, and Captain Mott went ashore with a boats' crew:

They scarce had landed when upwards of 50 Spaniards on horseback surrounded them, and it was with great difficulty they escaped being made prisoners, and it is possible they would not have escaped except had they not been well armed with muskets & cutlasses.

The situation was eased somewhat when they were able to buy a bullock and some limes. They also received permission to bring "some of our people ashore to bury them." This meant the desperate remedy of placing the scurvy-striken sailor in a bed of earth, with only the head exposed, the idea being that some elements in the soil would help clear up the dreaded disease.

Returning to London after an absence of two and one-half years, the

31. Head of a Single-fluke Harpoon from the Ship *Ospray*
Sailed for Rotch out of Dunkirk, France, in the eighteenth century, afterwards for the Rotch firm of New Bedford
[Photograph by Russell A. Fowler]
[Courtesy the Marine Historical Association, Mystic, Connecticut]

William had taken over 100 whales, filling her holds and completing a most successful voyage.

Seated in his comfortable library in the manor overlooking the lordly Hudson, Robert R. Livingston, one of the most esteemed men in America, sat deep in thought. On his desk was a letter—one of the most extraordinary he had ever read. It was late in the afternoon of October 28, 1797, and the letter was dated two days before, addressed to the "Chancellor of The State of New York," and had been written by Captain Alexander Coffin, then residing in the Town of Hudson, some twenty-five miles further up the river. (This letter is now preserved in the New York Historical Society collection.)

Captain Coffin, one of the founders of Hudson, was a man as unusual as the letter he had written. A staunch advocate of independence for the British Colonies, during the Revolution he had served as a dispatch bearer for John Adams in Amsterdam to the Continental Congress, in which capacity Mr. Livingston, a delegate, had probably first met him. After the war he had joined in the plan by the Nantucketers and their mainland friends to settle at Hudson, becoming one of the leaders in that community. It was Captain Coffin who had written Adams in 1785 of his fears for American whaling and of William Rotch's plans for a Nantucket colony in England.

Shortly before Captain Coffin wrote to Livingston, he had visited the Chancellor at the latter's home. On this occasion he had mentioned an idea he had "respecting the British whale fishery, and how important an object it would be to this country to have it annihilated, and return to the

United States, from whence . . . [it] first originated." The Captain contended that Spain would soon be at war with England, and that "the object of its [the whale fishery's] destruction will be the more easily accomplished."

Captain Coffin's plan was based on the knowledge he had obtained from the whalemen, as well as his own experience. He pointed out that one of the British fleets sailed from England to the Brazil Banks so as to arrive by November, to cruise until April, while another fleet usually left England in November "to cruise between the Cape of Good Hope and Madagascar for Spermaceti whales from March to June, then sail for Delagoa Bay," on the African southeastern coast. Still another fleet, described as the Cape Horn whaleships, "sails so as not to double the Cape in the middle of winter," and generally stayed from eighteen to twenty-four months in the Pacific.

"Therefore," he wrote, "these ships in that Fishery may be met with any month of the year, as they are on that Coast [Peru] at all seasons, for as winter approaches they sail into lower latitudes . . . as they stay on the Coast until they have filled their ships or consumed their provisions."

Basically, Captain Coffin's idea of "annihilating" the British whaleships was to be through the action of French and Spanish frigates sent to the whaling grounds. He stated that "one Spanish Frigate may destroy in one season" the fleet in the Pacific off the coast of Peru. As for the French participation, he advised:

One French Frigate to go & cruise on the Coast of Brazil from the Latt. of 36° to 48° South, and on the Banks, would almost be certain of meeting the greatest part of this fleet. One other Frigate to go round the Cape of Good Hope & enter Delagoa Bay about the 1st of July, would be morally certain to find all that Fleet there Snug at anchor; and one other Frigate to go into Walwich Bay about the same time might make an end to the Whale Business before their intentions were discovered and therefore prevent being counteracted.

It was pointed out that the British whalers, when they filled their ships, usually crossed the South Atlantic from Cape Horn to St. Helena, to join other whaleships at this Island from the Cape of Good Hope and Walwich Bay areas, and "there lay till there is a sufficient convoy made up for England," to proceed under the protection of British warships.

Captain Coffin touched upon the numbers of whaleships now (1797) comprising the American fleet—thirty-seven from Nantucket, twelve to fifteen from New Bedford, four from Hudson, and about twenty additional sailing from ports in New York, Connecticut, and Massachusetts, "making altogether 75 sail, which carry from 16 to 22 men. The voyages are performed according to the different routes, and take from 12 to 24 months."

Continuing, he stated:

The length of these voyages, and these [ships] being almost continually at sea, make one of the best Nurserys for our Seamen, who may in time to come be employed in vindicating their Country's Rights. Add to this the number of ships and men, which must be employed in exporting the suplus of what is consumed in this country to Foreign Markets—and as they diminish in their Fishery the United States will increase.

From 1793 the British–French War during its six years had reduced the British whaling fleet from "upwards of ninety sail" to twenty-five, Coffin declared, while the French whalers had been almost completely eliminated. If the small number of British whalers could be destroyed in the manner he had outlined, he wrote: "The whalemen would get scattered & return home, for they [the British ships] are mostly commanded by Americans; indeed the Head-men [harpooners] are altogether such."

Captain Coffin was certain that both France and England would attempt to revive their whale fishery upon the return of peace, "and that, in my opinion makes it the more necessary the few that remain should be destroyed. By that means it will be a difficult matter for them to get it going again."

When Adams and Franklin were in Paris, during the Revolution, a plan to destroy the British whaling fleet, similar to Coffin's, was proposed to the Continental Congress. Two years later (1778) Adams advanced the same plan to the State of Massachusetts. Both suggestions met with no action. It is more than coincidental that the ideas of 1778 and 1797 were fundamentally the same. Captain Coffin had been a courier for both Adams and Franklin during the Revolution, when he sailed out of Amsterdam as well as France.

As for Chancellor Livingston, had conditions been different it is certain he would have done something with the Coffin plan. As a man of perception, one of the authors of the Declaration of Independence, a jurist (and later the Minister to France who was to negotiate for the fabulous Louisiana Purchase, as well as the "angel" that later financed Fulton's successful steamboat *Clermont*), Livingston was well aware of the practicalities in Captain Coffin's daring idea. But he was also aware of the political scene. Adams' administration was dominated by the European wars, and his diplomatic experiences strengthened his desire for neutrality. The Republicans openly sided with France, while the Federalists favored England, but when the facts of French intrigue in this country were revealed both parties united against France. Thus, the timing of Coffin's plan made any diplomatic overtures impossible.

The long expected break between England and Spain finally materialized in 1797. As a result, the currents of naval action, like a back wash of the conflict, carried like a sea-tide around Cape Horn, making the west coast of South America and the distant Pacific feel the war's touch. Serving in the dual capacities of letter-of-marque and whaleship, the British armed whalers *Cornwall* and *Kingston* early in 1799 captured the Spanish ship *Nostra Senora de Bethlehem*, which was bound from Callao to Guyaquil, and took their prize to Sydney where she arrived in April. The Spanish ship *El Plumier*, loaded with wine and spirits, en route from California to Callao, was taken off the Maria Islands by three British whalers and brought across the Pacific to New South Wales. Early in 1800, the ship *Euphemia*, with a cargo of wines and spirits, was captured off the coast of Peru by the whaler *Betsey* and sailed to Sydney where the Spanish prize was re-named *Anna Josepha* in honor of the wife of Governor King.[5]

It was natural that the Spaniards should be suspicious of the American whalers, and regard them as British sympathizers because their crews spoke English. The Nantucket whaler *Washington* had been first to raise the Stars and Stripes in a South American port (when Captain George Bunker displayed the colors at Valparaiso in 1791), but even this new flag was regarded as that of a British ally. The *Beaver*, of Nantucket,

5. Cumpston, *Arrivals and Departures*, pp. 34–36.

first American whaler in the Pacific, had been ordered out of the port of Callao without obtaining the needed supplies. As the situation between Britain and Spain worsened, the American whalers had to be on guard.

In January 1799 Captain Amaziah Gardner and a boat's crew from the Nantucket whaleship *Commerce*, landed at St. Mary's (off the Chilean coast) for supplies and were "barbarously treated" before being released. Shortly after this incident, the sealer *Maryland*, Captain Liscomb, of New York, was captured with his boat's crew at the same island and roughly handled before being released. To add to his woes, Capt. Liscomb had his ship taken by a French privateer and two thousand skins stolen before he was allowed to continue his voyage.[6]

The *Tryal*, Capt. Thomas Coffin, was not so fortunate, being seized by the Spanish at Valparaiso and condemned in 1802. The ship *Miantonomoh*, of Norwich, Conn., was similarly taken by the Spanish authorities in that port. Shortly after, the London whaler *Redbridge* was trapped by one of the few Spanish armed vessels in this ocean and taken into Valparaiso. The *Hannah & Eliza*, and *Wareham* were also taken. On May 2, 1802, they were released and sailed to Callao under convoy.

It was natural enough for the American whalers to be sympathetic with the English, especially in view of the fact that so many Nantucketers were in command of the British South Seas Fleet. From the very nature of their calling they were all too aware of the dangers to be met within their normal voyaging and so were always to be depended upon for "lending a hand." An instance of this help took place in 1802. The brig *Anna Josepha*, the captured Spanish prize, sailed from Sydney in October 1801 with the second cargo of coal ever exported from New South Wales, together with timber for spars. Owned by Simeon Lord, the ex-convict who had become a shrewd trader, she was under a Captain McLean, but on board was one of the finest seamen of his time, Lieut. James Grant of the Royal Navy.

Grant, in the little brig *Lady Nelson*, had been the first man to explore the southern coast of Australia, "as far as Cape Schands," according to Matthew Flinders, mariner of equal ability. The *Anna Josepha* took a course through the South Pacific easterly for Cape Horn. Safely around

6. Starbuck, *History of The American Whale Fishery*, p. 195.

the Horn she reached the Falkland Islands where her crew were found to be in a desperate plight with scurvy, having been lacking provisions.[7]

Here was met the American sealing ship *Washington*, of Nantucket, under Captain Jedediah Fitch, who not only supplied them with fresh food but gave Lieut. Grant some information as to how the New England sealers and whalers used the Falklands to combat scurvy. Potatoes had been planted and eaten raw, were found to be anti-scurbutic. Captain Fitch also boiled the leaves from a variety of wild cabbage.

Her crew refreshed, the *Anna Josepha* continued her voyage and at length reached that lonely island of Tristan d'Acuhna, in the South Atlantic. Here, Lieut. Grant, dispirited and ill, had the good fortune to meet the American sealing vessel *Ocean*, Captain Dalton, out of Newburyport, Massachusetts. Bread, meat, and fresh water revived the Australian brig's crew, and the two vessels sailed for Table Bay and Cape Town with Lieut. Grant a guest on board the *Ocean*. Upon returning to England, Grant resumed his career in the Royal Navy, and during sea-fights off Holland distinguished himself in action.[8]

As if the elemental forces of the ocean storms, the regular combat with fractious whales, and the inroads of scurvy were not enough, the series of wars from 1793 to 1815 gave the whalemen a precarious time of it on their voyages to and from home. The mounting problems of neutral trade in the European theater were hazardous to shipping, as the successive decrees of France and England caught the whalemen as well as the merchantmen in a cross sea of troubles.

The journals and logbooks of London whaleships have virtually disappeared, only a scattered few remaining to tell the terse story of that remarkable period in British maritime history. Captain Paul West, of Nantucket, brought two of his logbooks home from London, and they are invaluable records of his voyages in the ship *Cyrus*. The ship itself had an unusual history. As a new whaleship out of Dunkirk, owned by the Rotch firm, the *Cyrus* had sailed on August 2, 1802, under Captain Archaelus Hammond, with Paul West as the first mate. A year later

7. Mrs. Charles Bruce Marriott, *The Logbooks of The Lady Nelson*, London, 1915, p. 76.
8. Ibid., p. 77.

(September 25, 1803), the *Cyrus* was captured by the British frigate *Scorpion* in Delagoa Bay, on Africa's east coast, and was taken to London. Purchased by the whaling merchant James Mather, she was placed under the command of Paul West. The fact that the Nantucketer had sailed under the French flag previously made little difference to either the new owner or the new captain—the *Cyrus* was a whaleship and her master was a whaleman.

For her first voyage from London, the *Cyrus* was first taken from her moorings at Limehouse Hole and warped into the dock at Ratcliffe Cross. Here she was literally taken off her keel, her garboard seams caulked, and her keel and keelson fastened with composition bolts. After being completely coppered, her decks were caulked, new hatch coamings installed, a new foremast stepped, and new topmasts raised. On June 28, 1804, provisions of salt beef, salt pork, bread, flour, peas, barley, oatmeal, and water were hoisted aboard; a gallows frame built between the mizzen mast and cabin campionway; new try-pots installed; and cables and anchors placed in position. A complement of cannon completed her deck gear, as it was clear that the ship was also to be letter-of-marque.[9]

On July 9, the *Cyrus* cleared from the Thames River, "bound for the South Seas," with Captain Paul West in command and his first mate Solomon Coffin, a fellow Nantucketer. Three days out they spoke the whaleship *Brothers*, returning to London under Captain Thomas Folger, with 150 tons of "spalm oil," taken in the Pacific. That Captain West was preparing for possible attacks by privateers was shown by his entry of August 27: "All hands to quarters; exercised our great guns and blew them off and prepared them again for action. At 6 p.m. sighted a strange sail standing to the westward." A few days later another strange sail standing to the westward. "Mustered all hands to quarters," records the log, "and prepared to engage her, but she kept off to the westward and we did not pursue her."

The potential dangers of the French–British war kept Captain West on guard. During a stay of a few days at Rio de Janeiro he requested that a French corsair be detained by the commandant of the port for twenty-four hours after the *Cyrus* cleared the port. While in Rio, getting ready

9. Logbook of the ship *Cyrus*, of London, Captain Paul West, Nicholson Collection, Providence Public Library, Providence, R.I.

to sail, the crew refused duty, "and behaved in a very mutinous manner . . . I sent a boat on board the Commandant and requested him to take hold of them, which he immediately did and took them on board his own ship . . ." On the day following, Captain West reported: "Went on board the Commandant and found my people a little Sobered and Glad to return, so I gave them permission to come back."

Proceeding on his voyage, Captain West fell in with the whaler *Samuel*, of Nantucket, under Captain Gideon Gardner, and the two vessels sailed in company to the Falkland Islands, where they stayed several days. Geese and ducks and penguin eggs made welcome additions to their daily menu. Some 400 eggs were gathered and stowed for future use. While in the Falklands, they found five men living ashore who were "without any ship or anything to live upon than what they caught daily. They applied to be taken off and set on shore in some other part of the world, but having a great number of Men on board . . . did not conclude to take them off."

Captain West afterwards relented and took two of the castaways, and the *Samuel* took the other three. A few days later the two men invited aboard the *Cyrus* were transferred to the ship *Hudson* of Nantucket, Captain Uriah Bunker, which had just arrived at the Falklands.

On November 30, 1804, Cape Horn was in sight. After a long struggle, the *Cyrus* managed to make the passage into the Pacific. On the last day of the year, December 31, 1804, they took their first sperm whale. Four days later, January 3, 1805, the logbook recorded:

Shipped a very heavy sea which stove the bulwarks all by the board, fore and aft, on the larboard side, but fortunately did not wreck the boats as might have been expected. This was the heaviest sea that ever struck or came aboard the Cyrus *since I have known her, which is nearly three years.*[10]

Well into the Pacific the *Cyrus* began whaling in earnest. Eight sperm whales were taken off the Island of Mocha, and on February 13 a fifty-barrel sperm was killed. While at the Island of St. Mary's a violent gale nearly drove the ship ashore, "the crew (drunk on wine) unable to perform their duty . . . was forced to cut our cable to escape." The ships

10. Ibid., entry for January 3, 1805.

Charles and *Winslow*, of Bedford, managed to escape also, and the *Maitland*, Captain Bunker, of Milford Haven, got under way before the gale struck. Other ships spoken at this time were the *Dianna*, Captain Waterman, of Bedford, and the *Adventure*, Captain Pinkham, of London. Later, the ships *Harlequin* and *Neutrality* of Nantucket, were spoken. Upon meeting the *Leo*, of Nantucket, Captain R. Gardner, Captain West learned he was bound for the coast of California.

The problems of obtaining supplies at the Spanish settlements on the coast was demonstrated by Captain West's experience at Coqimbo, Chili, on March 28, 1805:

Went on shore with one boat, well armed, and found reasons to suspect some mischief. Rec'd a letter wrote in Spanish from the Governor, which denied me any Refreshment, as the ports of Callao and Valparaiso were the only places for me to go into, and at the same time [stated] the Governor was in a dying condition and [needed] my surgeon. When I told the officer (who brought the letter) that I performed that office myself he seemed highly pleased (as he pretended) to find a surgeon and captain in one and the same person, and a horse was completed immediately for me to mount and go post-haste to the Governor. But without ceremony and little authority, I jumped into my boat and came on board (much to their astonishment), unmoored, took up my anchors, and stood out of the bay . . . and with but one bucket of onions and a basket of grapes from this place.[11]

Continuing up the coast the *Cyrus* spoke the brig *Polly*, of New Haven, with 13,900 seal skins, and on May 10 spoke the whalers *Pheobe Ann*, of New Bedford, and *Sukey*, of Nantucket, but found neither vessel had any "news political." There was good fortune awaiting them, however, as they took eleven whales, making 133 barrels on one day, always on the look-out for suspicious appearing craft, Captain West escaped a Spanish privateer which chased him, and on May 26 his log contained the following:

At 6 a.m. saw a strange sail bearing E by N 5 leagues . . . and drew up with her so fast that at 2p.m., the wind inclining E, got near enough to

11. Ibid., entry for March 28, 1805.

discover her hull. She appeared to be a large Spanish ship, mounting upwards of twenty guns, as she had one tier fore & aft run out of her black bulwarks, and hammocks stowed in a war-like manner, As the Cyrus *had much the advantage in sailing, and the wind steady, I passed under his lee . . . and hoisted American colors, which the Spaniard took no notice of. So kept steady on to the NE and at 4 p.m. lost sight of the Spanish ship.*

Proceeding to the north, the *Cyrus* reached the Island of Lobos, where Captain West careened the vessel to "pay the bends" with tar. It was early in June 1805 and as he had no news, "fearing bad will come first," he "thought it proper to meet it by mounting every gun." Upon leaving the Island he sighted six whaleships—the *Sterling*, *Fame*, *John Jay*, *Lima*, *Belvidere*, and *Maria*, all of Nantucket. Captain Wyer, in the *Fame*, provided Captain West with the news he expected. War had begun, declared between England and Spain, and several English ships had been taken, four at Concepcion, and a brig at Tumbez.

Captain West needed fresh supplies and during the first week in September he captured eight Peruvian fishing boats off Paita, taking the skippers on board and holding them until he had taken from the boats 23 pumpkins, 300 plantains, 100 lbs. of rice, some potatoes, 6 sheep and 2 goats. After releasing the captains and the vessels the *Cyrus* sailed north for the Galapagos Islands, where they finally dropped anchor in James Bay and took aboard seventy-eight tortoises.

During the next three months the *Cyrus* cruised for whales in the vicinity of the Galapagos Islands, during which time several Nantucket whaleships were spoken; Captain Folger, of the *Neutrality*, told of being boarded by a Spanish privateer at Tombez; the *Leo*, Captain Gardner, gave whaling news; the *Sukey*, Captain George W. Gardner, brought the sad facts of the murder of a fellow-Nantucketer, Captain Obed Cottle, of the *Minerva*, of London, who had been shot by a mutinous sailor.

Whaling proved excellent, and by the end of the year 1805 the *Cyrus* had a full ship, having taken sixty-three whales during her voyage. On January 1, 1806, she left the Coast of Peru off Callao and began her voyage home. After rounding Cape Horn, Captain West crossed the South Atlantic to St. Helena, where he joined a convoy being escorted home by British frigates. Off Beachey Head in the English Channel the second

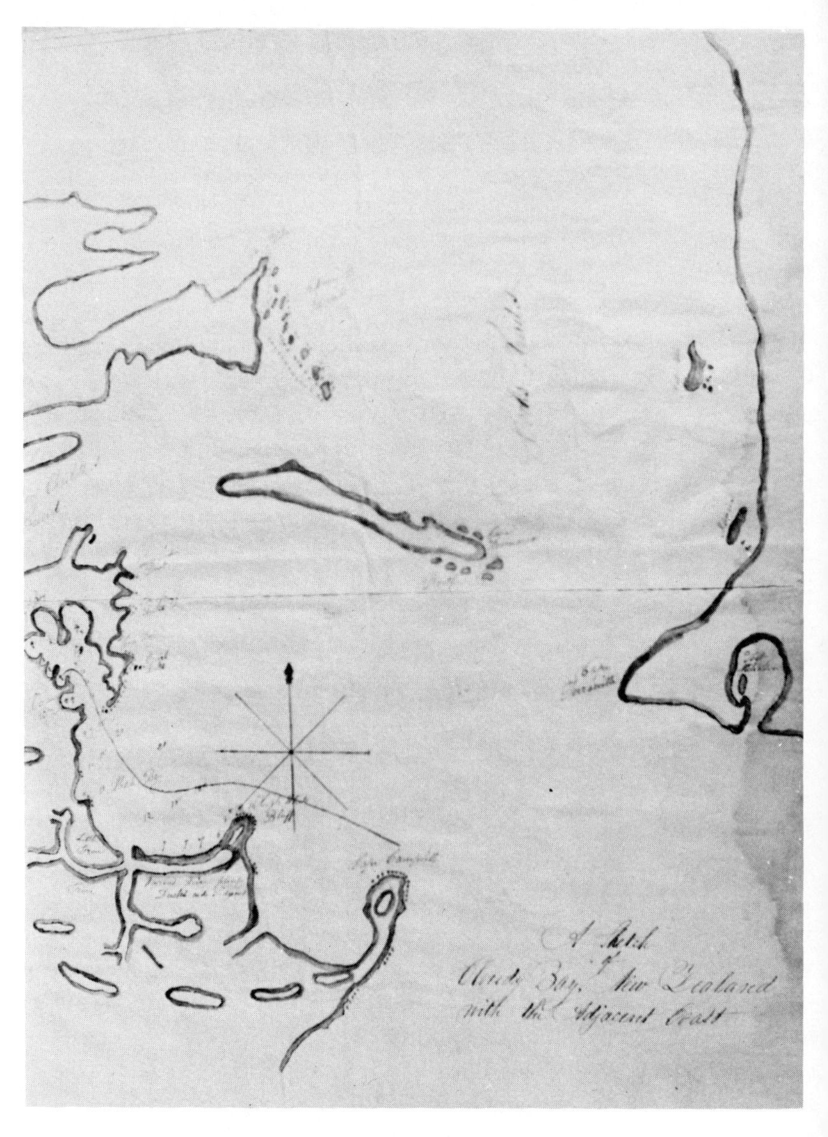

32. A Whaler's Chart of Cloudy Bay, New Zealand
One of the earliest charts of whaling area in Cook's Strait
[From Author's Whaling Collection]

mate, Alex Leith, and three other men who had been in trouble throughout the voyage, stole a boat and "ran away" to shore, thereby forfeiting their shares or "lays" in the voyage, but apparently fearing legal action by the Captain and owners for their conduct.

The *Cyrus* reached London on June 24, 1806, with 163 tuns of sperm oil on board. On his next voyage, Captain West took out the London whaler *Charlton*, returning in 1808. He then took the *Cyrus* for two more voyages to the Pacific—1808 to 1810, and 1810 through 1812. With the outbreak of the War of 1812 Captain Paul West returned to Nantucket, bringing with him as his bride Phoebe Hussey, daughter of Captain Benjamin Hussey of Dunkirk.

In the pattern of their cruisings in this part of the Pacific, the whaleships found the Galapagos Islands a favorite rendezvous. One of the most interesting whaling documents of this period has to do with an "agreement entered into for the mutual interest of the ships concerned," while cruising at the Islands, and signed by the following ship masters for their respective ships:

Seringapatam of London, Captain Wm. A. Day
Favourite of London, Captain James Keith
Aurora of Milford Haven, Captain Andrew Myrick
Leo of Nantucket, Captain Joseph Allen
Euphrates of London, Captain Ian Frazier
Neutrality of Nantucket, Captain Thaddeus Folger
Hannah & Eliza of Milford Haven, Captain Micajah Gardner

The "Articles of Agreement," dated 1801, contained several fundamental practices, and in this case give an excellent picture of whaling customs of that period. In this particular instance it is to be noted that four of the whaleships were commanded and manned by Nantucketers, and that the mate of the *Seringapatam* was David Joy, who became her master on her next voyage. No doubt, the other two London vessels also had Nantucketers aboard.

Because of the document, signed by these whaling masters, is the only one known to still exist, it is of interest to note the content of three "Articles":

1st—The ship which Discovers a School of Whales will immediately hoist his Ensign to the Miz. Topmast head . . .

2nd—If in a School of Whales our Lines get foul of each other we will use every endeavor to clear them and Not Cut each Other's lines without Mutual Consent.

3rd—If a boat is in Danger of losing her line the Nearest Boat is to Give his and without any Exception or Claim whatever.[12]

That the whaleships did well under such a set of articles is shown by reports by another London whaleship, the *Georgiana*. As of January 8, 1801, the *Leo*, Captain Allen, had 300 barrels of sperm, and the *Favourite* Captain Keith, had 200 barrels. The ships were then still at the Galapagos Islands.[13]

The easing of tensions in the European wars at this time (1802) was also apparent in the more frequent use of the Spanish ports of South America by the whaleships. In January 1802 the *Hannah & Eliza*, a Benjamin Rotch ship, then sailing from Milford Haven, was able to procure fresh beef and "soft bread," purchased at Concepcion, Chile, as well as five sheep, and fresh vegetables. Captain Micajah Gardner apparently spent some time ashore, as well, as his account shows charges for board and "horse hire." Later, in April of this year, he obtained fresh meat at Valparaiso.[14]

There was considerable difficulty encountered at Lima, Peru, however. Captain Gardner's account shows that he put in at Callao in June 1802 and had to pay £33 to have the ship's hatches sealed while the ship took on supplies, as well as paying an interpreter £10 for translating papers and a lawyer £18 for three petitions.

Captain Gardner had a whaling career which extended through a quarter century before he retired. During this time he sailed on vessels from four whaling ports: Dunkirk, Nantucket, Milford Haven, and Bedford. A granddaughter once described him as a man who enjoyed telling of his contemporaries—many of whom lived close by in Nantucket. Just up the short stretch of Broad Street from the Micajah Gardner house was the residence of Captain Jared Gardner, master of

12. "Articles of Agreement," from original manuscript in author's collection.
13. Logbook of the ship *Georgiana*, 1801–1803, Captain W. Mott, National Maritime Museum, Greenwich, England.
14. Papers of Captain Micajah Gardner, author's collection of manuscripts.

the *Diana* and *Enterprise*, whaleships and China-trade vessels; around the bend of Centre Street was the residence of Captain Coffin Whippey of the *Grand Sachem*, while next door was the new dwelling of Captain Archaelus Hammond, who aboard the *Emilia* was the first man to kill a sperm whale in the Pacific; and Captain George Rule of the *Sprin-grove*, was to build a home within a few ship's lengths away.

It is no wonder that these men, retired from such careers, should meet in the back room of a neighborhood store on Centre Street which they called the "London Club," in remembrance of that great port to which they had once brought fully laden ships after voyages to the most remote parts of the world.

The Nantucket merchants continued to send their vessels to the African coast as well as to the Pacific, and the Bedford owners followed this same pattern. New ships were built larger, especially for the Pacific voyages, and among the newcomers were the *Caesar* and *Minerva* of Nantucket; *American Hero* and *Oswego* of Hudson, N.Y.; *Juno*, *Lydia*, and *President* of New Bedford. All made good voyages, justifying the faith of the owner and judgment of their masters.

Throughout this period—with the last decade of the eighteenth century merging with the early years of the nineteenth—England's superiority in the Southern Whale Fishery continued. Although the numbers of the ships involved did not increase, the size of those engaged became larger. As a result, the amount of cargo totalled more, while the crews increased as well. The British Statistics of these years reflect this picture:

YEAR	VESSELS	TONS	MEN
1793	67	15,568	1,273
1794	55	13,431	1,320
1795	66	17,374	1,584
1796	61	15,416	1,464
1797	60	16,536	1,440
1798	58	17,194	1,392
1799	58	17,449	1,408
1800	64	18,980	1,536
1801	69	20,272	1,656[15]

15. Statistics from *Board of Trade Records*, B.T. 6/93, B.T. 6/95, and B.T. 6/230.

In 1795, the total value of the produce of the British Southern Whale Fishery was in excess of £210,000. In 1800 the total was raised to more than £260,000.

From 1800 through 1809, there was on an average per year some seventy-two whaleships registered in the British Southern Whale Fishery, and some thirty vessels were returning yearly, bringing an average of 1,634 tons of sperm oil (valued at £80 per ton), and 3,300 tons of right whale oil (valued at £32 per ton)—bringing an average total value of £233,400 per year over this period.

In France, the whalefishery at this time was undergoing a revival at Dunkirk and Le Havre. Encouraged by the apparent ability of the Directory in Paris to control the governing process of the country, William Rotch (now in New Bedford) continued his support of both Captain William Mooers and Captain Benjamin Hussey, who were keeping the whaling industry as a going concern at Dunkirk. A third American in France, Jeremiah Winslow, of New Bedford, was also a part of the business interests of Nantucket and New Bedford in both Dunkirk and Le Havre. Another Nantucketer active in the trade with France at this time was Dr. Benjamin Tupper. Writing from Le Havre in May 1796 Tupper mentioned being in residence at Paris with Mrs. Tupper and that his wife "enjoys France." He had purchased one-half ownership in a ship and chartered it to the French government for 10,000 guineas for a voyage to the West Indies. His letter continued:

She carries 500 negroes; and if she arrives safe I shall have money enough to come home and live with my friends, which I should like although I like France very much.[16]

In view of the Nantucket Quakers' strong aversion to slavery, this venture on Tupper's part was most unusual. The merchant–physician had been a controversial figure in Nantucket during the Revolution, and had been once arrested by the Continental as "a person dangerous to the liberties of the State," and later released on bail. Tupper was not a Quaker at this time, nor later, and was acting as a merchant entrepreneur while in Europe.

16. Dr. Benjamin Tupper, Letter from Le Havre, May 8, 1796, an author's manuscript collection.

In February 1797 three American Quakers visited Dunkirk from England, and one of them, thirty-year-old Benjamin Johnson, described his experience:

2mo.8, Dunkirk. Arrived at Dunkirk before dark and drove immediately to Benjamin Hussey's who with his family received us kindly. It was no small gratification, after six months travelling thro different governments, entirely among strangers and people of various customs & languages all different from our own, thus to find ourselves in a family such as this. We told them of our not wearing the cocade, and the various alarming reports we had heard. They said there might have been some difficulty in passing the frontiers and in conquered countries, but now we were in France we would find none. No such thing was ever required of them even in the severest times of the revolution; no people could be better treated than they had been during the whole time; that on occasions of public rejoinings and illuminations, instead of their windows being broken and their houses pillaged as had been the case in some other countries, the municipality of this place have ordered guards before their doors to protect them from insult. They were also exempted from military service and permitted to go before authorities with their hats on, which to others was not allowed.[17]

The young Quaker mentioned that there were as many as sixty Friends in Dunkirk "previous to the Revolution, nearly all from Nantucket," but that most of them had removed because of the war. As the leader of the Nantucket group, Captain Benjamin Hussey, with his wife and two younger children, and his eldest son Barzillai and family, were to continue a precarious existence at Dunkirk.

Boarding at the Widow Gardner's house, the visiting Quakers noted that her boarders were generally American captains, and that here were then fifteen American vessels in port. Commented Johnson:

They are among the finest in the place. Indeed, in every port I have seen they make a conspicuous appearance as strangers. They are mostly Yankees, who are at home wherever they can find freight for their vessels. These two have been employed for several months in bringing

17. Henry L. Cadberry, *The Dunkirk Colony of 1797,* in Annual "Proceedings" of Nantucket Historical Association, 1945.

salt from the south of France to this place, which french ships could not do without resque of capture. In this manner, coasting round about this quarter of the world a great many Americans were employed.

While in France, Captain Benjamin Hussey was active in programs of inoculating people against smallpox. For his contributions in this work he was presented a special silver medal by Napoleon. This did not help him later, however, when the French government confiscated his business and private property at Dunkirk. Soon after, Captain Hussey returned to Nantucket to continue his active life, joining old friends and business associates. A contemporary once said of him: "I shall never forget his huge head. When he took off his broad beaver hat, I could thing of nothing but that it contained a half-bushel of brains."

After the fall of Napoleon in 1814, Captain Hussey, at the age of seventy-eight, returned to Dunkirk, and once more organized a whaling venture. With the help of Jeremiah Winslow (later to be called by the French whalers "Pere Winslow"), Hussey fitted out a whaleship for the Greenland Fishery. During this voyage, while the ship was battling her way through heavy ice, the veteran whalemen met his end. A contemporary account reads:

Captain Hussey was at the wheel steering the ship, when the ice crushed against the rudder, throwing him against the wheel, breaking his ribs, from which wounds he soon died. He was then eighty years and five months old.[18]

18. Frederick C. Sanford, article in *The Boston Advertiser*, reprinted in *The Inquirer and Mirror*, Nantucket, Mass., September 15, 1877.

18. Meeting New Hazards and Challenges

WITH the end of the eighteenth and beginning of the nineteenth centuries the whalemen on both sides of the Atlantic found themselves sharing in uncertain fortunes. A major hazard to the Americans was the brief but damaging sea war with France, ironically called the Quasi-War. The underlying cause for this conflict was the commercial agreement between the United States and Britain which had been negotiated by John Jay in 1794. The Jay Treaty hardly satisfied most Americans, but to the French it stirred a quick and violent resentment. The Directory in Paris, controlling the affairs of that nation, declared the treaty an abrogation of the original 1778 treaty of alliance between France and the United States. When the Anglo-French war broke out in 1793 the French had expected the Americans to assume an active role in a common war against Britain. When this was not forthcoming the anger of the French smouldered and finally in 1797 broke out in actions at sea against American shipping.

Feeling against France in the United States had grown, with the activities of Citizen Genet, to effect a climax. But when the French Directory, in its attitude of revenge for the Jay Treaty, ordered her naval forces to prey on American commerce, matters assumed the proportions of open conflict. By June 1797 the French privateers and war-ships had captured nearly 200 American vessels. President Washington had recalled our Minister to France, James Monroe, and when John Adams became President he appointed Charles Pinckney as Ambassador—but the Directory repudiated the new Minister and asked him to leave.

Despite the fact that the Quasi-War with France was never officially declared it was a "shooting war," and the whaleships of Nantucket and Bedford were victims. The politicians spent hours in speeches and writings, and the conflict spread. While the Federalists, who favored England, and the Jeffersonian Republicans took the side of France, American shipping was badly mauled. Because of the naval situation in France, with the British blockade effective, the French had only a few frigates at sea, and these, with their privateers, were active in both the West Indies area and the English Channel. In the West Indies, following the recapture of Guadaloupe by the French, prize courts were established, in which the judges were many times part owners of the privateers.

Late in May 1798 Congress passed an act giving the President authority to instruct commanders of armed vessels of the United States to seize any armed vessel hovering near our coasts for the purpose of committing depredations, and also to retake any merchantman that might have been captured. It was inevitable that relations with France would quickly worsen and in June commercial intercourse between the United States and France was suspended. The French government was requested to disavow its preying on American merchantmen, and, of the most importance, to acknowledge the claim of the United States to be considered neutral. If this were done the United States agreed to discontinue its prohibitions on French trade.

So rapidly did the relations between the two countries deteriorate that on July 28, 1798, the Congress declared the nation was exonerated from the stipulations of the original French treaty. Thus, the American States were brought to the very brink of war—a condition marked by an existence of armed hostility without either country actually declaring a formal state of war.[1]

Returning homeward from long voyages, the whaleships were vulnerable to capture as they neared the West Indies, and many were taken in this area. The ship *Johanna*, Captain Zebdial Coffin, was within a few days sail of Nantucket with 2,000 barrels of oil when she was captured by a privateer schooner and brought into Cayenne, there condemned by a prize court. Her cargo was valued at $56,072.00. The *Active*, under the veteran Captain Micajah Gardner, was taken in 1799, and the ship and cargo were burned at sea. The owners, Mitchell and Gardner, suffered another loss when the *Minerva*, under Captain Obed Fitch, was taken by a French privateer out of Guadaloupe, but was retaken by an English sloop-of-war and brought into Antigua. Here the Admiralty Court or-

1. Harold U. Faulkner, *American Political and Social History* (New York, 1927), Chapter 9, p. 177.

dered the owners to pay fifty percent salvage and expenses, or the ship be placed at auction. Captain Fitch was forced to buy the ship and finally brought her home.

In an interesting affidavit, the master of the whaleship carefully itemized the "articuls robbed from on board of the *Minerva* by the French and English," listing among other things "1 pease of Rusha duck, 700 fathoms whale line, 12 whale irons, 2 tuns of wood hopp casks (shooks), I louer deck hatch hove overboard, and damage done our harsaw." A third Mitchell and Gardner vessel, the *Federal George*, was captured by a French frigate and condemned at Guadaloupe. These vessels were valued with their cargoes at $75,000 each.

After years of unsuccessful efforts to recover through the French Spoilation Claims, Aaron Mitchell died at the age of seventy-eight. Of this Nantucket Quaker merchant, a contemporary wrote:

For many years he was one of our most enterprising and successful merchants but, of late years, suffering under repeated and heavy losses. The frosts gathered white on his brow. He had been a long and anxious expectant of the payment of those just claims, hitherto so cruelly and unjustly disregarded by our Government in its petty scrambles of party strife.[2]

An interesting circumstance in connection with Paul Gardner, of the Mitchell firm, is that his corpse was the last ever to be attached for debt in Massachusetts. He died in 1830 and the statute permitting such an uncanny procedure was repealed shortly after.

The whaleships of New Bedford suffered similar disasters. On May 11, 1798, the ship *Fox* was captured by the French frigate *L'Importune* and brought into Guadaloupe to be condemned. She was commanded by Captain Coffin Whippey, a veteran of the Dunkirk whaling fleet, who had returned to the American fishery. In advancing claims for reparations, the owners of the *Fox* provide an excellent example of how the whaleships of the time were owned in shares, as follows:

Seth Russell & Sons	4/16
Daniel Ricketson & Sons	2/16
Cornelius Howland	2/16

2. *The Warder*, Nantucket, Mass., May 20, 1846.

Thomas Hazard, Jr.	2/16
William Rotch, Jr.	2/16
David Swain	2/16
Benjamin Taber, Jr.	1/16
William Handy	1/16

The claim was made up as follows:

Ship and outfit	$20,433.04
27 1/2 tuns sperm oil at 200	5,500.00
9 3/8 do. head matter at 300	2,812.50
188 do. whale oil at 73 1/3	13,795.83
17,500 lbs. whale bone	1,750.00
Charles in Guadaloupe	84.50
	$44,375.87[3]

Particularly ironic was the fate of the pioneer Cape Horn whaler *Rebecca*, out of New Bedford. Returning from her third voyage to the Pacific Ocean under Captain Andrew Gardner, she was captured by a French privateer, but was retaken by a British vessel and sent into Halifax, Nova Scotia. Before she was released, half the value of ship and cargo was awarded her last captors.

In a letter dated Nantucket, July 26, 1800, Silas Gardner wrote to his nephew: "Andrew Gardner from New Bedford on a whail voiage was taken on his return from 'round Cape Horn by the French and retaken by the Britons. He has got to Bedford."[4] A similar experience was shared by the ship *Nancy*, of Nantucket, Captain Zaccheus Swain, but had a far happier ending. While homeward bound early in 1800 the *Nancy* was captured by the French privateer *Reliance* and shortly after was retaken by the United States brig *Eagle* and so got into Bedford safely with a cargo of $50,000 worth of oil. Thomas and Abisha Delano were her owners, the latter recently returned from whaling out of London.

Sad, indeed, was the experience of the Nantucketer Captain Christopher Folger, in the London whaler *Perserverance*, captured by the French in 1801. With his oficers and crew, Folger was imprisoned at

3. French Spoilation Claims, *The Inquirer and Mirror*, Nantucket, Mass., March 14, 1891; also, *The Inquirer*, Nantucket, January 14, 1832.
4. Silas Gardner to Captain Micajah Gardner, July 6, 1800. Author's papers.

33. Captain Reuben Joy of the *Atlas*, Nantucket
From a tintype taken in his eighty-fourth year
[Courtesy the Nantucket Historical Association]

Sarre Libre, Department of the Moselle, and remained there several years. In one of his letters home, he wrote:

If nothing is done for my release I am fearful it may prove as hard with me as it has been for my fellow townsman, Henry Bunker, who departed this life the 20th of February, 1807, after a sickness of six weeks . . . I am well assured if nothing is done for me at home there will be nothing done in this country.[5]

Finally, the abortive government of France, The Directory, fell and Bonaparte came to power. When Napoleon, at the urging of Talleyrand, finally brought an end to the undeclared war between the United States and France, the French agreed to the abrogation of the terms in the original treaty of alliance, and promised to respect the rights of neutral traders. In the light of history the United States, as in the case of the treaty with England following the Revolution, had not fulfilled certain agreements stipulated in the original French treaty, and the quasi-war was a stiff price to pay for political waywardness.

Particularly unfortunate, also, was the treatment by the French government of the little whaling colony at Dunkirk. Following the crucial years of the French Revolution the industry had revived, but the advantage to the whalers was lost when Napoleon decreed the confiscation of the holdings of the "foreigners" in that port. However, the two Nantucket leaders, Captain William Mooers and Captain Benjamin Hussey, and Jeremiah Winslow managed to remain there, despite the loss of their property. Captain Hussey continued to serve as the port agent for the Rotch firm of New Bedford, while Winslow moved to Le Havre, where he established himself firmly in French history as "Pere Winslow," the father of the next revival in French whaling—from 1815 through 1840.

The short span of British–French peace—1801–1803—was but a preliminary to a resumption of the Napoleonic war. In October 1804 Spain became the ally of France, and once more the whaleships became prey of the privateers as they neared home ports. The *Kent*, Captain Christopher Bunker, the *Ganges*, Captain Obed Folger, and the *Falkland*,

5. Capt. Christopher Folger to William Coffin, October 16, 1806. In author's manuscript collection.

Captain Peleg Bunker, were captured by the French as they approached the English Channel.

The *Falkland* was a former Dunkirk whaler, captured by the British, and William Rotch had been unable to ransom her. Now, the tables were turned and she was again under the French flag. However, Captain Peleg Bunker, her master, and his men were treated as British subjects and were "marched like cattle" from the coast to a prison at Verdun.[6] During the next five years the British government, the Society of Friends in France and England, and private parties attempted to gain the Nantucketers' release—but to no avail. Finally, when the prisoners were released, Captain Bunker was so ill that he died from the emotional shock upon hearing he was to be freed. In his Nantucket home, he left a widow, Lydia Gardner Bunker, and seven children. He was fifty-nine years old.

There were ironic overtones to this incident. Captain Peleg Bunker had been one of the original shipmasters coming to Dunkirk with William Rotch, and among the French whalers he had commanded were the *Hope* (1789) and the *Ardent*. His oldest son, Captain Obed Bunker, had been master of the *Greyhound*, captured by a Dutch privateer in Delagoa Bay at the outbreak of the 1793 war between England and France. His second oldest son, Captain Tristram Bunker, had served his apprenticeship on the whaleship *Ruby* from Nantucket, and then had become master of the London whaleship *Scorpion*. On his second voyage to the Pacific in the *Scorpion* Captain Tristram Bunker had gone ashore on the South American coast at Chile for provisions when he was attacked "and barbarously murdered by the Spaniards."[7]

The Spanish colonies on South America's west coast were continuing to treat Americans as allies of Britain, creating many unfortunate incidents. In August 1809 the Nantucket ship *Atlas* was at Charles Island in the Galapagos Group, where Captain Reuben Joy careened the vessel in an effort to repair her bow and stop a dangerous leak. Finding it too rough a roadstead he sailed to Paita, in northern Peru, and managed to lighten the *Atlas* enough to at last check the leak. Out at sea again, Captain Joy cruised south and well off the coast, being "twenty-five leagues from

6. Captain Peleg Bunker, in a letter written from Verdun, France, December 22, 1805, and copied by his granddaughter, Lydia Bunker Gardner.
7. Thomas Sutcliffe, *Sixteen Years in Chili and Peru* (London, 1841), p. 96.

land" when he was captured by the *Vulture*, a privateer belonging to Lima, Peru and commanded by Don Juan Domingo Ameraza. Taken into the port of Callao, the *Atlas* was detained for a month, her provisions damaged and the ship ransacked. Finally, after weeks of hearings and interviews with the authorities at Lima, Captain Joy and the *Atlas* were set free, but he was unable to proceed because his crew were scattered and his provisions damaged.

"It is impossible for me to pursue my voyage," wrote Captain Joy in a formal protest to the Peruvian authorities, "I having no more than myself, two mates [Shubael Chase and John Harps] two men and a small boy [his nephew Samuel Joy] on board my ship, sixteen of my men being on board of the *Vulture*." The masters of three whaleships in Callao at this time (all of whom were fellow townsmen of Joy's) subscribed to the truth of the statement. Captain Joy's written protest was concluded as follows:

I must inform your Excellency that in case my Damages are not made good here, I shall immediately forward to the United States Papers and Documents substantiating the facts herein Related and refer [them] to my government for Redress.[8]

Fortunately, at this critical time, three whaleships commanded by fellow Nantucketers arrived at Callao—the *Hannah & Eliza*, of Milford Haven, under Captain Frederick Coffin, the *Maria*, of New Bedford, with Captain David Coffin, and the *Phoebe Ann*, under Captain Sylvanus Russell, also a New Bedford ship. Supplies were obtained, as well as some foremast hands, and Captain Joy was able to resume his voyage and return to Nantucket. It is not known whether the owners or Captain Joy were able to recover any damages.

When the *Atlas* reached home, however, she was repaired and again fitted out for a voyage, sailing under the command of Captain Obed Joy, brother of Captain Reuben Joy. She continued as a Nantucket whaler until sold in 1823.

Letters from the home port of the ships scattered throughout the world brought news of the varied fortunes of the Nantucket men who com-

8. Captain Reuben Joy, statement prepared at Callao, Peru, December 16, 1809. In author's manuscript collection.

manded ships from both Europe and America. Brief and terse, they still convey the international flavor of the times, with the pungency of a whale-oil flavor. Typical was a letter penned by Shubael Coffin, Jr., to his uncle, Captain Micajah Gardner, the veteran whaling master, the date being October 5, 1806:

Mother Pinkham is well and all the rest of our friends in general. Peter Coffin was marryed last Wednesday night to Joshua Coffin's daughter and is now awaiting a fare wind . . . his ship is partly loaded. My kinsman, Henry Folger, who had his ship taken from him by his own crew which mutinied and run his ship ashore on the coast of Chili, has arrived here two days since with Capt. Solomon Swain in the ship Lima, *with two hundred tuns of Sperm oil. The ship* Sukey, *Capt. Whippey, has got home with a Thousand barrels of Sperm oil. Give my respects to all friends. Peggy & Mary join us in respects.*[9]

The mention of the mutiny on the *Neutrality* is the only contemporary record of that misfortune, which occurred in February 1806. So far as is known it is the first mutiny on an American whaler, and that the crew ran the ship ashore after taking her by force from her captain and officers brings a further unusual character to the incident.

The necessity of arming the whaleships was recognized more by the British than the American whaleship owners, and there were many occasions when the practicality of such a practice was demonstrated. The London whaler *Argo*, under Captain Charles Gardner, sailed from England in July 1808, reaching Rio de Janeiro in September and there shifting his guns in preparation for arrival in the South Seas. Rounding Cape Horn a ship was sighted; as reported, "she set American colors and we did the same. . . . At 6 p.m. got some of the guns ready in case she proved an enemy."[10]

On the next occasion when a strange ship approached, the *Argo* came close enough to hail her, "but could not understand and soon we drew from him. He fired a large gun and the shot struck near us on the larboard quarter and a number of muskets. We steered WNW and very soon lost sight of him."

9. Shubael Coffin, Jr., to Captain Micajah Gardner, October 5, 1806. In author's manuscript collection.
10. Logbook of the *Argo*, owned by Isaac Sisson Riddell.

One of the most interesting entries in the *Argo*'s logbook presents an excellent picture of a whaling master's problems. On January 22, 1809, while the *Argo* was off the coast of Chile, sailing north, Captain Gardner reported in detail the following:

At 1 p.m. the hands at dinner and Grog was served to the crew, Orders was that there was none for John Morgan as he had cut his nee [knee] bad and much inflamed, not fit for him to drink Spirits. He sed if he could not have his Grog he would not come up to do any duty. I sent for him to come up and told him he was not to have Grog until his leg was well. He said he be Damed if he Caered if his leg was cut off, and many unbecoming words . . . I ordered him to turn to . . .[11]

A short time after this exchange a school of whales was sighted and the boats lowered to give chase. A sperm whale was killed and Captain Gardner returned to the ship, sending his boat back to help tow the dead whale alongside. When some hands were ordered aloft to furl the foresail the belligerent sailor Morgan joined them, only to be called back to the deck as Captain Gardner "thought it best for his knee to keep it still . . . but he told me he be Damed if he cared what he did, and that he did not care if his bloody leg was off. I saw he had hurt it on the foreyard and that it was bleeding. I sent the Steward for some brandy and lint and washed the wound and put it in dry lint and told him to be still."

But the man continued to confront Captain Gardner and "abuse me with Daming." The ship was now coming down to the dead whale with the ship's boats clustered around the carcass, and Captain Gardner noted "the fish was worth from 800 to 1000 pounds to the ship, and this man Morgan was going on with his damning on the quarterdeck." The logbook entry continued:

The man at the wheel could not hear my orders, and we were in danger of running the ship by the fish. I told Morgan I would make him still. He was about to make a pass at me and I fended off his arm and he clinched me with both hands. I now found it was time to Defend myself. I took hold of him and in a minute the Chief Mate (Barney) came to my assistance and took hold of him. Whilst he had hold of me he tore the

11. Ibid., January 22, 1809.

fore part of my waistcoat nearly off. I then tied his hands. Then we got the fish alongside, which was not completed until dark. All the time he would keep talking in a abusive manner, saying . . . he would have satisfaction of me, as was knowing of many Disaffeted words being Spoak in the Ship by Several. I thought it would be most proper to put him in irons, which I ordered to be done for the General Good and Safety of the Ship and our lives. Latter part cut in the Fish and began boiling.

Two days later, January 24, 1809, The *Argo*'s logbook noted:

At 8 p.m. when all the officers were below, the Steward and Cabin Boy and James Prumus a Boatsteerer on Deck, and he (Prumus) said to the Steward that it would not be long before there would be a Mutiny in the Ship. He would wait a while but he would see it out; if he did begin some would be Killed—and further said hand-spikes was as good as guns. This the Steward John Mead and Wm. Grant heard him say and many other mutinous words, and came and told me the above." [12]

The course of the *Argo* was towards the Galapagos Islands, and on January 28 the New Bedford whaleship *Maria*, Captain David Coffin, was sighted. The ships then sailed in company, with the two Nantucket captains enjoying each other's cabins. The incipient mutiny was now well blocked. Two weeks later Captain Gardner reported that Morgan came aft and requested he be freed. "I told him If he would do his duty and behave himself I would release him on tryal and ordered his hands loosed." Fresh provisions were obtained from the *Maria* and the *Argo*'s log mentioned "the hands that had the Scurvy getting better." The ship had not dared put into any port from the time she had sailed from Rio the previous October—six months before—and the potatoes and pumpkins from the *Maria* were welcomed. Arrival at the Galapagos Islands, early in May, was of great importance as here the land turtle, or terrapin, provided fresh meat for a food-starved crew.

Here was a favorite area for whaleships to be refreshed, as well as a prime region for catching whales. On the 13th of May, the *Argo* sailed to an anchorage at Charles Island, putting men ashore for terrapin and pumpkins. Two other whaleships were anchored close by, and when

12. Ibid., January 24, 1809.

they left the London whaler *Cyrus*, under Captain Paul West, came in and moored close by the *Argo*. While ashore the crew brought seventy of the land tortoise to the boats. Two men did not return, escaping into "the woods, and we could not find them, & hollered and looked . . . Night coming on we were obliged to get to the ship. Left some vittles and water for them." The next day the younger of the two escapees came down to the shore and shouted to be taken back to the ship. The other runaway, Joseph Pollard, remained in hiding.

The place on Charles Island where the boats landed was known as "Post Office Bay." On another side of the Island was a section of the shore which was called "Paddy's Beach." The man known as "Paddy" was a castaway who lived in a wooded section of the Island. The desperate character of the man was shown by the following incident:

At 5 a.m. went to N.E. part of the island known as Paddy's Beach to get pumpkins, with two boats from each ship, and arrived at 6 a.m.. Went up near to Paddy's Plantation and found pumpkins, according to agreement, and we took down to the beach 60—and came up and got 60 more, and went up the third time . . . Made the rest of our way down to the boats, where we found two of the Cyrus' *stove, one of the* Argo's *whole and one gone, together with the oars and sails and everything we had aboard the boats—excepting two steering oars and one boat sail that was up in the bushes, with which we reached the ships, and sent off boats and tarpaulines for the boats wrecked, and then took them on board. Brought off Joseph Pollard that had been with Paddy, and he informed me . . . that Paddy double-loaded 3 muskets and 2 pistols and got his cutlasses in order, so that it was premedatated business. Getting ready to leave the island.*

In company with the *Cyrus* the *Argo* sailed to James Island, where they sent boats ashore for water, and "found very little to run at the rate of 10 gallons an hour." Here they saw another self-exiled whaler, who had come ashore from the London whaleship *Thames*. A wild cotton plant was picked to serve as tea. Several of the men who had shown little improvement from attacks by the scurvy were taken ashore "to bury in the ground," in an attempt to arrest the progress of the dreaded affliction.

Getting underway the *Argo* sailed to Albemarle Island, and off this high island spoke the whaleships *Charlton*, Captain Folger, of London, *Sally*, Captain Edward Clark, *Winslow*, Captain Prince Coleman, of New Bedford, and the *Lady Adams*, Captain Elisha Folger, of Nantucket. On May 31 a strange sail was sighted. The *Argo* attempted to escape but the newcomer outsailed her and, upon firing her cannon, forced the whaleship to heave to. Captain Gardner sent his boat and ship's papers to the armed vessel, which proved to be a Spanish government ship. The logbook recorded: "They kept the officer and crew and came on board, several officers and 14 men, and searched the ship in every part, and was well satisfied and used us well."

After another year of cruising, spending most of the time in the vicinity of the Galapagos Islands, the *Argo* sailed for home, arriving at London on February 26, 1811, having been absent twenty months and bringing home a full cargo of twenty-five-hundred barrels of oil.

The *Cyrus*, Captain Paul West, was boarded (June 2, 1809) by the same Spanish armed vessel which had forced the *Argo* to heave to and be inspected. In reporting the incident, Captain West described the same procedures as those that took place on the other whaleship. The government ship was the *Hero*, and the *Cyrus*' logbook recorded: "The Captain gave me a certificate of examination and permitted me to depart, but his crew immediately demanded that the *Cyrus* be sent to Lima for further examination, saying that some of the crew of the *Cyrus* had given them every reason to consider her a good prize."[13]

Fortunately, the original orders of the *Hero*'s captain prevailed and the *Cyrus* was permitted to proceed. Upon returning to his ship (having been detained on board the *Hero* during the examination), Captain West reported his decks, steerage, cabin and "every other part, stores, provisions, etc., all in heaps, looking as if some powerful hand with a strong instrument stirred them up in a manner similar to a cook stirring a dish of Pease, but having got my ship and crew without any considerable damage or loss, I plyed to windward of the *Hero*."

The voyage of the *Cyrus* was one of the most successful of Captain Paul West's career, and his logbook reflects, as few such contemporary writings do, the full flavor of such extended voyagings. Among the wealth of

13. Logbook of the *Cyrus*, Captain Paul West, July 4, 1808, to July 24, 1810. Peter Foulger Museum Library, Nantucket Historical Association.

detailed information is an account of an encounter with a bull sperm off the coast of Peru. One of the boats "fastened" (plunged the harpoon deeply into the whale) at 6:30 one morning and three hours later the whale was still towing the boat, having carried it some twenty miles. At that moment the harpoon broke, freeing the whale, which swam away before the crew could again attack. All during this three-hour period the boat was unable to get close enough to the whale to "dart" a second harpoon at the animal, so vigorously did he keep the boat at tow behind him.

Another interesting entry was that of May 26, 1809, when Captain West listed the names of ten vessels spoken by the *Cyrus* in a single day off Narborough Island in the Galapagos Group. These were the *Lion*, Capt. Peter Paddack, Nantucket; *Winslow*, Capt. Prince Coleman, New Bedford: *Lady Adams*, Capt. Elisha Folger, Nantucket; *Maria*, Capt. David Coffin, New Bedford; *Sally*, Capt. Edward Clark, New Bedford; *Danube*, Capt. James Mosher, London; *Hannah & Eliza*, Capt. Frederick Coffin, Milford Haven; *Charleton*, Capt. Christopher Folger, London; *Volunteer*, Capt. Uriel Bunker, Milford Haven; and *Walker*, Capt. Stephen West, of New Bedford, a brother of Captain Paul West. All but one of these vessels were commanded by Nantucket men, and most of the officers were probably natives of that island, the exception being Captain Mosher.

As was the customary experience of whaling masters the problems of maintaining discipline aboard a whaleship required a constant vigilance on the part of the officers. In one instance, Captain West had to intervene when a mutinous sailor "clinched "with one of the mates and then produced a knife, which the shipmaster laconically mentioned, "I took from him." On another occasion he subdued another rebellious man and "seized him in the main rigging and gave him a dozen strokes of the cat, to show him the error of his ways."

Even while homeward bound, crew members threatened mutiny and Captain West reported (December 12, 1809): "I find it highly necessary to continue keeping the deck myself half of the night, with loaded pistols, and a sharp cutlass at the hands of my officers as well as myself."

The master of the *Cyrus* made one particular entry which showed him to be a compassionate man. During the first few months at sea he ordered two of the apprentice boys aloft to set the main royal, the topmost sail on the mainmast. His full record of what then transpired is worthy of being quoted in full:

They had not been long aloft when one of them call's 'hoist the royal', and the helmsman steped along the halyards and hoisted a pull without looking aloft. At the same instant the boy on the Mn t M crosstrees called out loudly, which caused me to start round, and observe one of them, Jos Jakcson, flying as a bird in the air, coming from the Royal-crosstrees down upon deck. It appears that he had loosed the royal, and was making fast the gaskets, when the yard was hoisted suddenly and threw him backwards, clear of everything, until he came down below the Main top. Then failling into the bight of the signal Halyd that happened to have loose turns round a backstay, and rendering as he descended, broke his fall wonderfully until as low as the leading blocks, then the signal halyards parted and, of course, the boy fell upon deck, striking upon his side and back flat upon the deck. I took him up and put him in an easy posture, and before I could get a Lancet to him (which was done in less than one minute) he recovered so well as to make any such assistance unnecessary; and I was extremely happy and very agreeably disappointed to see the boy, five minutes after his fall from the Main Royal yard down upon deck, walk away as deliberately as usualy, and perform his Sunday's duty as if nothing particular had happened.[14]

The eras that marked the greatest expansion in the geographical extent of the Southern Whale Fishery were those which found the ships from three nations opening to the world previously little-known parts. First, to the Brazilian Banks and down the Atlantic to the Falklands; then along the African coast to Good Hope, to round that promontory and reach the Delagoa Bay area, on to Madagascar, the East Indies and Australia; then around Cape Horn and along the west coast of South America; to proceed northward to California and the Northwest Coast of North America: through four of the world's great oceans the whaleships made their bluff-bowed way in their relentless chase of the whale.

Contributions which the whalemen and sealers made to the growth of Australia and New Zealand were such as to represent vital roles in the prosperity of these remote regions. Sydney and Hobart Town were

14. Ibid., October 16, 1808.

34. Ship *Atlas* of Nantucket, Captain Reuben Joy (1809)
[Photograph by Frederick G. S. Clow]
[Courtesy the Nantucket Historical Association]

greatly aided in their formative years by the whaler and sealer. Following the pioneer voyages of Eber Bunker and Thomas Melville in British whalers the ubiquitous Yankee whaler soon made his appearance in Australian waters. With the turn of the nineteenth century the sealing vessels from New England were playing equally important parts in this area, and New South Wales, despite the monopoly of the East India Company, which profited from and was sustained by these vessels. American ships were of particular importance as they were not bound by the restrictions of the Honorable Company. When in 1803 Governor King established a settlement in Van Diemen's Land (Tasmania), another center of the whaler-sealer activities was created.[15]

The first American vessel into Sydney (Port Jackson) was the brigantine *Philadelphia*, arriving on November 1, 1792, a trading craft from that city to China (Canton), under Captain Thomas Patrickson, with

15. "Romance and Thrills of Whaling," *The Mercury*, Hobart, Tasmania, July 5, 1924, p. 57.

beef, pork, and "spirits." As already noted, by that time several British whalers had reached New South Wales, having been pressed into service as transports for carrying convicts. A month later, December 24, the Rhode Island ship *Hope*, Captain Benjamin Page, anchored in Sydney, bound for Canton with seal skins picked up in the Falkland Islands. The third American craft to arrive (October 26, 1792) was the snow *Fairy*, a sealer, under Captain Rogers. Having rounded the Cape of Good Hope and collected her seal skins at St. Paul's Island, she was to sail to Canton, thence bound for the Northwest Coast of North America.[16]

Two more trading ships from Providence, R.I., the *Halcyon*, with Captain Benjamin Page paying his second visit, and the *Hope*, commanded by his uncle, came in during June and July 1794, both heading for China. The second American sealer, the brig *Mercury*, of Rhode Island, put in late in October of this same year, also bound for Canton, her destination the Northwest Coast of America. The next American sealer to put in at New South Wales was the brig *Otter*, of Boston, under that adventurous Captain Ebenezer Dorr, which arrived on January 24, 1796, after having taken seal pelts at Amsterdam Island.

The first American whaleship to reach Australian waters was the *Diana*, of New Bedford, with Captain Jared Gardner in command. Owned by William Rotch and Samuel Rodman, she had been one of the Rotch firm's ships originally sailing out of Dunkirk. Rounding the Cape of Good Hope she reached Sydney on November 20, 1800. She was bound for China with some trading goods, and cruised in the western Pacific en route, stopping at Naru, or Pleasant Island, where Captain Gardner reported the natives had never before bartered with white men.[17]

The arrival of the *General Boyd*, of London, under Captain George Hales, in June 1801 marked the initial appearance of this British whaler which was owned by Watson and Co. Upon returning to England she was again sent to the Australian area under Captain Owen Bunker, arriving at Sydney in June 1802. Captain Bunker was a cousin of Captain Eber Bunker, whose career, as we have seen, made whaling history. Captain Owen was back in these waters in 1805 in command of the whaler *Honduras Packet*, one of the Spanish vessels captured in the war. The

16. Cumpston, *Arrivals and Departures*, pp. 27–28.
17. Ibid., p. 37.

first Nantucket whaler to arrive at Sydney (July 1805), was the *Brothers* under Captain Benjamin Worth, a ship which took whales off the coast of New Zealand. But she had been preceded by the *Rose*, Captain Cary, a China-bound merchantman of Nantucket, and the sealers *Criterion* and *Favourite*.

The *Brothers* provided an interesting footnote to Australian history by participating in a unique event in Sydney harbor, when a boat's crew from the Nantucket ship won a whale boat race with a boat from the London whaler *Honduras Packet*, on August 3, 1805. The boats began the race at Benelong Point and raced around Shark Island—a distance of seven miles. The *Sydney Gazette* reported: "They started at 9:15 a.m., and at 10:20 the Brothers' crew regained their ship, closely followed by their antagonists." Captain Benjamin Worth, of the *Brothers*, and Captain Owen Bunker, of the *Honduras Packet*, were from Nantucket.

After staying in port for an extended period of four months, the *Brothers* sailed on November 1, 1805. Late in December she was spoken by Captain Gardner in the *Hannah & Eliza*, trying out oil near Barrier Island, New Zealand. The two ships probably sailed in company as on January 20, 1806, Captain Worth took a whale the *Hannah & Eliza* had lost, and both shared the blubber. The *Brothers* returned to Sydney on July 22, 1806, but this time remained only a month, then sailed for home.

The advantage which the Americans gained over the British in these waters was a matter of deep concern to Governor King of New South Wales. China was the chief market for seal skins, and yet the Colony could not ship to Canton because of the East India Company's monopoly. Appeals to government were denied, and even an effort to trade with Canton by way of London proved abortive. Problems with American vessels carrying trade goods and entering Sydney had been a problem, also, and Governor King's predecessor, Captain Hunter, was instructed to "ascertain the number of and description of all such vessels as may arrive at or proceed from Port Jackson," and "no vessel is to be allowed to land any articles or break bulk before the return is filled up." One of the reasons for checks so proscribed was to prevent the "importation of spirits without your license first obtained."[18]

18. *Historical Records of Australia, Series I*, Vol. 2, p. 341.

The sealers' voyages were managed shrewdly. They would bring in a cargo of trading goods to the Colony, proceed to the sealing localities to obtain fur pelts, bring them to China for sale and put aboard a return cargo of China goods. Often they would buy seal pelts at Sydney, since by 1799 local vessels were bringing in large numbers from the newly discovered sealing areas in Bass Straits. Between January 1803 and December 1804 a total of 47,402 seal skins were brought into Sydney by the Colonial sealers.[19]

But the American sealers were ranging far: the *Union*, Captain Isaac Pendleton (a Stonington man), came into Sydney in January 1804, with 12,000 skins he had taken in Bass Straits. Pendleton is thought to have been the pioneer in the region around Kangaroo Island in the Straits, but recent researches by J. R. Cumpston, of Australia, indicate that the New England sealer *Fairy*, Captain Rogers, may have touched at Kangaroo Island on a voyage from St. Paul's Island to Sydney, where she arrived in October 1793.

Further research by Cumpston reveals that when Captain Matthew Flinders made his historic voyage along the southern coast of Australia early in 1802, he found on Kangaroo Island a piece of sheet copper, on which was inscribed: "August 27, 1800. Chr. Dixson—ship *Elligood*." The remains of the wreck were later found by other mariners on the other side of the Island. The *Elligood* was a London whaleship, owned by Daniel Bennett, and had made one voyage to the African coast (1798–1799). An entry in a Cape Town newspaper indicates the *Elligood* returned to that port in May 1801, reporting the death of her master and nine men by scurvy.[20]

The French exploring ships *Le Geographe* and *Le Naturaliste*, under command of Captain Nicholas Baudin, sailing from Le Havre on October 19, 1800, reached Tasmania in January 1802. They followed the route which Captain John Hunter, commanding the convoy of convict-transports and store ships, had first used in 1787–1788 in sailing to Botany Bay, traveling south of 50° latitude to reach his destination in New South

19. Gordon Greenwood, *Early American–Australian Relations To 1830* (Melbourne University Press, Australia, 1944), p. 82.
20. J. S. Cumpston, *Kangaroo Island, 1800–1836* (Canberra, Australia, 1970), p. 3.

Wales. The French expedition eventually arrived in King George's Sound on Australia's southwest coast in February 1802, where they met the American sealing brig *Union*, Captain Pendleton, who had sailed from New York the same month (October) that the French had left Le Havre.

The *Union* sailed east to winter at Kangaroo Island, where the bulk of her cargo of seal skins was taken. Here Captain Pendleton and Mate Owen Folger Smith built a small schooner, named the *Independence*. On January 6, 1804, the *Union* reached Sydney, and Governor King immediately chartered the brig for a cruise to Norfolk Island to bring back some pork, bringing tea, sugar and other supplies to that Island, which was an offshoot of the larger penal colony. Upon her return from this assignment, Captain Pendleton again sailed for Bass Straits and Kangaroo Island, to return in June with the *Independence* and 12,000 seal skins.[21]

At this point the figure of merchant Simeon Lord, of Sydney, enters the story. An agreement was made between Captain Pendleton and Lord, whereby the seal pelts would be purchased by Lord and payment provided by a cargo of the sandalwood to be procured at the Fijis through the agency of a certain John Boston. Both ships sailed for China with their skins on August 29, 1804. The *Union* stopped at Tongatabu in the Tonga Islands, and Captain Pendleton and John Boston went ashore. It was a most unfortunate decision, as both Pendleton and Boston and the boat's crew were killed by natives.

When the boat did not return from shore Mate Wright was alarmed by the appearance of many canoes. He would have hesitated in getting under way, expecting Captain Pendleton's return, but an amazing scene developed. In one of the canoes a white woman suddenly stood erect to shout a warning in English that the boat's occupants had been massacred and the vessel was in danger of being attacked. Mate Wright, with considerable presence of mind, called upon the woman to leap into the sea and be rescued. Before the startled natives realized what was happening the woman quickly obeyed, and was rescued under the cover of musket fire that drove the canoes off. The white woman was the sole survivor of

21. Ibid., p. 27.

an American vessel, the *Duke of Portland*, her master and the boat's crew massacred upon visiting the shore. The woman's name was Elizabeth Morey.[22]

The *Union* got under way immediately, to return to Sydney. After conferring with Lord it was decided the brig should go back to the Fijis to obtain the cargo of sandalwood so much in demand on the Chinese market. But the expedition was doomed when the *Union* became trapped among uncharted reefs and wrecked. Those who managed to escape drowning by swimming ashore were killed by the natives.

The *Independence*, under Captain Isaiah Townsend, did not sail for China, despite the declaration in the Sydney port record. Using a typical "dodge," she sailed "in ballast," and actually headed for New Zealand where she conducted an exploratory cruise through Foveaux Straits, that passage between South Island, New Zealand, and Stewart Island. A chart prepared by the mate, Owen Folger Smith, has been preserved, showing that this American navigator was the first to chart the strait now called Foveaux. Continuing on to the Antipodes Islands, the *Independence* landed a shore gang for sealing, and then returned to Sydney on April 21, 1805. On June 11, the little schooner sailed again, in company with the ship *Favourite*, Captain Jonathan Paddock, a Nantucket vessel owned by Paul Gardner, Jr., and Co. She had sailed from Nantucket in October 1804, rounded Good Hope and gone to the Crozet Islands for seal pelts, becoming one of the first sealers in those remote islands. The *Favourite* had reached Sydney on March 11, 1805, and Simeon Lord had arranged with Captain Paddack and Daniel Whitney, the supercargo, for the *Favourite*'s sailing with the *Independence*.[23]

Both Captain Townsend and Owen Folger Smith were Americans, and Smith, who had incurred the displeasure of Governor King for alleged smuggling of wine to Norfolk Island, was probably a relative of Captain Paddack's fellow townsman, Owen Folger, who had married Eunice Smith. In his historic chart of Foveaux Straits (the original now in the Alexander Turnbull Library in Wellington, New Zealand), Smith had

22. Greenwood, pp. 95–96.
23. Papers of the Ship *Favourite*, Captain Jonathan Paddack, Nantucket Historical Association.

noted the place-names of the discoveries by the sealers, all of which are significant. One harbor, called "Port Honduras," probably was named for the vessel commanded by Captain Owen Bunker, another Nantucket man in the British whaler-sealer *Honduras Packet.* The mate of the *Favourite* was Thaddeus Bunker, a cousin of both Owen and Eber Bunker. Captain Owen Bunker had married Mary Fenn, in London, while another Captain (Tristram) Bunker had married the sister, Cecilia Fenn.

The *Favourite* began her voyage to the sealing islands from Sydney on June 11, 1805. Once out to sea, Owen F. Smith and eleven men were transferred from the schooner to the ship, and both craft headed for the South Antipodes. When the *Favourite* arrived she took aboard 60,000 skins from the shore gang which the *Union* had placed on the Islands. Captain Paddack waited in vain for the *Independence*, and finally sailed back to Sydney, where the schooner was reported as missing. All trace of her vanished, and her fate remains a mystery. Arriving at Sydney, March 10, 1806, Captain Paddack completed his dealings with Simeon Lord and on July 29 sailed for Canton.

The ship's papers, filed when the *Favourite* returned to Nantucket under date of November 27, 1807, show that 87,080 skins were sold at Canton in January 1807. Captain Edmund Fanning, representing the owners of the *Union* declared that Lord never paid as much as "one farthing" for the skins left by the *Union* in his care, or for a share of the skins taken from the Antipodes. How many of the latter number there were is not listed in the *Favourite*'s accounts, but Obed Macy, the agent for the owners, noted:

Whitney and Paddock gave me an account of the skins obtained by the ship's crew, of which they were to draw their shares, valuing the skins at the price in Canton. After deducting customary charges the whole number of skins they furnished me with were 34,356 . . . valuing the skins at 80 cents each, which they said was the price in Canton, amounting to $27,484.[24]

It would thus appear that the owners of the *Union* had a just case against Lord. Edmund Fanning, in his account of the affair, as printed

24. Ibid.

in *Voyages and Discoveries In The South Seas,* declared that 14,000 skins had been landed at Sydney in Lord's care before the *Union* began her last voyage. The bulk of the skins from the Antipodes, he stated, totaled "rising 60,000." The New York owners, through Willet Coles, issued a writ of attachment against Lord claiming the Sydney man was indebted to them "in the sum of one thousand dollars and upwards of lawful money of the United States over the above all discounts." This does not appear to be an amount consistent for the sale of 53,000 skins, the number that would result after deducting the Nantucket ship's total. The actual facts of the undisclosed agreement between Captain Paddack, Supercargo Whitney, and Simeon Lord will appear later in this chapter.

Captain Jonathan Paddack was no newcomer to sealing. From 1801 through 1803 he had sailed in the ship *Sally*, of New Haven, and had obtained over 60,000 skins, which he sold at Canton in January 1803. Among the owners of the *Sally* appear the names of three members of the Coles family.

It is of interest to note that while in Sydney in March 1806, Captain Paddack and Daniel Whitney offered Governor King a charter for the *Favourite* to bring food supplies to the port, possibly from India. Governor King declined, but offered an alternative plan to bring rice within four months at £30 a ton. The Nantucketers asked for an advance of 10,000 Spanish dollars, and the right to bring in 5,000 gallons of spirits. The reply of the Governor was prompt:

> *Mr. Harris [the port officer] will have the goodness to inform the Supercargo and Master of the American ship* Favourite *that Government has no dollars, nor do I choose to give 7 shillings a piece for such Dollars as I might be inclined to take from the few Individuals who possess them. . . . To the request for bringing Spirits I have every objection.*[25]

Simeon Lord, that somewhat devious entrepreneur, had realized enough profit from the sandalwood trade at Canton, to again make an attempt at chartering an American vessel. Following the abortive contract with the *Union*, he arranged a similar charter with Captain Peter

25. Ibid.

Chase, of the Nantucket sealing ship *Criterion*, that had arrived at Sydney on April 21, 1805, with 1,300 seal pelts procured at the Crozet Islands. Among her cargo of trade goods were sperm candles and tobacco. To perform the charter-voyage to the Fijis, Captain Chase needed a pilot, but the Governor refused to allow James Aicken, an English pilot, to go aboard the American ship. To circumvent this Aicken and another British sailor shipped aboard the London whaleship *Harriet*, which had arrived in Sydney under Captain Thaddeus Coffin, another Nantucketer. The whaleship, bound home to England with a full cargo, sailed from Sydney on May 28, 1805, but the *Criterion* had gotten under way a few hours before. Out in the Tasman Sea, Pilot Aicken and James Bailey trans-shipped from the *Harriet* to the *Criterion*.

On May 25, 1806 (a year later), the *Criterion* returned to Sydney. She had obtained her sandalwood from the Fijis, no doubt, and brought a cargo of Chinese goods, intended for America. However, before sailing, a "Letter of Attorney" was drawn up between Simeon Lord and Captain Chase, dated July 22, 1806, in which Captain Chase was appointed Simeon Lord's "true and lawful attorney, to ask, sue for, demand, recover and receive for me," any money due Lord in the United States. At the same time an "Article of Agreement" was written and "entered into," between Chase and Lord: "Witness, that the said Simeon Lord has a set of Bills of Exchange at twelve months after date, drawn upon Paul Gardner, Esquire, of Nantucket, for $30,000, and the said Lord is also entitled to a certain part of moiety of what the present cargo of the said ship *Criterion* may sell in America or elsewhere."[26]

Upon arrival in America, Captain Chase agreed to do everything necessary to realize the money to which Lord claimed, "and with the said property purchase a Vessel and Cargo and dispatch her to this Colony with such goods as may be thought most likely to be beneficial to the parties concerned." Among the articles listed as "best calculated for the market of New South Wales," were 4,000 gallons of rum, as well as 2,000 gallons each of Hollands, wine and porter, barrels of flour, beef, pork (salt), and five tons of tobacco.

Of particular significance in this document is the set of bills of ex-

26. Article of Agreement, Captain Peter Chase and Simeon Lord, July 22, 1806. In author's manuscript collection.

35. The Henry Coffin (Carlisle) House, Nantucket
[Photograph by Cortlandt V. B. Hubbard]
[Courtesy Historic American Buildings Survey]

change in the amount of $30,000 drawn against Paul Gardner, of Nantucket. The agent and principal owner of the *Favourite* was Paul Gardner, and it is natural to assume this was Lord's share of the 53,000 seal skins that ship had brought to the China market.

The *Criterion* sailed from Sydney on July 29, 1806, going first to the Derwent River in Tasmania. On August 31, Lieut-Governor Collins wrote to Viscount Castlereaugh:

I understood from the master, Mr. Peter Chace, that having incurred the displeasure of Governor King by receiving on board a British Subject whom he had expressly forbidden him to ship anywhere to the eastward of Cape Horn and proceeded with him to Canton, where they laid in an extensive Cargo of Tea and China goods with which he returned to Port Jackson; that he would not allow him to land a single article from his ship, not even a trifling present. I discovered that he came in here with a view of disposing of part of her Cargo, which was to find its way back to Port Jackson, but considering the transaction in the same light as the

Governor and conceiving myself equally bound to discourage any clandestine communications with the possessions of the East India Company as well as the coast of China, I refused my consent; and although the Article of Tea was one of those comforts of which we had been for some months wholly destitute, I would not permit even the smallest quantity to be landed from her but directed him to quit the Port forthwith.[27]

It should be mentioned that Captain Peter Chase was a nephew of Owen Folger, and a cousin of Owen Folger Smith. The relationship of the Nantucket mariners was surmised by Surgeon James Thomson when he wrote on June 28, 1804: "Ever since I have been in the service of New South Wales I have observed most of the commanders of the South Sea ships to be Nantucket men."

The *Criterion* reached Rio de Janeiro in November 1806 and here Captain Chase shipped a cargo of teas, carpeting, chinaware, window blinds, silks, candles, tools, and "one bag of whales' teeth," on board the brig *Hannah & Sally*, of Philadelphia, Captain Nathaniel Cogswell, consigned to Port Jackson and "Simeon Lord, Merchant." The brig arrived in New South Wales on April 5, 1807. Upon arrival in Nantucket in January 1807 Captain Chase insured the cargo on board the *Hannah & Sally* for Simeon Lord with the Nantucket Union Marine Insurance Company in the amount of $10,000, "the risk to commence upon her leaving Rio de Janeiro and to cease upon her safe arrival at Port Jackson."[28]

The whaleships were not the problem to the authorities of New South Wales as were the sealing vessels. Battles between rival gangs of American and Australian sealers were reported from Bass Straits in 1803 and 1804. The Port Jackson authorities were always alert to illicit trade. When Owen Folger Smith petitioned to go ashore there, from the *Union*, as an agent for the sealers, Governor King refused him on the grounds that:

27. Thomas Dunbadin, *Nantucket Vessels at Sydney*, Enclosure in letter to E. A. Stackpole, August 10, 1948.
28. Insurance Policy issued by the Nantucket Union Marine Insurance Company, January 22, 1807.

Clearing such vessels out for the purpose of Skinning and Oiling, or with a view of their returning here and making this place depot for their Trade in Skins and Oil, is of manifest injury to his Majesty's subjects in this Territory . . . It is therefore . . . directed that no vessel under Foreign colors . . . be cleared from the Port for any Sealing voyage within the limits of this Territory.

Among the American sealers arriving earlier on the scene, whose activities ultimately caused Governor King to issue such stringent orders, were Captain Amasa Delano and his brother Samuel, who had come into these territorial seas in the ship *Perseverance* and schooner *Pilgrim*. No mariner left a more revealing document of his adventures than Captain Amasa. In *A Narrative of Voyages and Travels*, he recounted adventures in all parts of the Pacific Ocean, giving some idea of the variety of dangers the American whaleman and sealer encountered on his voyages into these uncharted seas.

When the Colonial cutter *Integrity* became disabled by a broken rudder while en route from Sydney to the Derwent in March 1804, Captain Delano and his brother Samuel came to the aid of Lieutenant Bowen. The cutter was towed to safety, and the passengers, freight, and company were then brought to their destination. The situation developed unfortunately, however, when the parties involved quarreled over the prices charged for the tasks.

In Bass Straits, Delano had an experience which only a truly honest man could report. It is a chilling and remarkable incident: he had set out for shore three-quarters of a mile away in the "moses boat" with five men, and half way across the boat swamped in a heavy sea and the men found themselves in the cold water. Amasa's story continues:

While moving my legs in the water . . . I hit a bundle of small sticks . . . and placed it under my breast . . . it prevented me from sinking . . . I was just heading for the land when I saw one of my faithful sailors . . . John Forstram, making towards me with all possible exertion. I turned my head from him and made every possible exertion to prevent his reaching me . . . the poor fellow, finding his attempts fail, relinquished the oar he had grasped . . . until his strength being entire exhausted he gave up and sank. I never until then experienced any satisfaction of seeing a man die, but so great is the regard we have for ourselves in danger than we would sooner see the whole human race perish than die ourselves . . . the oar he relinquished . . . I immediately seized and headed for the land. Very soon after I observed another poor, distressed sailor . . . making towards me on the right hand . . . I pulled from him . . . the poor fellow soon met his fate . . . I likewise made slift to procure his oar . . . then once more headed for the land.[29]

This frank narrative might indicate that Delano was entirely cold-blooded and calculating, but another incident he recounted gives another side of his character. While at the island of St. Mary's off the coast of Chile, he saw a vessel behaving strangely and upon investigating found it had been captured by a mutinous crew. Knowing full well the desperate situation of the mutineers, Delano nevertheless attacked and re-took the ship, named the *Tryal*, restoring it to the one of the few officers left alive after an orgy of murder.

Delano's daring action restored the ship to its owner, Don Benito Cereno. Many years later Herman Melville immortalized the incident in a story now famed for the content and the literary form. While in Peru, Captain Delano was responsible for also securing the release of a number of British and American mariners held captive at Lima.

All through the world of the Pacific the mysterious voyages of the whalers and sealers added to the geographical knowledge of their times. The Colonial sealers and whalers of New South Wales contributed a remarkable story in their own right. Mariners such as George Bass, in the *Venus*; Joseph Murrell, in the *Endeavor*; William Stewart in the *Fly*, among others, out of Sydney; Frederick Hasselburgh, master of the *Perseverance*, discoverer of Macquarie Island; Captain Richard Siddons, in the *Lynx*—all were excellent shipmasters and leaders. Such American sealers as Captain Samuel Rodman Chase, in the cutter *King George*, and Captain Robert Johnson, who disappeared on a voyage "south of New Zealand," in the *Jane Maria*, also made history.

An almost obscure British whaling master, Captain Abraham Bristow, was the discoverer of the Auckland Islands, well south of Tasmania, in late 1805, while on his third voyage in the Enderby ship *Ocean*. "This

29. Amasa Delano, *A Narrative of Voyages and Travels in the Pacific Ocean*, (Boston, 1817), p. 468.

island or islands, as being the first discoverer, I shall call Lord Auckland's (my friend through my father)," wrote Bristow in his log. "This place I should suppose abounds with seals, and sorry I am that the time and the lumbered state of my ship do not allow me to examine it." On his next voyage in another Enderby ship. the *Sarah*, he landed on the Aucklands in July 1807, and took possession for Britain. The harbor where he anchored he appropriately named "Sarah's Bosom."[30]

In the maritime world the most bold, resilient, and persevering body of seafarers were the whaler-sealers, who sailed in a world where charts were unknown and calculated risks were always a part of the voyage.

30. Samuel Enderby, Jr., to George Chalmers, July 21, 1809. *Banks Papers*, Vol. 20, (A.83) Mitchell Library, Canberra, Australia.

19. The Pioneer Port Resurges: And Another Crisis Looms

THE industrious character of Nantucket, which sustained men throughout these critical years, now enabled them to revive their basic business. From the time it originated Colonial deep-sea whaling, the Island was like a mother ship, and the ocean was as much the home of its people as was the land. The decades marking the end of the eighteenth and the beginnings of the nineteenth centuries were to find the wandering mariners returning from the international scene. The resurgence of whaling at the old home port was a feature of this period.

Having passed through the post-war crisis, the resiliency of the people now slowly transformed the depression of the 1780's into the opportunities of the 1790's. Accustomed to depending on their own resources, they set about to recoup the industry which had given them a tradition. In the first years of the recovery period several new firms were launched. Shubael Coffin, back from Europe, Richard Mitchell, Francis Joy, John Elkins, and Micajah Coffin and Sons came into the field, sending out two to three ships each, some purchased, others newly built. One of the new ships was the *Ranger*, constructed in the North River at Pembroke, and owned by Francis Joy and Co., which sailed from Nantucket in September 1788 under the command of Captain William Swain.

Word of new whaling grounds, especially Walvis Bay on the southwest coast of Africa, had reached Nantucket from fellow-Islanders with the Rotch fleet, and the *Ranger* went directly to that locality. In June 1789 she returned with 1,150 barrels of oil, becoming the first American whaler to bring back a cargo in excess of 1,000 bbls. It is of significance that out of the *Ranger*'s crew on that voyage, numbering fourteen, nine eventually became shipmasters in their own right. This demonstrates the Nantucket tradition of applied learning.[1]

1. "Reminiscence of A Whaling Voyage." *The Inquirer*, Nantucket, February 7, 1838.

Walvis Bay proved to be a prolific whaling area. Called Woolwich Bay or Walwich Bay by the American and British whalers, located just above the Tropic of Capricorn on the African coast, it was a wide indentation of some length, protected from all but northerly gales (which seldom blew), and its western side formed a peninsula four miles long. With its three-mile-wide entrance, and good holding ground in most of its six-mile length, the Bay became an excellent rendezvous for the whaleships. Large numbers of right whales frequented the coast, particularly during the early summer months.

Besides the *Ranger* there were ten additional Nantucket ships and brigs sailing to Walvis Bay during the 1789 season, and all made good voyages. The first New Bedford ship, the *Eliza*, under Captain Benjamin Coleman went there in 1794. The Dunkirk fleet, organized by William Rotch, made particular good use of this area. One of the few Dunkirk logbooks extant, that of the ship *Ann*, Captain Thaddeus Coffin, records whaleships in the area in the 1792 season. The *Ann* left Dunkirk November 10, 1791, and arrived at the Isle of May, Cape Verdes, on January 16, 1792. While at anchor here, Captain Coffin greeted the *Speedy*, of London, Captain Lock, "bound 'round Cape Horn," and taking some salt aboard for curing seal pelts. Reaching the African coast in March the *Ann* cruised a while with the *Polly* and *Union* of Bedford. They went to Cape Town in April, where they found three London whalers bound home, one of them the *Rasper*, Captain Gage, who had been also sealing at St. Paul's Island.[2]

Arriving at Walvis Bay in May, the *Ann* spent two and one-half months in this vicinity, with the *Polly*, Captain Cottle, of New Bedford, in company. Joining them were the *Brothers*, Captain David Swain, the *Alexander*, Captain Starbuck, and the *Canton*, Captain Coffin Whippey, all three out of Dunkirk. From the coast of Brazil the ship *Leo*, of Nantucket, Captain William Barnard, sailed in to complete his voyage here before returning home. The *Ann* returned to Dunkirk in September 1792.

Sailing again from France on January 26, 1793, the *Ann* (commanded this time by Captain Jonathan Barney), made another voyage to Walvis

2. Logbook of the Ship *Ann*, Peter Foulger Museum Library, Nantucket Historical Association, Nantucket, Mass.

Bay, during which she spoke the *Canton*, now out of New Bedford, and the *Lydia* and *Columbia*, of Nantucket. Learning of the outbreak of the Anglo-French War, the *Ann* did not return to Dunkirk but crossed the Atlantic for Nantucket.

As an indication of the growth of the Nantucket whaling fleet over this period there was in 1798 a total of thirty vessels representing that port, an increase of some twenty vessels in over a decade. It was a true gauge of promise for American whaling's future.

The resurgence of Nantucket's single industry was doubly important. She had not only returned to the role of leader in American whaling but was now to challenge the world leader of the Southern Whale Fishery—London. New Bedford's steady growth was also manifested. In addition, the town on the Acushnet had developed another string to her bow—trade with Europe—a business in which the Quaker Islanders, being wedded to sperm whaling, did not become too deeply involved. This "neutral trading" brought in considerable wealth to New Bedford, especially for the Howlands, Hazards, Russells and Rotches. Several of these firms had representatives in New York City, where New Bedford merchants established headquarters for the European trade.

Associated with the revival of American whaling were such factors as the build-up of an American market for whale oil; the return of Nantucket whalemen from France and England, with both money and experience; and the rebirth of the indomitable spirit of the Islanders as they applied their unique system to the whaling industry. It was this unusual combination of enterprise and cooperation which had given the Nantucketers their leadership.

To illustrate, through their own words, this characteristic, the fitting out of the Micajah and Zenas Coffin ships *Lydia* and *John Jay* serves as an example of the energy and planning of these industrious Islanders. The diary of one of the owners reports:

1801—7 mo., 21st: Ship John Jay *has arrived at Nantucket Bar.*
23d: At about 9 o'clock in the evening we got the John Jay *into the wharf.*
30th: In the forenoon put Gardner and Mitchell's money in the bank on deposit for them and took their receipt for $2400 [*investment in forthcoming voyage of ship* Lydia.] *Also deposited* $64 *to Micajah Coffin & Sons. After dinner, Gilbert, Micajah, Paul, Walter and*

Thomas Coffin went up with Gilbert's mare and finished hilling the piece of corn. Isaiah and Zenas [Coffin, sons of Micajah] at the wharf to see to finishing the graving of ship Lydia.

8th Mo. 1st: This forenoon weather cleared off. The carpenter acaulking [Lydia's] *decks and others a-fitting the standard knees between decks on board the* John Jay. *Our company bought at Vendue of Wm. Coleman six barrels of Rye flour, 3 bbls. of which we carried to Ben'j Walcott, baker . . . took out ship* Lydia's *pumps and sent to Paul Coffin, block maker. After dinner, myself, Gilbert and Paul Coffin went up into the field plowed and hilled one piece of the other piece of corn . . . also Zenas Coffin with his company went at the same time and plowed & hoed among his corn.*

6th: Carpenters and caulkers at work . . . putting on the trussel trees on the foremast and set the foremast a little before sun setting and also caulking the Quick work on starboard side . . . bought 3/4 cord of pine wood . . . carted it alongside ship Lydia; *from the pump . . . bought 28 lbs. of oakum of D. Pinkham for the* John Jay.[3]

Throughout the days of August 1801 the shipowners worked at fitting out their ships. While the carpenters and riggers were busy on the outer perimeters of the ship, the owners put down the water casks on the lower tier, took aboard casks filled with potatoes, beans, flour, cornmeal, and "Boston bread" (hard tack, baked by Benjamin Wolcott), and fetched barrels of beef and pork, salted down. The tryworks were set up, sheathing and cedar boards and shooks of casks hoisted over the gangway and carefully lowered through the main hatch, and casks of sail canvas and whale line, double and single tackles, blocks and straps packed below. Among the final stores placed aboard were codfish tongues and cranberries, Captain Micajah even went into the swamps and "brought down some brush to make brooms for the ships."

On the morning of August 22, 1801, the *Lydia*'s mooring lines were taken aboard and she left the South Wharf and rounded Brant Point, to drop anchor "in the channel off Cliff Shoal and in the evening they got over the Barr." After putting her final supplies on board from a lighter, Captain Reuben Starbuck came aboard with his stores, "they got the

3. Transcript of the Zenas Coffin Diary, Peter Foulger Library.

Lydia's anchors on her bows and between one and two o'clock the ship squared away for the Point being bound . . . on a whaling voyage and at sundown she was out of sight behind the high land at the head of the harbor."

On August 24, 1801, the diarist recorded the arrival at Nantucket of the ships *Boston Packet* and *Hudson*, "full, from the Gallapogas Islands." These were two veteran whaleships, once sailing out of Dunkirk.

The *John Jay* sailed September 16, 1801. Her final days of preparation were filled with almost continuous activity insofar as her owners were concerned. Captain Clark lost an anchor (recovered before he sailed) while riding out a September gale in back of Nantucket bar. Both the *Lydia* and *John Jay* had successful voyages, returning full of oil, most of which was taken in the South Atlantic, their voyaging ranging from the Brazil Banks off the South American coast to "Woolwich Bay" on the Coast of Africa. The *Jay* was always a productive investment and made twelve consecutive voyages from 1801 through 1830, when she was broken up at Nantucket.

The *Lydia* had an equally busy career, embracing voyages from 1790 to 1818, her end coming when she was broken up at Nantucket. Her adventures were many. In 1797 she was caught in a hurricane off the West Indies and limped into Antigua for repairs; in 1800 one of her crew was discovered to have been a female who had successfully disguised herself during one voyage and part of another; in 1811 her master, David Swain, was killed by a whale. Following her last voyage, the Coffins built a second *Lydia* in 1822, which did not last as long as the original vessel, her end coming in 1835, when she was set afire by a mutinous crew.[4]

The *Lydia* and *John Jay* lay the foundation for one of the most successful firms in the whaling industry—Micajah Coffin and Sons. Captain Micajah was one of ten sons of two marriages of Benjamin Coffin, a Quaker schoolmaster. His half-brother, Captain Thomas Coffin, was the father of Lucretia Mott, famous as a teacher and abolitionist. Micajah founded the firm in 1793, joined by his three sons Isaiah, Gilbert and Zenas. In the ships *Fame*, *Lydia*, *John Jay*, *Brothers 2d*, and *Alligator* the firm acquired a considerable fortune by judicious processing and

4. Starbuck, *History of Nantucket*, pp. 410–411.

marketing of the oil and candles. The *Alligator* was named for the Royal Navy frigate commanded by a Coffin kinsman, Captain (later Admiral) Sir Isaac Coffin. As an ironic twist, the *Alligator* was captured by the British during the War of 1812 and confiscated as a prize!

Captain Zenas Coffin had a busy and most successful career. Even after his retirement as a whaling master to devote all his time to the firm's business ashore, he often served as a pilot for his ships, inward and outward bound. Other ships coming under the management of the firm included some famous Nantucket whaling craft such as the *Weymouth*, *Golden Farmer*, *Independence*, and *Dauphin*—this latter not to be confused with an earlier vessel of the same name. On Washington Street in Nantucket still stands the brick warehouse of the Coffin firm, and on Main Street are the two brick houses which Charles and Henry Coffin, Zenas' sons, built with money derived from whaling fortunes. Zenas' home remains today on Pine Street, much as he built it—a square, strong wooden structure—as sturdy as the man himself.

During this same time a reverse turn in the tide of Nantucket whalemen going to Britain had already set in. Among the first to return home was Captain Elisha Pinkham, who had been absent in England for fourteen years; then Captain James Gwinn from Milford Haven; Captain William Swain, Jr., from London; Captain Francis Baxter from France, and Elijah Coffin and Uriel Bunker, both of whom had been whaling masters in England.

The restrictions of American commerce imposed by the questionable Jeffersonian policies had a depressing effect on all New England ports carrying on the lucrative "neutral trade"—a trade made possible by the embargoes which both England and France had decreed on their respective ports, but which provided American vessels opportunity to risk voyages to both countries. Jefferson attempted an economic boycott by stopping American trade with England. This was not feasible. Again politics and free trade could not be mixed.

In August 1808 the inhabitants of New Bedford held a special town meeting to petition for relief "from the great and extensive Calamity" of the Federal government's restraint of trade. One paragraph in the petition is highly significant, as it reflects a forthrightness that is characteristic of the writings of William Rotch, Sr. As he was then a leading

36. Captain Eber Bunker
Father of Australian Whaling
[Miniature in the collection of the Historical Museum at Melbourne, Australia]

citizen of New Bedford, it is possible he was called upon to phrase the petition, which, in part, reads as follows:

Shipping and commodities for maritime trade left on their hands without value either for sale or employment—are rendered totally useless to their country . . . by depriving seamen of their usual means of subsistence, the continuance of this measure must either reduce them to idleness and beggary, or compel them to seek employment in a foreign country and thus deprive the nation of their services.[5]

But the embargo continued—and the incoming Madison administration held little or no promise of a more practical policy.

As a graphic illustration of the feeling in New Bedford and Nantucket at the prospect of Mr. Madison's handling of commercial affairs, the New Bedford *Mercury*'s issue of January 20, 1809, appeared with a black border around each page. The editorial stated "Our Constitution Dead!" and a notice in the obituary column so declared that it had "died in Washington on the 9th inst.," with the New England states listed as the chief mourners.

A number of sidelights on the extent of the sale of oil and general trade in Europe were reflected in the newspapers of the day, but none were more interesting than a voyage reported by the New Bedford *Mercury*. The little merchantman *Ann Alexander*, New Bedford (later to become a famous whaleship), owned by the Howlands, was captured by a Spanish privateer on January 8, 1807, and a prize crew being put aboard the ship was ordered to proceed to Vigo, Spain. The next day, however, the *Ann Alexander* was recaptured by an English man-of-war, who took out the crew and put her own prize-crew of nine men aboard and ordered the ship to Gibraltar.

When in the Straits, in sight of that port, another Spanish privateer boarded the American and carried her into Algeciras. The astute master of the *Ann Alexander*, Captain Loum Snow, now tried strategy. Before dropping anchor in that Spanish port, he persuaded the English prize crew to enroll as his own crew, and, upon entry through customs, presented a new crew list. He was allowed by the authorities to again take possession of his ship, and proceeded on his voyage to Leghorn, his original destination.

5. *The New Bedford Mercury*, New Bedford, Mass., August 17, 1808.

It was this same *Ann Alexander* which, a year and a half before, while bound to the Mediterranean late in October 1805, fell in with the British fleet only a few days after the famous battle off Cape Trafalgar. A seventeen-year-old crew member on board the American ship, John Aiken, recalled that day:

The different ships were repairing damages . . . we had on board a deckload of lumber . . . An English officer boarded us and informed our Captain that Lord Nelson had been shot through the shoulder and spine, and had died on board the Victory . . . *We could readily see the effects of the enemy's fire upon the English ships. The English officer returned to the* Victory, *Lord Collingwood's flagship, and soon after came back with a request that we let him have our lumber, a quantity of flour, and some apples. Our Captain agreed . . . and was paid for these goods in English gold . . . We squared away for the Straits of Gibraltar and on the following day came up with the new 74-gun frigate* United States . . . *We gave them the news and sent the commander two barrels of apples.*[6]

On April 21, 1811, the ship *Frederick*, of New Bedford, arrived at that port with a passenger on board—Captain David Jenkins who had escaped from France, where he had been a prisoner for over a year. He was one of the fortunate few to be exchanged, as a number of whalemen captured off the islands of Dominique and Guadaloupe by French privateers, never returned. This was in direct violation to Article XVII of the Treaty of 1800, between France and the United States, which guaranteed that the whale fishery would "be free to all nations in all parts of the world."

The French consul admitted that the starving conditions in the West India port of Jacmel led to the capture of certain whalers in 1808. In reporting these captures, the *Baltimore Whig* made the following observation:

This would perhaps be the place to say something about the particular character which distinguished the seamen of Nantucket. But those who have seen them will probably consider as a fiction that which is no more than a faithful picture of the perseverence and fortitude which, in the pursuit of their industry, supports them against the privations and dangers of every kind.

6. Letter of Captain John Aiken, owned by Rodman S. Moeller, Cincinnati, Ohio.

We will only say that, indefatigable, and inaccessible to fear, they navigate during whole years in almost open boats, exposed to all the solitude and perils of the sea. When we see such men lose at once the object of their labours and their personal freedom; when we reflect that their misfortune has also reduced to misery their wives and children, whose sole subsistence depends upon the bold industry of the chief of the family—there is not a feeling heart that will not be interested in their lot, and sympathize with their evils and sufferings.[7]

One of the most obnoxious of the governmental laws was called the "Force Act," which allowed officials of the Federal government to seize certain goods under suspicion of foreign destination. There was danger of a civil war, as New England bitterly opposed the Act. Even the repeal of the Embargo Act itself failed to lessen the resentment. Jefferson's political sagacity and sincere liberal philosophy remain unquestioned, but his inability to admit to himself the limitations of his policies should also be recognized.

His successor, James Madison, a thorough scholar but an impractical administrator, lacked Jefferson's leadership, and knowledge of international affairs. Trade conditions with the British and French had changed but little in his first years in office, and could have been improved had Britain not repudiated the plan advanced by her Minister Erskine. It was the Indian wars along the frontiers, and the election of a "War-Hawk" Congress in 1810, that forced the indecisive Madison to send his war message to Congress on June 1, 1812.

Diplomacy on the part of both countries had been badly handled. Ironically, concessions regarding the resumption of trade between them had been decided upon in Britain even while President Madison was declaring that a state of war existed. The United States was divided in its support of such a war, with the agricultural and land-conscious west and maritime-trade-minded New England bitterly opposed. Thus, the War of 1812 became a sharply debated national issue.

Just as in the Revolution, Nantucket expected that many of their young whalemen, balked by the War of 1812, would go to England and France to take out the European whalers. But most of those who did go

7. *Memoir* of D. B. Warden, American Consul, addressed to the Council of Prizes, in Paris, January 25, 1811.

left before the actual outbreak of the war, as Captain Andrew Pinkham wrote early in 1812 to Walter Folger, Nantucket's Representative to Congress in Washington:

England has held out such fair prospects for carrying on the whale-fishing, some of our young men have gone there and sent for their families; in fact our situation at the moment appears alarming, in case of war, we are accessible on all points, and have not the means of defending ourselves against an enemy which has reduced the value of our real estate to almost nothing . . . should a war ensue, a few ships of war of either of the belligerents would convert all our whaleships and take us all away together.[8]

Such a grim prophecy was to prove all too true. Added to the melancholy cast of events a number of families unwillingly moved away, not wishing to face the experience of the revolution, and many of these, including Captain Pinkham, went to the Ohio country—seafarers who became farmers.

The British custom of arming her whaleships was a protective method to off-set the numbers of French privateers, but such conditions were bound to cause trouble in other ways. In August 1810 the New Bedford *Mercury* reported an incident, which was ironic at best in that it involved the whaleship *Mary*, of London, and a New Bedford ship, the *Sally*, commanded by Captain Obed Clark—who had at one time also been the master of London whaleships. In January 1811 the marine column of the *Mercury* noted:

Arrived yesterday, ship Sally, *Captain Clark, from Coast of Peru, full of oil, for Messrs. Rotch and Hazard. On July 20, in latitude 6° 54′ and longitude 33° 30′, she was brought to by the whaleship* Mary *of London, and the colors set; but the* Mary *still continued firing with swivel and musket until the colors were hauled down. She then came alongside. Captain Clark was called a damned rascal for not setting his colors; the* Mary's *Captain sent his Lieutenant aboard and detained her an hour.*[9]

8. Letter of Captain Andrew Pinkham to Walter Folger, Jr., in author's manuscript collection of the "Pinkham Papers."
9. Marine Column, *The New Bedford Mercury*, January 20, 1811.

On her next voyage, the *Sally* was not so fortunate. On July 20, 1812, she was captured off Burmuda by H.M.S. *Recruit* and sent into that port. Value of vessel and cargo was $40,000. She was owned by the Rotch firm of New Bedford.

The all too familiar pattern was now established, and the whaleships found themselves trapped. Despite the high feelings of the times the men at sea were unprepared for an actual state of war between the two nations. While illegal impressment of American seamen by English frigates, the virtual blockade of our coasts, and seizure of our vessels in British waters were contributing factors they, in themselves, could not have brought on the conflict.

It took a long time for the news of the war to reach the Pacific Ocean and the west coast of South America. With the whaleships scattered from Cape Horn to the Galapagos Islands, it was to take even longer. In many instances the news was but an immediate prelude to disaster.

One account of a whaleman who participated in the events of that period presents a good picture of the experiences of the whaleships. Off the coast of Chile, on April 22, 1813, the ship *President*, Captain Solomon Folger, was hailed by a British whaler, and received the first word of the war. On board the *President*, a young harpooner named Nathaniel Fitzgerald described the event:

We discovered a ship under our lee, and, the wind being light, she lowered one of her boats and boarded us. It proved to be Capt. Joy of ship Atlas *of Nantucket. He informed us that he had spoken with ship* Atlantic *Capt. Obed Wyer, ninety days from London, who informed him that there was war between England and America.*[10]

There was an ironic twist to these circumstances. Captain Wyer of the British whaler, who had warned his fellow whalemen of the war's outbreak, was one of the many Nantucketers remaining in London and still taking out whaleships from England. His command, the *Atlantic*, was herself an armed whaler. The *President* and *Atlas* immediately put away for the Chilean port of Talcahuano. Upon arrival at their harbor,

10. Reminiscences of Captain Nathaniel Fitzgerald, *The Inquirer and Mirror*, September 7 and 14, 1872.

Captain Folger found a fleet of whalers at anchor—mostly Nantucket ships. Young Fitzgerald recorded them:

There we found the Criterion, *Capt. William Clark*, Lion, *Albert Clark*, Sukey, *John Macy*, Gardner, *Capt. Ray*, Mary Ann, *Capt. Russel*, Perseveranda, *Capt. Paddock*, Monticello, *Capt. Coffin*, Chili, *Capt. R. Gardner*, John and James, *Capt. R. Clasby*, Lima, *Capt. S. Swain*—President *and* Atlas. *The* Perseveranda *and* Sukey *brought in part of the crew of ship* Renown, *Capt. Barnard and they were distributed among the ships. The* Renown *had been taken by an English armed whaler and I think the crew had been landed on Massafuera.*

Ashore the entire coast was in Revolution. Taking advantage of the European war, which more than occupied the Spanish army and navy, the Revolutionists, or Patriot forces (so-called), of Chile and Peru had become well organized and were forcing the Royalist armies to give ground. The Patriots announced they were allies of England and sent their raiders toward Concepcion and Talcahuano, and lay seige to these ports. Like sitting ducks, the American whaleships were helpless to protect themselves once the combined Patriot privateers and armed British whaleships closed in upon them.

One of the British armed whalers was the *Comet* from Hull, under Captain Abel Scurr. She carried ten carronades and two long "fours" or six-pounders, and had left England on September 13, 1812, with another armed whaler, the *Indispensable.* Joining a convoy guarded by H.M.S. *Nimrod* and H.M.S. *Impregnable*, the two whalers were escorted as far as Madeira, where they set their own course for Cape Horn and the Pacific. In the South Atlantic they spoke the *Leo* and *Rebecca*, both Nantucket whalers, and soon after rounded Cape Horn. On February 2, 1813, the *Comet* reached the island of Mocha, off the coast of Chile, a favored rendezvous for whaleships, where she met the Nantucket whalers *George*, *John & James*, and *Lima.*[11]

Reaching Talcahuano on March 6, the *Comet* found herself anchored with a fleet of seven American whalers, one British whaler, the *Atlantic*, and three Spanish coasters. During the next two months five more Nantucket whaleships arrived—all as listed by young Fitzgerald, of the

11. Logbook of the Whaleship *Comet*, of Hull, England, Manuscript Collections, National Maritime Museum, Greenwich, England.

President. As was the custom in those days the British had two captains on these armed whalers, and aboard the *Comet* Captain Scurr was the "Fighting Captain" and Captain Dunn, the "Whaling Captain." The mate was Simeon Long, of Nantucket.

It was a time of frustration and apprehension. Completely caught between these two contending forces, and well aware of the armed British whalers off shore, the American whalers—eleven of the twelve being Nantucket vessels—decided to remain together. They could only hope for some propitious moment to escape and head for Cape Horn—and home.

With the Patriot forces announcing themselves as allies of England, and with two British armed whalers in the very midst of the fleet, the American whalers might have taken the initiative at this time. But as the master of the *Atlantic*, Captain Obed Wyer, and the mate of the *Comet*, Simeon Long, were Nantucketers, there was not the advantage expected by the Chilean officers ashore. This was more apparent when Captain Scurr went ashore, due to "ill health." Ten days later five ships from Lima (Callao) came in with Royalist troops who were landed and proceeded to recapture both Talcahuano and Concepcion. Now the situation was more dangerous than before. It was known that the *Comet* wished to put to sea to cruise for whales, leaving Captain Scurr ashore. This, temporarily, would remove one threat—but the Royalist commander could not guarantee permission for the fleet to sail.[12]

The whaleships remained unmolested for several days. Then, one morning, a boat from one of the Spanish vessels, on a trip ashore, stopped aboard the *President* to trade for tobacco. Observers ashore became suspicious and when the boat reached the land the crew were seized and the Royalist officers from the fort came aboard the *President* and searched her. In Captain Folger's quarters, a box with $800 in gold was found, and the Peruvian officer immediately lay claim to the sum. Captain Folger protested, and was placed under arrest and taken ashore. Later in the day, the entire crew were also made prisoners, being brought to the fort, leaving only the mate and two men aboard. Realizing there might be an attempt to recover the ship through action by the other whalemen, the Royalists attempted to unhang the rudder but were

12. Ibid.

unable to do so, not knowing how to complete the task. Apprehensive at this incident, the whaling masters held a council of war.[13]

At this moment, an impressive figure made his appearance in the person of Joel R. Poinsett, the American consul from Buenos Aires, who had crossed overland from Argentine. Joining the Patriot forces, under General Serenna, he waited until the contending South American forces had met in a stalemate at the Battle of Carlos. Organizing a party of four hundred volunteers, Poinsett retook Talcahuano, and released the whaleships. Some of the whaling masters had suffered loss of their papers by theft and Poinsett issued consular certificates to replace them. It was learned that Captain Scurr of the *Comet* had actually agreed to become a privateer to help the Peruvian revolutionary cause. Scurr was in danger of losing his vessel but cooler heads among the American whaleship captains saved the *Comet*, although they knew her commander's actions were unacceptable. The British whaler was allowed to sail on March 13, 1813, but after cruising two months returned to Talcahuano on May 27. With her came the *Thomas*, an armed whaler chartered by the Royalist forces, which anchored close to the harbor mouth while the *Comet* came into the anchorage basin.

Consul Poinsett and General Serena boarded the *Comet*. The ship's log recorded:

A Spanish boat came alongside and a complement of 18 soldiers took the crew away as prisoners, except Mr. Long (the Nantucket mate) and 8 men. They likewise stole from the ship 12 cutlasses, 20 muskets, bayonets, 6 pistols, a carbine and 19 full cartouch boxes, and dismantled our steering wheel. A guard of soldiers was left behind.[14]

The next day, the logbook recorded: "the Rebels took out all the cannon . . . and all the ammunition."

Boatsteerer Fitzgerald, of the *President*, had a different account of the incident:

An English whale-ship came in and anchored, and it was noised about that she intended to "cut us out;" as she was well armed. Her name was the Comet*; Captain Skerr was the fighting captain, and Captain Dunn*

13. Reminiscences, Captain Fitzgerald.
14. Logbook of whaleship *Comet*, op. cit.

was the whaling captain; Simeon Long, of Nantucket was mate. General Poinsett called on Captain Dunn for his papers, and he also called Captain Skerr for his papers, and was refused. General Poinsett got volunteers, manned and sent a gunboat to take out all the arms and ammunition. At first the ship's crew thought of resisting, and there was hard talk about fighting; at last the officer told Mr. Long, if he did not hold his peace, he would have him put in irons. So we took out arms and ammunition until we landed it all on shore; then Poinsett told the Captain he was on the same footing as the American whalers.[15] [*The* Comet *was detained at Talcahuano for a year, but finally resumed her whaling career, and returned home in December, 1815.*]

Two days later, June 8, 1813, Poinsett decided the Royalist gunboat *Thomas* should be captured and thus eliminate another threat to the fleet. The energetic American Consul called for volunteers to man one of the small gunboats. A number of the whalemen eagerly joined and, under cover of the darkness, they boarded the *Thomas*—ex-British whaler—and took her without a shot being fired. The ship's documents, however, were destroyed before they could be confiscated. In reporting the incident, young Fitzgerald stated, as a sequel: "The *Thomas* had on board Royalist officers who planned to take command of the place and I understood they were taken to Conception and shot!"

On September 18, 1813, the fleet of Nantucket whaleships left Talcahuano bound round Cape Horn for home. Having left the frying pan of the South American coast, they were heading for the fire off the British navy blockade of the New England coast, and only a few managed to reach the safety of an American port. The *Criterion*, Capt. Clark, slipped through the British cordon and got safely into Newport in December 1813. The *President* was also one of the fortunate ones, coming in between Block Island and Vineyard Sound and arriving at Newport late in December 1813. A few days before them the *Criterion* had reached the haven of that harbor with a full cargo of oil, and Captain William Clark brought the information that Captain Porter and the frigate *Essex* were in the Pacific to protect American whalemen, but this news had already reached the United States, being reported by *Niles Register* in August.

15. Reminiscences of Fitzgerald, op. cit.

Of the other craft in the fleet of twelve, gathered in Talcahuano Harbor, April–May 1813, the *Atlas* got safely into Old Town, Martha's Vineyard on December 8, 1813. Captain Obed Joy was fortunate to get home, as four days later, as the diary of Obed Macy recorded: "A British frigate is standing off and on the south side, fresh wind at N.E., supposed she is cruising for remainder of fleet."

On December 14, 1813, a small schooner approached the islands south shore off Siasconset and landed forty prisoners—all whalemen. Among them were Captain Isaiah Ray and his crew of the *Gardner*, captured by the British frigate *Loire*, within a few days of safety. He reported speaking the *Monticello*, Captain Coffin, which had been taken a few days before Captain Ray's ship—the British brig *Albion* having brought her to off the Delaware Capes, only a dozen miles from haven.

The *Chili*, Captain Robert Gardner, Jr., came within sight of the safe anchorage of Nantucket. After leaving the coast of South America on September 15 with 550 bbls. of sperm, and slipping through the blockade, he ran into Vineyard Sound on December 6, 1813, and anchored at Tarpaulin Cove for the night.

The ship had only been here for two hours when the British armed brig *Nimrod* came in and captured the *Chili*. Captain Gardner and his crew were put aboard a schooner and arrived at Nantucket with the sad news the following day. He brought news of the sailing from Talcahuano with the *Lima*, *George*, *William Penn*, and *Charles*, of Nantucket, and *Barclay*, of New Bedford. Later, five boats with sixty men rowed out to the *Chili*, lying off Gay Head, and fired muskets, but did not board the ship. Had they done so it would have been easy to re-take her as there were only nine British on board.[16]

Obed Macy recorded in his Diary: "The public mind is much agitated, such great dependence is made upon the ships now daily looked for that very little is done but talking in company about them, flocking to every boat that arrives to hear something."

This diarist gives further detail:

Finding all the whaleships in the Pacific except the Charles *has put away for home and the danger of falling in with the enemy on this coast, the owners met and agreed to pay $500 to any Vineyard pilot who would*

16. Obed Macy, Diary. Peter Foulger Library, Nantucket Historical Association.

37. The "London Club" Centre Street, Nantucket
Captain Coffin Whippey's house on left; Captain Archaelus Hammond's on the right. Store removed from land in 1897
[Courtesy the Nantucket Historical Association]

get one of the ships into port, so that the property become secured to who it belongs, and so for as many as they should get in. Charles Cartwright went to the Vineyard to publish the same.[17]

Of all the cruel fates which overtook the returning whalemen perhaps the most devastating was that which saw the *Perseveranda*, under Captain Thomas Paddock, come all the way from around Cape Horn, through the British blockade, and into Nantucket Sound. Just as the ship "made the turn" at Tuckernuck Shoal to head for the harbor of Nantucket, a British seventy-four came into view from the eastward, fired a few long shots and brought the whaleship to—capturing her within ten miles of harbor.

As for the *Charles*, which had continued whaling to fill the ship before returning home, her commander, Captain Grafton Gardner was a lucky man. He not only reached home safely on February 27, 1814, but brought the "most valuable cargo of oil ever brought here."

17. Obed Macy, *A History of Nantucket*, p. 164.

Thus, of the dozen American whalers who fled from the Chilean port in September 1813, only four managed to reach safety on New England ports. Nantucket lost by capture nineteen out of her fleet of thirty-seven ships, and New Bedford lost five out of nine.

Of the two American ports carrying on deep-sea whaling, Nantucket and Bedford, Nantucket was the most vulnerable to the British blockade, and her inevitable suffering was yet to come. The islanders knew, from the experience of the Revolution, what was in store. Obed Macy, the Nantucket Quaker historian, summed it up:

There seemed no alternative .... but to submit to the calamaties which [the War] would probably bring among them. The reflection on the situation ... was appalling. Nearly the whole amount of the trading capital was in the Pacific Ocean, the greatest part of which was not likely to return in less than one year and perhaps not in two ... led many to devise measures to ward off impending ruin.

But what of the Nantucket and New Bedford whalers still in the remote Pacific? Unknown to them, or to the home ports, a grim story was already beginning in that part of the whaling world, marking another unusual chapter in the history of the Southern Whale Fishery.

20. The Whaleships Go to War

THERE was no more dramatic an episode in the drab War of 1812 than that developing from the voyage of the United States frigate *Essex*, Captain David Porter, and her cruisings in the Pacific Ocean. On the great stage of the Pacific seascape the whaleships of both America and Britain played supporting roles. The play had three acts, a variety of scenes, and a gripping climax. The motivation was the war but the action in all but the final scene was directly connected with the whaling industry.

It was in October 1812 that Captain Porter received orders from Commodore Bainbridge to prepare the *Essex* for a long cruise. When the frigate put out to sea she had a complement of 319 men. Proceeding first to the Cape de Verdes, near the African coast, the *Essex* then re-crossed the Atlantic to the island of Fernando de Noronha, off Brazil, where a rendezvous with Bainbridge had been planned. But the Commodore's frigate did not appear—either at this island or at Cape Frio—and Captain Porter, aware of the British warships in this area, decided to head offshore again and to the south. His next landfall was the whaler's provisioning spot at the island of St. Catherine between Rio and the mouth of the River Plate.

It was at this time that Porter determined to put into effect a novel plan for a naval cruise. The plan was as daring as it was unprecedented, insofar as the U.S. Navy was concerned, but it was one to which the imaginative officer had given considerable thought over a period of months. He had decided to sail around Cape Horn.

When war's outbreak was imminent, Porter had written to the Secretary of the Navy and proposed just such a plan of a voyage into the Pacific to harass the enemy's commerce in that ocean. This, of course, meant the British whaling fleet off the west coast of South America. Both the Secretary and Commodore Bainbridge approved. Had Porter remained in the Atlantic, seeking Bainbridge and the *Constitution*, it is possible that he would have been chased into port, and probably been bottled up as were the other American frigates blockaded by the numerically superior British squadrons. The South Seas was a logical and, as it proved, a perfect theatre of operations for the *Essex*.

The purpose of this extraordinary adventure is best expressed in Porter's own words:

Before the declaration of war, I wrote a letter to the Secretary [of the Navy] containing a plan for annoying the enemy's commerce in the Pacific Ocean, which was approved by him, and prior to my sailing, Com. Bainbridge requested my opinion as to the best mode of annoying the enemy. I laid before him the same plan and received his answer approving the same.[1]

During his passage south from St. Catherine's, Porter announced to his men his plan, promising his crew that "the girls of the Sandwich Islands will reward you for your sufferings during the passage round Cape Horn." Proceeding through the Straits of Le Maire, the *Essex* came into the region of "Cape Stiff," where piercing cold, violent squalls of rain and hail, and a leaky ship added to the problems of navigation. It was early February 1813, late summer in this part of the world.

No better description of this experience has been given than that in Porter's *Journal*. As he wrote, he must have gained a new respect for those pioneer whalemen who had first dared the passage through the Straits of Le Maire and 'round the Horn a quarter of a century before.

I am induced to believe no part of the world presents a more horrible aspect than Staten Land [the island just south and east of the tip end of South America]. The breakers appeared to lie about half a mile from the shore. While we were standing off, the whole sea, from the violence of the current, appeared in a foam of breakers, and nothing but the apprehension of immediate distruction could have induced me to have ventured through it, but thanks to the excellent quality of the ship we received no material injury, although we are putting our forecastle under with a heavy press of sail, and the violence of the sea was such that

1. Captain David Porter, *Journal of a Cruise Made In The Pacific Ocean In The U.S. Frigate "Essex"* (Philadephia, 1815), Vol. I, p. 61.

it was impossible for any man to stand without grasping something to support himself.[2]

It was not until February 24, 1813, that Porter deemed himself in far enough west longitude. He had sailed as far south as latitude 60°, then began his cruise north. A few days later, March 3, the *Essex* was struck by a fierce gale that caused the shingle ballast to shift and clog the pumps. A great sea burst into the ports of the gun-deck, stove both quarterboats, washed his spare spars from their chains, broke the hammock stanchions, and deluged the ship with water. Although the crew was thrown into consternation, Porter reported that he promoted the men at the wheel for their coolness.

When the *Essex* gradually got into warmer latitudes the men were cheered. On March 6, 1813, off the coast of Chile, the green-clothed hills of Mocha Island came into view. Its fresh and beautiful appearance and the smell of the land were like food and drink to the crew and all hands welcomed shore-leave. Now these Navy men could fully appreciate the perils and hardships of the whalemen's existence, the long months of cruising, far away from land and civilized ports!

Porter then proceeded to St. Mary's island, a favored rendezvous of American whalers and sealers. He recorded sighting a great number of whales, "which gave us strong hopes of soon meeting those vessels engaged in catching them."[3]

In running up the coast of Chile, the commander of the *Essex* had much in common with the whalemen in respect to locating his position. He wrote just as a whaling master might have written:

I . . . had only one chart of the whole coast of America, and that on so small a scale as to be relied on but for the direction of the coast, projections or headlands, etc., and on that the island of St. Mary's merely marked as a point.[4]

On March 13, the *Essex* came closer to the rock-bound coast of Chile, "at the back of which the eternally snow-capped mountains of the Andes

2. Ibid., p. 73.
3. Ibid., p. 98.
4. Ibid., p. 99.

reared their lofty heads, and altogether presented to us a scene of gloomy solitude." They were now in the vicinity of the port of Valparaiso, and while Porter was tempted to sail into the wide bay he did not do so, wishing not to give away the fact that an American frigate was now in the South Seas. Instead he hoisted English colors and kept to sea, past the harbor, at the same time getting a good look at the shipping there through his glass. One of the craft so observed he believed to be a British whaler, and he determined to remain in the vicinity in hopes of intercepting her. Changing course, he once again approached the harbor entrance, lowered a boat and sent Lieutenant John Downes ashore to interview the Spanish Governor. It was the first time one of the U.S. Marines participated in a diplomatic mission in the Pacific.

Lieut. Downes was not to see the Royal Governor, however. Instead he met Don Francisco Lastre, a young Chilean leader of the Patriot forces, who informed him that Chile had decided to throw off the yoke of Spain and become an independent nation. Quickly, Downes returned to the *Essex* with the astounding news. Lastre offered the facilities of his port, announcing that Chile had shaken off allegiance to Spain and was looking for help from the United States. Porter was highly relieved by this news. The *Essex* then came into Valparaiso and dropped her anchor.

A round of "socials" greeted the men of the *Essex*, and after their long voyage, this was doubly appreciated. Porter was informed that the Peruvians, revolting against the Spanish authorities and allying themselves with England, were sending privateers out of Lima. He did not comfortably receive the news that some of these privateers had captured a number of American whalers. Lastre, the leader of the Revolutionists in Valparaiso, assured Porter of his friendship, and of the availability of supplies. He also permitted the dispatch of a courier to Joel Poinsett, the American Consul, who was at the capital Santiago up in the hills. As already noted, Poinsett had come overland to Chile from Buenos Aires and was now preparing the attack on Talcahuano.

After the customary parties and receptions had taken place ashore and aboard ship, Porter prepared to leave port. But that same day, as if ordained by fortune, a whaleship came into Valparaiso and anchored. She proved to be the American ship *George*, out of Nantucket, commanded by the veteran Captain Benjamin Worth, one of the finest whalemen who ever sailed from New England. The information which

Worth gave Porter was such as to determine the future course of the *Essex* and by the same token, change the course of whaling history.

Captain Worth informed Captain Porter that the *George* had been liberated a few days before at Callao, after paying the Peruvians what was virtually a ransom. Of more vital concern was the whaling master's account of having spoken two armed British whaleships only a few days before, and these had given him his first knowledge of the declaration of war by Congress. As a number of these vessels had Nantucket Captains, it was natural they should warn their kinfolk of the danger, just as Captain Obed Wyer, of the London whaleship *Atlantic*, had warned the *Atlas* of Nantucket.

Porter's question was obvious: "Why didn't the British ships, armed for such action, capture the Americans?" Captain Worth replied that the British had no orders for proceeding in this manner, but were in daily expectation of such orders.

It was at this time that Captain Worth suggested that the *Essex* cruise to the Galapagos Islands, which was a favorite rendezvous for the whalers at the coming season. This advice changed the whole picture and made possible the initial success of the cruise of the frigate. Wrote Porter:

> *He* [*Captain Worth*] *also gave me such information as would render my falling in with them probable while running along the coast. He represented our whale-fishers, which were very numerous, as in a helpless and unprotected state, entirely exposed to attack and capture by the armed English whaleships in those seas carrying from 14 to 20 guns and well manned. He stated that as our whaleships sometimes kept the sea for 6 months at a time, most of them were ignorant of the war and would fall on easy prey to the British.*[5]

Such information and advice from an experienced whaleman convinced Porter that he could strike the enemy's commerce a crippling blow by destroying her fleet of whaleships in this ocean. If this was a departure from any previous plan, it proved to be a wise and justified one throughout.

Two days out of Valparaiso, on March 25, 1813, the *Essex* sighted her second whaleship. This was the *Charles*, another Nantucket vessel, under Captain Grafton Gardner. He had just been released from Peruvian

5. Ibid., p. 111.

capture by paying certain costs at Callao, the port for Lima, as had the *George*. Captain Gardner gave the news that two New Bedford whaleships, the *Barclay* and *Walker*, had been captured off Coquimbo by a Peruvian armed vessel.

Porter now decided on a course of strategic action. Heading up the coast for the port of Coquimbo, he sent the *Charles* on ahead to act as a decoy. The plan was an immediate success. Off the harbor of that port a Peruvian privateer, the *Nereyda*, came into view, and she swooped down on the *Charles* like a hawk on a luckless fowl. The *Essex* hoisted British colors, with the whaleship following suit.

After approaching to within hailing distance, the *Nereyda* hove to, sending her lieutenant to visit the *Essex*, which he naturally believed to be a Royal Navy frigate. The Peruvian lieutenant told Captain Porter of the capture of two whaleships, the *Walker* and *Barclay*, but went on to report that the *Walker* had been retaken by the British armed whaler *Nimrod*, and that the *Nereyda* was now intending to recapture what was considered her rightful prize. On board the privateer were the *Walker*'s master and the crew of the *Barclay*.

At the curt request of Captain Porter, the Peruvian lieutenant returned to the privateer and brought back to the *Essex* Captain Stephen West of the *Walker* and Isaac Bly, one of the *Barclay*'s mates. When informed of the true identity of the frigate, Captain West, a Nantucket veteran of sealing and whaling, offered his services on board the *Essex*. A short time later the twenty-three captured whalemen on the privateer were equally delighted to learn the news. The *Nereyda* was ordered to haul down her colors.

All the guns of the *Nereyda* were thrown overboard. The Americans noted with surprise that the round shot, star shot, and bar shot were all made of copper, a metal cheaper than iron on this coast. Porter then announced he would return the privateer to the Peruvians, but put a prize crew aboard to keep the craft under his guns for a few more days.

Before the *Nereyda* was finally released, Porter lowered a boat, and Lieutenant John Downes, together with Captain West and a boat's crew, rowed into the harbor of Coquimbo to reconnoitre. By eleven o'clock that night they returned with the news that the British armed whaler *Nimrod* had sailed. Captain West was put aboard the *Charles* and sailed for Callao. Porter advised West to go overland to San Jago to claim damages.

The *Nereyda* was then released, with a letter addressed to the commanding officer recommending, among other things, that the privateer's officers and crew be punished for capturing and plundering the vessels of a neutral nation, the United States of America.

When Porter headed north on March 28, he had a complete list (prepared by Captains Worth and West) of all the whaleships—both American and British—which were in this part of the Pacific. This list gave Porter a detailed description of the chief features of the British whalers—the *Nimrod* had no figurehead; the *Perseverance* had an excellent one, as did the *Seringapatam* and the *Charlton*; the *Sirius* was a low ship, with a wide billet head; the *Catharine* would show the figurehead of a woman, and the *Thames*, *Rose*, and *Greenwich* each possessed certain characteristics, all carefully noted. Other British whalers were expected to be in the area, also, so that there would be probably twenty vessels, all fine ships each of 400 tons burthen, carrying cargoes worth $200,000—the whole fleet valued at about $4,000,000.[6]

Also listed for Porter's information were the twenty-one Nantucket and two New Bedford whaleships known to be in the Pacific. As Porter was well aware, two of the American whalers had already been taken by the *Nimrod*. How many more of the American whalers had been captured by the British armed whaleships was not then known, but Porter expected (correctly) this was now being accomplished. As a matter of fact, the first Nantucket whaleship taken was the *Renown*, Captain Zaccheus Barnard, whose crew was put ashore on Massafuero Island. They were later rescued by their fellow Nantucket ships *Perseverenda* and *Sukey*, and brought into Talcahuano when those vessels joined the beleagued whaling fleet in that port.

Porter's plan for "annoying the enemy" was now crystallized:

> *Besides capture or destruction of these [British] vessels, I had another object in view, of no less importance, which was the protection of the American whale-ships . . . and that effecting this object alone would be a sufficient compensation for the hardships and dangers we have experienced.*[7]

Porter had been well advised by Captain Worth and Captain West (the latter had once served as a lieutenant on an armed British ship), to steer

6. Ibid., p. 123.
7. Ibid., p. 129.

38. The Thomas Macy House, Nantucket
Herman Melville was a guest of Macy in July 1852 and here received a copy of Obed Macy's *History*
[Photograph by Cortlandt V. B. Hubbard]

north for the Galapagos Islands. On March 29, 1813, he was off Callao, where he sighted three sail standing out of the harbor and maneuvred to cut them off. Fleeing before him the three ships headed back to the safety of the harbor. They were aided by the wind failing off San Lorenzo island. Two of the three got into Callao—but the third was captured by boats lowered by the *Essex* in pursuit, and these boats towed the captive craft out to sea and under the guns of the frigate, which was flying British colors.

The ship proved to be the captured New Bedford whaler *Barclay*, with her deposed master, Captain Gideon Randall, still on board. This was the old Rotch ship which had carried William, Sr., home from England in 1794. Then twenty years old, this ship was still only one-fourth through her long career—and her good fortune in being recaptured was another reason for her being known as a lucky ship.

When the boat from the frigate came alongside the whaleship the officer in charge was none other than one of the *Barclay*'s former mates, Isaac Bly, and when Captain Randall saw him he could only think that the man had been impressed aboard this (as he believed) British frigate. Knowing how the whaling master was viewing the situation, the former mate could not resist a bit of fun.

"Come, Captain Randall," he called, "don't look so down in the mouth. What have you got to drink?"

"You know as well as I do," was the reply of the astonished whaling master.[8]

Captain Randall was transferred to the *Essex*. His mixed emotions increased as he mounted the quarterdeck of the frigate, being not a little surprised by the degree of formality that seemed more in accordance with a reception prepared for one of equal rank than for a prisoner.

"Welcome, Captain Randall," said Captain Porter, gravely, "Welcome on board the United States frigate *Essex*."

"Would to God I was aboard her!" was the prompt and despairing reply of the veteran whaleman.

When Porter, in a few words, explained the import of the situation, Randall was overwhelmed. Under the hatches of his ship was a full cargo of sperm oil—twice supposed lost and twice restored to him. No wonder he was unnerved.

It was decided that the *Barclay* would sail in company with the *Essex*. Midshipman Cowan and eight men from the frigate went aboard the whaleship with Captain Randall. Porter's intention of proceeding to the Galapagos Islands was approved by the old whaleman.

The two ships separated for a time, to cruise and then rejoin each other at Paita. On the way up the coast they saw ten large rafts, drifting and sailing, on their way from Guayaquil to Lima. A mast, a rude keel, a stone for an anchor, a long paddle for a rudder—all these were features of these primitive craft, which annually made the six-hundred-mile trip down the coast.

Porter, with the *Essex* and *Barclay*, arrived at the Galapagos Islands in mid-April. At Hood's (Espanola) Island, a whaleboat taken from the *Barclay* was sent into a small bay and Lieutenant Downes reconnoitred, finding no signs of whaleships but a good anchorage for future use. They next sailed for Charles (Floriana) Island to the west, a favorite place for

8. "Leaves From The Log of The *Barclay*," *The Nantucket Journal*, Nantucket, Mass., May 26, 1887.

the whaleships to stop for refreshment. Here the whalers obtained the giant terrapin for welcomed fresh meat.

At Charles Island was located the famous "Post Office Bay" of the whaling fleet. From the landing place, a shore party went up to where a half cask was nailed to a section of yardarm, set into the ground as a post. Here were found letters which gave Porter the important news that the British whaleships *Charlton*, *Nimrod*, *Hector*, *Atlantic* and *Cyrus* had put in here during June 1812, and were then on their way to Albemarle (Isabela) Island, the largest in the group. They generally cruised here a year at a time, according to advice from Captains Worth, West, and Randall.

Also in the "Post Office" were found letters from *Perseverenda*, Captain Thomas Paddock, and the *Sukey*, Captain John Macy, both out of Nantucket. The letter from the *Sukey*'s commander was considered a "rare specimen" by Porter, who preserved it, reading as follows:

Ship Sukey, *John Macy, 7 1/2 months out 150 barrels. 75 days from Lima. No oil since leaving that port, Spanyards Very Savage. Lost in the Brazil Banks John Sealin Apprentice to Capt. Benjamin Worth, fell from fore-top sail yard in a Gale of Wind. Left* Diana *Captain (David) Paddock 14 days Since 250 barrels. I leave this port this day with 250 Turpen, 8 Boat Loads Wood. Yesterday went up to Pat's Landing. East Side to the Starboard hand of the Landing 1 1/4 Miles Saw 100 Turpen. 20 Rods Apart Road Very Bad. Yours Forever—John Macy.*[9]

Nearby the "Post Office" was a rude shelter containing a cask of water, clothes, tinder box, and a barrel of bread. This was for any unfortunates who might be shipwrecked on the island or nearby, and who might find their way to this place.

Porter than sailed for Narborough Island, where he found an excellent anchorage at Elizabeth Bay. Here the high cliffs formed a protected basin where Narborough nestled close to the long flank of the larger island of Albemarle.

On April 29, 1813, the *Essex* took her first British whaleship, the *Montezuma*, under Captain David Baxter, a Nantucket man. Seeing

9. Porter, *Journal*, p. 139.

British colors at the *Essex*'s masthead, Captain Baxter went on board and repaired to the cabin with Captain Porter. When he returned to the deck, the whaling master was bewildered to find his boat's crew prisoners, a prize crew on board his own ship, and snapping in the breeze overhead the flag of the United States of America.

This, in the light of whaling history, was an ironic capture. The *Montezuma*, one of Benjamin Rotch's ships, had sailed out of Milford Haven, Wales. Her crew were, for the most part, Nantucketers, fellow townsmen of Captain Baxter. The *Barclay*, of New Bedford, close at hand, was owned by Benjamin's father, William Rotch, Sr.[10]

Later that same day the British whalers *Georgiana* and *Policy* were sighted northeast of Albemarle Island: both were armed, the former with six 18-pounders and the latter with ten 6-pounders. The wind had dropped to a dead calm, but Porter ordered his boat away under First Lieutenant John Downes. The whaleships were boarded with alacrity. The *Georgiana* was a former East Indiaman. Seeing her possibilities as a consort, Porter placed Downes in command with a prize crew of forty-one men.

But it was with the capture of another armed British whaler—the *Atlantic*—that the tragic overtones of the situation are best revealed. After a month of lying in wait, Captain Porter was rewarded by the appearance of this whaleship, coming in toward Albermarle Island. Again lured by the *Essex* flying British colors, the unsuspecting whaleship hove to close by, and her captain came aboard. The Nantucket master of the British *Atlantic* was Captain Obed Wyer, who only a few months before had hailed the Nantucket whaleship *Atlas* and warned Captain Obed Joy, his fellow townsman, of the outbreak of war.

Captain Porter once more went through the pretence of welcoming an English captain aboard an English naval vessel. It was then that he learned Captain Wyer was a Nantucket man, and that his wife and children were then residing on the Island.

"I asked him," wrote Porter, "how he reconciled it to himself to sail from England under the British flag, and in an armed ship, after hostilities had broken out between the two countries. He said he found no difficulties in reconciling it to himself for, although born in America

10. "Leaves From The Log of The *Barclay*,"

he was an Englishman at heart . . . I permitted him to remain in error some time, but at length introduced him to the Captains of the *Montezuma* and *Georgiana*, who soon undeceived him with respect to our being an English frigate . . . he endeavored to apologize . . . explaining his conduct . . . by artfully putting the case to me."[11] Porter failed to mention that Captain Wyer had met several American whaleships but had made no attempt to capture them and had warned one.

By using the same tactics, Porter was able to chase and capture the whaler *Greenwich*, and again found the British vessel well armed. Had the two whaleships joined forces, Porter believed they might have beaten the *Essex* off, as many of his officers and men were absent.

Porter now had five of the British whaling fleet in the little bay off Narborough Island, as well as the recaptured *Barclay*, now restored to her former master, Captain Gideon Randall. Leaving Lieut. Downes in the *Georgiana* to cruise in the Galapagos group, Porter placed the remaining craft together and convoyed them towards the South American coast, arrived at his destination in the bay of Guayaquil on June 19, 1813.

While the British were now aware that the *Essex* was in the South Seas they had no idea where she was cruising. In April 1813 Captain Peter Heywood, of H.M.S. *Nereus*, at Buenos Aires wrote Admiral Dixon and the British Naval Chief at Rio de Janeiro that the *Essex* had rounded the Horn and was at Valparaiso. Several weeks later Heywood wrote Dixon that the American frigate might sail into the East Indies and Australian waters, disrupting trade routes. This caused Dixon to send to the Pacific the veteran Captain James Hillyar, in the frigate *Phoebe*, accompanied by the sloop-of-war *Cherub*.

The two vessels arrived at Valparaiso on June 24, 1813, five days after the *Essex* reached Guayaquil. The *Georgiana*, now an American armed prize, sailed into Tumbez and Lieut. Downes presented Porter with three more captured British whaleships—the *Catharine*, Captain Thomas Folger, another Nantucket man; the *Rose*, Captain Monroe, and the *Hector*. The latter had refused to strike and Lieutenant Downes put five broadsides into the whaleship before she surrendered, two men on board being killed and six wounded. Here was war in all its ugliness.

11. Porter, *Journal*, pp. 184–185.

Captain Randall, of the *Barclay*, learned of the capture of the *Rose* with mixed emotions—his own son was first mate on the ship. Porter selected the *Rose* as a cartel, sending her around the Horn to St. Helena with nearly one hundred British whalemen as prisoners. But George Randall was allowed to go aboard the *Barclay* to join his father.[12]

Now Porter transferred some of the armament from the *Georgiana* and put it aboard the *Atlantic*, a faster ship, and gave Lieutenant Downes the new command. The ship was rechristened the *Essex, Jr.*, and was dispatched for Valparaiso with five vessels, four of which were to be sold as prizes. Sailing in the convoy were the *Hector*, *Policy*, *Montezuma*, *Catharine*, and *Barclay*. The last named still had Captain Randall on board with his son as his mate. Captain Randall was told he could act as navigator on the voyage but that the protégé of Captain Porter, a young midshipman named David Glasgow Farragut, was in charge as prize master. After discharging Porter's supplies (taken from captured ships) at the Chilean port, the *Barclay* was to be released.[13]

There have been numerous accounts of the trouble between Captain Randall and twelve-year-old Midshipman Farragut, whom Porter had placed in command of the *Barclay*. Most of the stories point up Randall as a bitter old man, angered by the fact that he had to take orders from a mere stripling. He is said to have refused to steer his whaleship into Valparaiso in the wake of the *Essex, Jr.*, declaring he would shoot any man who dared to touch a rope without his orders, and that he was going below to get his pistols. Farragut then gave the command to keep the course into Valparaiso and sent word down that if Captain Randall came on deck he would have him "thrown overboard."

Considering the subsequent fame of Farragut and the comparative obscurity of Randall, such incidents give color only to the career of Admiral Farragut. But it must not be overlooked that Gideon Randall was a veteran whaleman who had commanded the *Barclay* since 1801. In this particular episode he naturally resented Porter's action in placing a young midshipman in command of the ship of which he had been master until her double capture. Downes reported that Captain Randall told

12. Ibid., p. 211.
13. "Leaves From The Log of The *Barclay*," *The Nantucket Journal*. Porter, *Journal*, p. 213.

him afterwards his threats were meant to "browbeat" Farragut—but that the youngster refused to "scare."[14]

The whaling master proved his own courage when he brought the *Barclay* back to New Bedford, through the British blockade.

As soon as Lieut. Downes in the *Essex, Jr.* left with his improvised fleet for Valparaiso, Captain Porter, and his own fleet, again headed for the Galapagos Islands. With the *Essex*, well off shore from Guayaquil Bay on July 9, 1813, were, as consorts, the *Georgiana* and the *Greenwich*. Four days later, off Banks Bay, in the Islands, three British whaleships were sighted. With the American frigate in chase, the quarry separated. One of these, observing the *Essex* had laid a course after one of her companions, came about and headed for the *Greenwich*. Lieutenant John Gamble, in charge of this prize, took several experienced men out of the *Georgiana*, and stood boldly for the stranger.

While Porter watched with his glass from the deck of the *Essex*, the stranger—the British armed whaler *Seringapatam* of London, carrying fourteen guns and manned by forty men—engaged the *Greenwich*. A few broadsides settled the matter; Gamble's experienced navy men could handle the cannon better and fired so rapidly that the British craft hauled down her colors.[15]

The *Essex* soon came up to her quarry and took her; she was the *Charlton*. In the meantime, the *Serringapatam* suddenly squared her yards and tried to escape. In the running fight which ensued, the Britisher found his course would bring him close to the *Essex* and she suddenly came about, ending her dash for freedom as quickly as she began it. The third ship, the *New Zealander*, was taken a short time later.

In the *Seringapatam*, Porter had a fine ship of 357 tons. Her master, Captain William Stavers, had no letter-of-marque, although he had already captured one American whaler, the *Edward*, of Nantucket, and would probably have taken others. Porter held him as a pirate, and ordered him aboard the *Georgiana*. This latter ship was manned by a mixed crew composed of men from the *Essex*, whose enlistments had

14. David Glasgow Farragut, quoted by Betty Shepard in *Bound For Battle*, (New York, 1967), p. 103.
15. Porter, *Journal*, p. 216.

run out, and American whalemen. Loaded with oil, she was dispatched for America on July 25. Several British whalemen joined her crew, preferring America to impressment, their expected ultimate fate.[16]

The *Charlton* was a dull sailor, and so the prisoners were put aboard and she was sent to Rio as a cartel. To utilize the capable *Seringapatam*, Porter had her refitted, putting the guns from the *New Zealander* aboard, thus giving her a battery of twenty-two guns. This actually gave her only the advantage of the first broadside as she had a small crew. With this new craft and the *Greenwich*, the *Essex* once more put away for Albemarle Island.

The next whaleship sighted (on July 29) managed to escape. Being in a region of baffling calms, the boats of the *Essex* and the whaler engaged in a towing match, with brisk volleys of musket-fire at long range. However, just as the Britisher appeared to heave to, a breeze sprang up on the almost motionless sea, and taking advantage of it, the British craft squared her yards and escaped into the quickly falling dusk. She was the *Sir Andrew Hammond* of London. Porter was nettled. This was the first whaler to escape.

During the next month, the *Essex* and her consorts scoured the sea in the vicinity of Albemarle and Narborough Islands and off Redondo Rock, known to be a whaler's haunt. At length their patience was amply rewarded. On September 15 the lookouts sighted, at a considerable distance, a whaler's smoking trypots.

Sending down his topyards to give the *Essex* more of a whaleman's appearance, Porter stood for the smoke. He succeeded in getting within four miles of the unsuspecting craft. She was thirty-five miles south of Albemarle Island, her decks full of blubber, trying out her oil. She made sail—but too late; the *Essex* overhauled her easily and she came about without firing a shot. This was the escaped British whaler *Sir Andrew Hammond*, an excellent craft, of twelve guns and thirty-five men. Thus, another potential menace was taken.[17]

Returning to Banks Bay, Captain Porter's fleet was joined in mid-September by Lieutenant Downes in the *Essex, Jr.* Downes brought important news from Valparaiso. Word had reached there of President

16. Ibid., p. 220.
17. Ibid., p. 249.

Madison's reelection—but more important to them was the information (conveyed overland from Buenos Aires) that the British frigates *Racoon*, *Phoebe*, and sloop-of-war *Cherub*, had sailed to the Pacific to track down the American frigate *Essex*.

Porter now made his decision to sail to the Marquesas Island and there overhaul the *Essex* in comparative safety. He reached Nukahiva with his squadron on October 23, 1813, built a fort to command the harbor, and took possession formally of the island itself. When Porter finally left the Marquesas, there was a series of incidents aboard the prize vessels, and the *Seringapatam* consequently escaped to reach Sydney in safety.

The subsequent history of Porter, his men, and his cruises is well known to history—an exciting page in this nation's naval annals. The final scene in the drama was the disastrous engagement in the neutral waters of Valparaiso harbor on March 28, 1814, when the British frigate *Phoebe* and sloop-of-war *Cherub*, at long range, hammered the *Essex* into a wreck.

Porter's contribution to the annals of the War of 1812 was one of the outstanding achievements of the entire period. As the representative of the U.S. Navy in the Pacific, he had taken a dozen British whalers, valued with their cargoes at $2,500,000, he had utterly destroyed British whaling in this part of the South Seas, and by the same token he had salvaged the American whaling fleet in the Pacific. It is worthy of note, that British naval historians designate Porter a pirate, although that nation armed the British whalers with the same design in mind which the *Essex* successfully accomplished. As Alexander Starbuck, whaling historian, commented in 1876, the British whalers "went out to shear and returned shorn."

With their whaling fleet destroyed and their tradesmen and craftsmen thrown out of employment, Nantucket again was at a turning point in its existence. The leaders in the community, as in the Revolutionary War, were determined to save the town from possible pillage. Town meetings were convened, and on September 26, 1813, one of these meetings voted:

That this town will not pay any Direct Tax or Internal duties during the present war between the United States of America and the Government of Great Britain.

That there be a committee appointed to Carry Into Effect the Neutrality which is agreed on with Admiral Henry Hotham . . .[18]

The committee sailed the sloop *Hawk* in search of the British admiral commanding the section of the coast and, after weeks of negotiations, certain conditions were agreed upon. Thus, Nantucket entered into a tacit, if unofficial, treaty of neutrality with Admiral Hotham, representing His Britannic Majesty's Navy in North American waters, being the "party of the second part." It was a situation unprecedented.

Notwithstanding, one of the most sanguinary sea-fights of the entire war took place off the island's south shore, October 10, 1814, when the American privateer *Prince de Neufchatel* defeated barges of would-be boarders from H.M.S. *Endymion.* Among those killed was Hilburn, the Nantucket Pilot of the privateer.

An islander, who heard the sound of the guns reaching the town, reported the fight in her diary:

The Barges attackted the Privateer—there was great Havvock made among them. They were repulsed and some taken—numbers killed and wounded—2 was dead in the Barge which was buried on the shoar—Waggons went down and brought them up . . . some were badly wounded —many were killed & thrown overboard . . . the Americans put 14 of their wounded on shoar.[19]

Thus, the war came close to the shores of the Quaker whaling kingdom.

But the tide of the conflict was now ebbing. When news of the peace reached Nantucket on February 16, 1815, it was greeted with uncontrolled joy—"bells began to ring . . . shouting by the boys, etc., through the streets, the town offices were lighted . . ."[20] The island was then ringed with ice, the harbor frozen; there was great suffering from cold

18. Obed Macy, *A History of Nantucket*, pp. 176–177. Original copies of documents in author's collections.
19. Diary of Kezia Coffin Fanning, Nantucket, October 11, 1814. Printed in *Historic Nantucket*, Nantucket Historical Association.
20. Obed Macy, *Diary*, February 16, 1815. Peter Foulger Library, Nantucket Historical Association.

and hunger, many houses stood forlorn and empty, their former inhabitants having moved off to the mainland; the wharves held only a few ships.

Yet, there was joy—and that quiet, Quaker faith. With the war ended, Nantucket, the island headquarters of American whaling, was ready to assert herself again.

21. Careers and Adventures of the Migrant Whalemen

THE aftermath of the War of 1812 brought recognition on the part of Britain that the young United States was potentially a nation with which to reckon. The quick resumption of commercial relations between the two countries was a favorable sign. As for the people of the American nation there was a renewed awareness of their destiny. As Albert Gallatin observed the war had "reinstated the national feeling . . . the people are more American . . . the permanency of the Union is thus better secured."[1]

This feeling of permanency was also felt by the many Nantucket whalemen who were returning home after years in British and French whaleships. Some of these have already been mentioned—men like Captain Shubael Coffin and Captain Thomas Delano, who came directly back to their Island home. Others went to New Bedford to continue their association with the Rotches and Russells. The number included mates, harpooners and foremast hands, as well as master mariners.

The return of these self-exiled seafarers gave American whaling valuable assets. First, there was the experience they had gained in their voyages to new whaling grounds. They had adapted themselves to other countries because their world was unchanged; the sea was their home, and the common risks they shared unified them and made them more aware of their backgrounds; they drew their endurance and faith from the tradition that had become a proud heritage. Secondly, they usually returned with money hard-earned and carefully saved. This was reflected in new homes in the town and in purchase of shares in new whaleships. A number came home to retire as comparatively young men.

Among this latter group in Nantucket were such veterans as Captains Archaelus Hammond, Coffin Whippey, and Paul West. All three had

1. Alfred T. Mahan, *Sea Power and Its Relation To The War of 1812*, (Boston, 1905), Vol. II, p. 436.

39. Obed Macy, Nantucket Historian
Photograph of portrait owned by the Macy Heirs by Bill Haddon
[Courtesy the Nantucket Historical Association]

served in both London and Dunkirk whaleships. Captain Hammond, it will be remembered, was the first man to harpoon a sperm whale in the Pacific, being then (1789) the mate of the London whaler *Emilia*. After this experience, he had gone to Dunkirk to take command of William Rotch's new ship *Cyrus*, and also serve as mate on other Dunkirk ships. Hammond came back to Nantucket to build a new house on Centre Street in 1802. Captain Coffin Whippey, his contemporary, who had commanded the *Canton* out of Dunkirk, and then gone to Milford Haven to take out the Benjamin Rotch ship *Grand Sachem*, came home in 1810 to live next door to his friend. Captain Paul West returned to Nantucket in 1815, following his last voyage in the *Cyrus*—being one of the few British whalers to escape the *Essex* and her consorts.[2]

Others of the first group to return home from their European adventures were Captain Uriah Swain, who had been a pioneer whaler from Dunkirk to the Falkland Islands with Captain Hussey in the *United States*; Captain Silas Jones, who had made successful voyages from that same French port in the *Swan*, whose brick house on Orange Street was the first of that material ever built in this island town of wooden dwellings, and who upon his return to Nantucket, took command of the ship *Favorite* for one voyage, then retired; Captain Benjamin Paddock, commander of the *Royal Bounty*, out of London, who had reached Nantucket in 1800.

Captain William Swain who had gone to sea with his father on the *Ranger* of Nantucket in 1789, had also gone to London to command Enderby ships, one of which was the *Cumberland*. His brother, Captain George Swain, had remained at home to sail in Nantucket ships. Upon his return to America, Captain William removed to Auburn, New York, where he lived to the ripe old age of ninety-three.

Among the first to return, after experiences in London ships, were the cousins, Captain Elisha and Matthew Pinkham, the former once commanding the *Ark* and the latter the *Romulus*. In 1790, when Captain Elisha came to Sherborn for a visit, Keziah Fanning wrote in her diary that he had been absent for fourteen years, during which time he had been "Captain of a whaling ship out of England." Captain

2. Data relating to the return of the whalemen compiled from newspapers of Nantucket, the Sanford Papers, Nantucket Atheneum, and author's files.

Matthew had first gone to Dartmouth, Nova Scotia, but then, instead of joining the group to Milford Haven, had taken the *Romulus* directly to London when that ship was sold to London merchants. With the advent of the 1812 War, he had returned to Nantucket. During the Revolution he had served on the privateer *Hound* (being thus disowned by the Society of Friends for "going to sea on an armed vessel"), and it was this experience which no doubt dissuaded him from serving on British armed whalers.

In November 1814 Captain James Gwinn arrived in Nantucket, via Halifax, together with Captain Jedediah Fitch—both of whom "had been absent for many years whaling out of England." Gwinn had been with William Rotch in Dunkirk, and had then commanded the Benjamin Rotch ship *Wareham* out of Milford Haven, Wales. Soon after returning home he took out the *Reaper*, of Nantucket, but died during the voyage.[3]

When Captain Shubael Coffin came home he had all his savings from voyages in the *Brothers*, of Dunkirk, the *Prince of Wales* and *William* of London. He now invested his money in several Nantucket ships, owning the majority of shares in the *Swan*, *Favorite* and *America*.

But there were a number of Nantucket men who remained in England, enjoying prosperous careers in London. Of these, Captain Henry Delano was perhaps the most successful. He was one of three sons of Thomas Delano, Sr., all of whom went to sea at an early age, and eventually removed to London—apparently among the first to be employed by the Enderby firm. Captain Thomas Delano, Jr., had command of the *Hercules*, Captain Abishai Delano took out the *Sea Horse*, and the younger brother Captain Henry Delano was master of the *Kingston*, the *Rasper*, and the very successful *Atlantic*—all Enderby ships.

Recognizing the opportunities in the rapid growth of the British Southern Whale Fishery, Captain Henry Delano eventually stayed ashore and invested his money in a South Seas whaler owned jointly with his brother, Captain Abishai Delano, and three others. Appropriately enough, they named her the *Columbia*. She was a brig, built in Trinity, Newfoundland, in 1786, and first registered in Poole, England, but in

3. Obed Macy, *Diary*, November 8, 1814. Mss. Collections, Peter Foulger Museum Library, Nantucket Historical Association.

1791 she was purchased by the Delanos and Andrew Swain, a fellow Nantucketer, together with James and Ann Osborn, of Middlesex—into which family Henry Delano had recently married. Soon after the brig *Columbia* was registered in London (Aug. 30, 1791), she sailed for the Pacific under Captain Abishai Delano.[4]

In 1790 Captain Henry Delano was listed as of "Edmonton, Middlesex, Gentleman," and had married Miss Francis Osborne, sister of James who was a share owner of the *Columbia*. Some of his descendants were known as Delano and Delano-Osborne. The young Nantucketer must have felt a close affinity to England. It was his ancestor, originally known as De la Noye, who, immigrating from that nation to America a century and a half before, eventually settled in New England. The children of that immigrant, in Dartmouth, Massachusetts, soon adopted the name as "Delano." The American President Franklin Delano Roosevelt was a descendant.

The older of the three brothers, Captain Thomas Delano, Jr., was one of the pioneer whaling masters who went to South Georgia, that remote island in the South Atlantic, for sea elephants. In 1786 he took the *Lord Hawkesbury*, owned by Alex and Benjamin Champion, sailed to South Georgia, and returned with a full cargo. On a visit home to Nantucket, Captain Delano often told of his adventures to his young nephew Joseph Gardner Swift, son of his sister Deborah and Dr. Swift. General Joseph Gardner Swift was born in the Delano homestead then standing on the corner of Main and Orange Streets. The youngster, who became the first graduate of West Point and the second Commandant of that school, often commented on the spirit of adventure that the stories of his Delano uncle always conveyed and which inspired him.[5]

Without doubt, it was the example set by the father, Captain Thomas Delano, Sr., which influenced the sons. As early as 1745, Captain Paul Paddack, a Nantucketer, had built a ship named the *Royal Bounty* for parties in London and taken her to Greenland on a whaling voyage. He

4. London Ship Registers, National Maritime Museum, Greenwich, England; List of Nantucket Shipmasters in British Whaleships, compiled by Hussey and Robinson, Nantucket, Mass., 1876; Correspondence with Mrs. Cecily Wallers, Kent, England, and Mr. Robert Craig, London, England.
5. Gen. Joseph Gardner Swift, *Memoirs*, U.S. Military Academy, West Point, Library, West Point, New York.

was the first American to actually command a London whaler, and in the crew of mostly Nantucket men were several boys among which were Thomas Delano, Sr., and Charles West, the latter who was to marry Hepsabeth Paddack, daughter of Captain Paddack. From the latter union were born three boys who became outstanding shipmasters.

These grandsons of Captain Paul Paddack had their share of adventures in French and English whaleships. Captain Stephen West's adventures while master of the New Bedford whaler *Walker*, during the War of 1812, have been described in the previous chapter. Upon reaching the age of twelve, he had shipped on the Rotch ship *Speedwell*, out of Dunkirk. Before his long sea career was completed, he had commanded ships out of four different ports, including whaleships, sealing vessels, packet ships; then returned to whaling rather than command the early clippers in the China trade. A close friend of Captain Stephen West once wrote:

It was at the knee of his grandfather, Captain Paul Paddack, that Stephen first imbibed a taste for sailor life and in which he became as famed as his ancestors had been.[6]

Still standing in Nantucket is the ancient dwelling on Sunset Hill (built in 1686) where the Delanos and Wests heard Grandfather Paddack recall his adventures. The youngest of the West brothers, Captain Silas West, was killed by a sperm whale while in command of the ship *Indian*, of London. The news of the tragic death was reported by the whaler *Woodlark*, Captain Moore:

The unfortunate Captain West was heading his beat as usual, off New Zealand, when the frail craft was stove in by a blow from the whale to which they were fast, and the animal, almost at the same instant, seized Captain West by the middle of the body and nearly tore him asunder.[7]

In contrast, the second oldest of the three West brothers, Captain Paul West, was one on whom fortune smiled. At the age of thirteen he was a greenhorn on the *Maria* (the same vessel which had carried William Rotch to England). Later, aboard this whaleship, then sailing from New Bedford under Captain Benjamin Paddack, the boy first

6. William Hussey Macy, in *The Mirror*, March 12, 1862.
7. *The Sydney Gazette*, Sydney, Australia, May 24, 1822.

crossed the equator. After two more years at sea he made his first voyage as a boatsteerer on the ship *Alliance* in 1797, out of Nantucket.[8]

On his next voyage, Paul West shipped as mate of the *Cyrus*, a Rotch ship out of Dunkirk, commanded by Archaelus Hammond, sailing in 1803. The capture of that whaler by the *Scorpion* in Delagoa Bay (September 1803) and Paul West's appointment as Captain of the vessel, following her arrival and subsequent sale in London, have already been recorded.

Late in the year 1812, Captain Paul West came back to Nantucket to retire after twenty years at sea. He was then in his thirty-fourth year, and a well-to-do mariner. For the next fifty years he lived a busy life as a director of the Pacific Bank and Union Insurance Company. A contemporary wrote of him:

He has been a most methodical man. Everything connected with his affairs has been exact to a fault. His house and grounds the very patterns of neatness, and all things about him arranged with perfect system. He never filled a check away from his checkbook at home without cutting it out after the fashion of a 'Mediterranean Pass,' so that it could be fitted on if necessary when returned. His integrity has always been remarkable.[9]

Many years later a family controversy arose as to how Captain West accumulated his fortune at such an early age. His great-grandson and namesake, Paul West, wrote a brief account of the Captain's life, commenting that, as the *Cyrus* was an armed whaler, she may have taken part in privateering as did others in this category as a "letter of Marque" vessel. Another grandson, Alfred Bunker, took his cousin to task for suggesting their ancestor was a "privateer," dismissing it as "romance," to which the younger Paul West replied:

It may interest you to know how I discovered that our respected progenitor, of whose history I am no less proud than you are, did not devote his entire sea-going life to the peaceable pursuit of the whale. It all came about by accident, without which I might have lived on, as you have done, in ignorance of the interesting facts, since I had never heard any of them from the family.

8. Obituary, Captain Paul West, *The Mirror*, Nantucket, March 12, 1859.
9. William Hussey Macy.

The following summer, visiting Nantucket, I asked my grandfather about this, and he said that he had heard some such thing. I asked him—with a fleeting hope that it was so—if Captain Paul West had been one of the number. As I recall our conversation he made no direct admission, but, before I went home, handed me a little book, telling me it might interest me.

This book, which I still have and cherish, was in the writing of his father, and was a list of the private signals used among ships engaged in the whaling business during the years of the French–Spanish–British wars, between 1800 and 1812.

It contained, in well painted flags and lanterns, a full set of signal arrangements for all contingencies, many of them far from being what were the usual requirements for whalers . . . The Cyrus *is known to have sailed from London heavily gunned . . . I have found a picture of the old whaler with a lot of cannon—nine on her starboard side and, I presume, an equal number on her port. It shows the* Cyrus *going into St. Helena, under the British flag . . . I think these facts give me the right to request you to withdraw your implication that I am a romancer. If my story distresses you, it does not distress me. I am proud of it.*[10]

Of that Nantucket family of Gardners who entered the European whaling scene, there were eight who became masters of French whalers and twelve who commanded British ships. In the latter number, Captain Calvin Gardner married a London girl. At his death in 1812, his widow requested that her brother-in-law Henry Gardner be appointed administrator of her husband's estate. It is worthy of note that in the legal documents she referred to Henry as "my friend and brother-in-law," despite the fact that the Atlantic Ocean lay between their relationship and precluded personal meetings. This was another indication of the close association of the Nantucketers, and the feeling of confidence in their integrity on the part of distant "in-laws."

Three of the Bunker captains married English girls, as did two of the Delanos and Clarks. Those remaining in Britain following their marriages were usually in favorable circumstances, both as to the marital

10. Paul West, "My Ancestor, Captain Paul West," *The Inquirer and Mirror*, Nantucket, Mass., December 16, 1916.

state and business prospects. Of those who married English and French girls a number returned with wives and families.

Two of the Gardner clan coming back to their Island homes were Captain Micajah Gardner (already mentioned), and Captain Amaziah Gardner. Both became well-known citizens in retirement in their Island homes. Another family, the Clarks, resumed their old business of whaling upon return to Nantucket, as did the Husseys, Whippeys, and Rawsons. In most cases those who had acquired a "nest egg" built substantial homes, such as Jared Gardner and David Baxter. One member of the latter family, who remained in France, became so prominent in French whaling as to be knighted by King Louis and became Sir Francis Baxter.

The full account of the subsequent accomplishments of these returning Nantucket mariners would fill a volume in itself. Captain Nathaniel E. Cary, who sailed out of Le Havre in the French whaler *George*, afterwards commanded ships from Nantucket, New Bedford and Mattapoisett, and, upon "coming ashore," had established a record for the third highest total of oil ever taken by any whaler. Near his end, while in his eighty-sixth year he sometimes suffered lapses of memory and often believed he was still aboard ship, so that when his call for "all hands on deck" failed to bring prompt answer he would man-handle the supposed "mutinous sailors," much to the embarrassment of the household—especially if there were guests.

Another of these long-lived master mariners, Captain Alexander Macy, transferred his activities to Boston, taking out the Israel Thorndike whaler *Palladium* in 1822 and on his voyage to the Sandwich (Hawaiian) Islands he took out the first printing press to be used at Honolulu. During his retiring years on Nantucket he became a bank president and teacher of navigation.[11]

Captain Noah Pease Folger's return to Nantucket had some unusual overtones. He had taken out the ship *Partridge* for James Mellish, the affluent London whaling merchant, and the voyage had been a financial failure. Mellish, on the Exchange, made some uncomplimentary remarks about the ability of Captain Folger, whereby the mariner asked for a retraction and, when Mellish refused, the Nantucket man threatened to shoot him. This brought a disdainful remark from the merchant,

11. Logbook of the ship *Palladium*, Captain Alexander Macy, in the author's collections.

at which moment his former shipmaster fired, the bullet coming close to the astonished Londoner's head. Captain Folger was arrested and sentenced to Newgate Prison. The young man's uncle, Captain Noah Pease, was still residing in London and he appealed to none other than William IV—the friend of all mariners—who saw to it that Captain Folger was pardoned, on the condition that he leave England immediately.[12]

Among other Nantucket shipmasters who made names for themselves in whaling history were Captain Frederick Coffin, in the London whaler *Syren*, and Captain Joseph Allen in the *Maro*, of Nantucket, who shared in the honor of being the first whaleships to take sperm in the new whaling area known as the "Japan Grounds." While off the Palau Islands, the *Syren* was boarded by natives who suddenly decided to capture the ship. During the fighting the mate was killed and several of the crew wounded before the South Sea Islanders were repelled. The *Syren* returned on April 21, 1822, with one of the largest cargoes of sperm brought into London for many years—2,768 barrels of sperm oil.[13]

Before the Milford Haven whaling families finally disappeared from the maritime scene, the grandson of Samuel Starbuck, Sr., played the leading role in a most unusual adventure. This was Captain Valentine Starbuck who had been most successful in his first voyage out of London in the ship *L'Aigle*, and in 1822 he was again in command of the whaleship, owned by Hill and Bouliott. Proceeding into the Pacific he became one of the first of the British whalers to reach the Sandwich Islands, although the English sealers had long been visitors there.

The Hawaiian King Kamehameha became friendly with Captain Starbuck. After a series of pleasant visits the King confided to the whaling master his great curiosity respecting King George and the British kingdom. He had been impressed by the visit of H.M.S. *Cherub*, under Captain Tucker, during that vessel's stay at Oahu, but was also well aware of the influence of the growing number of American whaleships which were now visiting the Sandwich Islands and bringing considerable trade.

12. Frederick C. Sanford, *Sanford Papers*, Nantucket Atheneum Library.
13. Thomas Beale, *The Natural History of The Sperm Whale* (London, 1839), p. 150.

Captain Starbuck was anxious to secure the advantages of these Islands for the British Southern Whale Fishery and decided on a bold move. He proposed Kamehameha that he and his Queen accompany him on board *L'Aigle* and sail directly to London. After some weeks of persuasion, the Hawaiian King decided to accept the offer, and late in November 1823, the ship sailed with her royal passengers.[14]

On her passage toward Cape Horn, the *L'Aigle* sighted a low, uninhabited island some 5 degrees below the equator, and Captain Starbuck believed it a new discovery. However, it was first charted by another whaleman, Captain Obed Starbuck of Nantucket, who named it "Starbuck's Island." He was a close relative of the Milford Haven Starbuck.

In March 1824 the British whaleship put in at Rio de Janeiro, and the British Admiral George Eyre wrote to London that the *L'Aigle* had arrived with the King and Queen of the Sandwich Islands and suite on board, "for a passage, to pay his respects to His Britannic Majesty and to put his islands, he says, under His Majesty's protection. I avail myself of this opportunity . . . to acquaint their Lordships with the circumstances."[15]

Captain Valentine Starbuck sailed his ship up the Thames in May 1824, and King Kamehameha and his Queen and their suite were duly installed in appropriate lodgings in London. Just when it appeared the young whaling master had accomplished a remarkable coup—and the coveted Sandwich Islands, the veritable crossroads of the Pacific, might be placed under Britain's control—the royal visitors became ill with the measles.

The Queen of the Sandwich Islands died first on July 2, 1824, followed six days later by King Kamehameha. Although they had been vaccinated against smallpox they were victims of a disease of milder concern to Europeans but fatal to a race which had no immunity against it.

All the assistance which medical knowledge could devise was administered the sufferers, by his majesty's own physicians accompanied by others eminent in the profession.[16]

14. Captain Valentine Starbuck, *Memorial*, Foreign Office Papers, Public Record Office, London, F.O. 58—3.
15. Rear Admiral George Eyre, H.M.S. *Spartrake*, March 5, 1824, at Rio de Janeiro, to Foreign Office, "In-Letters," F.O. 58.
16. Captain Valentine Starbuck, *Memorial*.

The bodies of the unfortunate royal couple were brought back to Honolulu by H.M.S. *Blonde* under Captain John Byron, the news of their deaths having previously been brought to the islands by Richard Charlton, the first British representative to the Sandwich Islands.

After the Hawaiian King's death, King Kamehameha's secretary presented Captain Starbuck with one of the famous Hawaiian feathered capes, the mark of royalty of the native kings—often called a cloak. Geometrically designed in a checkerboard pattern of concentric rows of rectangles, the cape is made of the feathers of the then very rare, and now nearly extinct, Iiwi bird. These birds lived high up among the crags of the mountains in the Hawaiian Islands and were very difficult to capture. Only two or three golden feathers grew under the wings of this black-feathered bird. After these were plucked for use on the capes, the birds were released so that the feathers could grow back and be plundered again.

A century later (in 1927) the granddaughters of Captain Starbuck, Eva and Lucretia Starbuck of Aylesbury, England, presented the feathered cape to the people of Hawaii through Mr. Robert P. Lewis, of Honolulu, whose hobby was Hawaiian feather-work. The cape is now in the Bernice P. Bishop Museum, in Honolulu.

Captain Starbuck, so close to becoming a hero in British diplomatic history, became highly criticized by the ship's owners, who declared: "Captain Starbuck took it upon himself to deviate from his instructions . . . we have a serious demand against him for deficiency of cargo." The whaleship master wrote a Memorial directly to His Majesty, King George IV, stating that, "during his voyages to the South Seas, he has been in the habit of calling at the Sandwich Islands, the King of which had invariably displayed the utmost veneration for the name of your Majesty . . . that during the last voyage . . . the King and Queen had expressed the utmost anxiety to see, and enter into a solemn treaty with your Majesty, that your Majesty might in future possess an influence in these Islands and your Majesty's subjects might therefore derive peculiar immunities and privileges above that of all other nations."

The Memorial went on to describe the growing trade of the United States with the Sandwich Islands, and that:

Considering the great importance to your Majesty's subjects as place of call and refreshment, it was highly essential to do everything reasonable

40. Captain Zophar Haden
Master in the Nantucket–Dunkirk trade
[Courtesy the Nantucket Historical Association]

for the purpose of insuring the continuance of the amicable feeling, coupled with the circumstances of these Islands producing immense quantities of sandalwood and other valuable articles—important to the commerce of this country . . . as a lucrative trade with China . . . your Memorialist presumed to think . . . it would be in the interests of your Majesty's government . . . by conveying to Great Britain the King and Queen of said Islands.[17]

Thus, the difference between a secure place in history for Captain Starbuck's diplomatic coup, and bitter criticisms and censure, was the intervention of what was then considered a mild children's disease—the measles.

But there were more pleasant experiences in the careers of these transplanted Nantucket whaling masters, such as those enjoyed by Captain Laban Russell. Joining the Milford Haven migration the young man became master of the *Charles* from that port, and made several successful voyages. He married Mary Hayden, of Nantucket, and she went to sea with him, on the ship *Hydra*, of Plymouth, England, probably the first of the whaling masters out of England to bring his wife on a whaling voyage. In 1822, after visiting her family in Nantucket, Mrs. Russell returned to England with her younger son Charles, and boarded her husband's new ship the *Emily*, then bound for the New Holland (Australia) grounds. William, the Russell's oldest son, had signed on as a boatsteerer (or harpooner) for this voyage.[18]

Young William Russell was well qualified for the demanding role as a harpooner on his father's ship. On the previous voyage he had accompanied his parents on the *Hydra*, out of Plymouth, which sailed on February 25, 1817, on a three-year voyage. William was then twelve years old and kept a journal of the voyage, which is still in the hands of his descendants in the United States. Despite his youth, he kept an excellent record, no doubt carefully supervised by his father, Captain Laban, and the account is one of the very few whaling logs of that period still extant—especially of a British whaleship.

The *Hydra* had more than the usual share of adventures during her

17. Ibid.
18. Correspondence with Mrs. Katherine Tristan, Mrs. Margaret Cook Tristan, and Dr. Theodore Tristan, and E. A. Stackpole, January–April, 1968.

three-year voyage. At the "Tres Maria" Islands, off the coast of Mexico, Captain Russell helped a fellow British Captain quell a mutiny on the *Shakespeare*. Shortly after this incident the *Hydra* was held captive by the Spaniards in Callao for six months and thus unable to obtain her first 100 barrels of sperm until a year had elapsed. Cruising to the favored Galapagos Islands, Captain Russell spoke to several Nantucket whalers, including the *Lady Adams*, *Thomas*, and *Hero*, and had a pleasant gam with his cousin, Captain James Russell aboard the latter ship. Little did either shipmaster realize that the *Hero*'s Captain was to meet disaster a few months later off Arauco, Chile, when his ship was captured by Spanish revolutionists and James Russell was murdered by the pirate Benevedes.

In October 1818 the *Hydra* spoke the famous Nantucket whaleship *Maro*, under Captain Joseph Allen, first American whaler on the "Japan Grounds." Captain Allen, at this time, told Captain Russell of the unusually large number of whales the *Maro* had come upon in the area of the Maria Islands, and the *Hydra* immediately sailed north. The cruise was successful, and by June 1819 the ship had taken its seventy-first whale. Off the California coast, at the island now called Cedros Island, the crew killed sea elephants and stripped off the blubber as in whaling, making 334 bbls of oil. When the *Hydra* returned home in May 1820 she had 2,000 bbls. of oil—a fine total.[19]

On his next whaling voyage, Captain Russell was given command of the new whaleship *Emily*, of London. When the ship sailed from the Thames, Mrs. Russell was on board along with the older son William and the younger son Charles. The *Emily* rounded the Cape of Good Hope and was headed for Australia and eventually the "Japan Grounds."

A most unusual situation faced Captain Russell when son Charles fell in the cabin companionway and broke his arm just above the wrist. As was the custom the Captain of a whaleship also served as doctor on board, and the master of the *Emily* was now faced with setting his son's arm. Mrs. Russell described the situation:

Such an accident on the land would have been distressing but what were my feelings when I saw the child writhing in agony with no

19. The logbook of the *Hydra*, in *The Rochester Times–Union*, Rochester, N.Y., November 12, 1927.

surgeon on board . . . His dear father, with that fortitude and presence of mind that seldom forsakes him, took him below, and, with a man to steady the arm, set it and splintered it up. The dear fellow bore the operation with courage that would have done credit to a man.[20]

The *Emily* made whaling history for the British Southern whalers in that Mrs. Russell was the first wife of a whaling master to go ashore in the Molucca Islands in the East Indies. She wrote that her appearance attracted much attention: "As whaleships usually touch for refreshments, a white man is no novelty, but the female created a wonderful commotion."

Mary Hayden Russell was of help to her husband in the care of the sick. "I have often had reason since I left to bless the little knowledge I had of medicine, as it had contributed to take a great care off the mind of my husband. He examines the cases and reports them to me . . . the medicines I have administered have never failed of their desired effect."

Perhaps the outstanding entry of Mrs. Russell's journal describes her emotions upon watching her husband and son in boats approaching a school of whales:

My husband was soon fast to one, and in lancing another he had his boat's head completely broke off and his wrist joint badly sprained. The whales were so near I could distinctly view the scene with a [spy] glass. My terror was extreme, and as he set down low in the stern to keep the boat balanced, I concluded him more hurt than he really was. Under this impression I was with difficulty kept from fainting. To my great relief he was soon on board.[21]

There was no lack of humorous incidents, and one Mrs. Russell reported has such a twist. Off the coast of New Guinea an English ship, bound to Calcutta from Manila, hove to for a visit. Her master Captain Baker, came aboard the *Emily*, and remarked that the morning previous he had spoken another British whaler from London, the *Cape Packet*, and that her master could not be distinguished from the sailors, while

20. Letters of Mary H. Russell to her daughter, Mrs. Mary Mount, typescript in author's collections.
21. Ibid.

Captain Russell "seemed what he really was." Mrs. Russell wrote to her daughter:

"This is easily accounted for," said your Father, "as that captain is a Yankee."
"Indeed," replied Captain Baker, "Had you been a Yankee, Captain Russell, You would not have seen me on board here,—for I detest and despise those Yankees."
Little did he think at that moment he was conversing with two of those detestable things!

While cruising in this part of the East Indies the *Emily* spoke two other London whaleships, the *Ann*, Captain Kemp, and the *Catharine*, Captain Younger. Joining forces, the three whaleships dared to approach the New Guinea coast with the hopes of bartering for fresh vegetables offered by the natives in their proas, but keeping a close eye for any hostile action which might signal a native attack. Mrs. Russell's description follows:

The ships were hove to to wait the proas coming. In the head proa was the chief, seated on the roof of the house and paddled by about sixty men. The length of this royal proa was about 80 feet. Some of the others were double canoes, manned as they are in time of war . . . Care was taken not to allow more than one proa to come alongside at a time . . . Commodities traded were cocoanuts, plantains, and Paradise birds in high preservation, for which we exchanged, tin pots, iron hoops, handkerchiefs, etc.[22]

One of these Birds of Paradise is still on display at the Whaling Museum in Nantucket, but its donor is unknown. It would be most significant if he were Captain Russell—or his son William. The latter became a shipmaster in his own right. Upon returning to Nantucket, he went into the merchant service and became one of the packet ship commanders in the New York–Liverpool trade. Still in the possession of his family is a letter of appreciation for his ship-board courtesy written by Nathaniel Hawthorne who sailed with Captain Russell on the voyage to Liverpool when Hawthorne assumed his duties as American Consul.

22. Ibid.

Before leaving the letters of Mary Hayden Russell there is one passage which epitomizes not only dedication to her family but that of the other "whaling wives," who made the long voyages into the South Seas. She wrote, on November 18, 1824, as follows:

It was 5:00 o'clock in the afternoon when the boats left the ship to pursue a school of whales. The Weather was squally and threatening, and with anxious heart I followed their motions with the glass (telescope). So distant did they become that to the naked eye they would only be discovered on the horizon as they rose high on the swell.

Directly, one of the boats, which proved to be William's, rapidly approached the ship, as the whale to which he was fastened was running (in our direction) with all his strength. As they drew near, exhausted by loss of blood, the whale slackened his pace and, in a few minutes, died close to the ship.

By this time sunset had set in, and the ship's other boats were as far off as the eye to discern. William Russell came on board and after lashing the whale alongside, put sail on the *Emily* and headed for the distant boats. Due to the drag of the whale their way was slow and a thunder squall with sharp lightning broke over the sea. It was some four hours after the first lowerings that the ship and her boats finally came together again. Mrs. Russell's letter to her daughter completed the episode with these significant words:

Think, my dear Mary Ann, how anxious I must have been, and how happy I was to see your dear Father once more . . . He had not a dry thread in his clothes. I thought: "this is the way these sons of the ocean earn their money—that is so thoughtlessly spent at home."

Could some of the ladies whose husband are occupied in this dangerous business have been here these few hours past, I think it would be a lesson they would not forget. It would teach them prudence and economy more powerfully than all the books ever written on the subject since the invention of printing.[23]

23. Ibid.

22. New England Regains the Ascendancy in the Southern Whale Fishery

IN THAT bustling decade of 1815 to 1825 America's economic fortunes became vigorous and steady. It was during this time that New England at length wrested from Old England control of the Southern Whale Fishery. The remarkable recovery of Nantucket and the rapid growth of New Bedford were the factors leading to this whaling ascendancy. The success of these two ports encouraged other seafaring towns to enter the industry, with Fairhaven, Westport, Sag Harbor, and Newport having companies formed. Hudson, that offshoot of Nantucket in New York State, in 1817 sent two ships to the Pacific, the *Diana* and the *Eliza Barker*, and Edgartown on Martha's Vineyard, dispatched its first whaler to the Pacific, the *Apollo*, under Captain Daggett. Salem was represented by the *Britannia*, in 1818, while New London sent the ship *Carrier* whaling in the next year.

Nantucket's swift resumption of the industry in 1815 was phenomenal. Despite the disasters that overcame her fleet in the War of 1812, the pioneer port assembled a fleet of forty-eight vessels in the 1815–1816 year, with sixteen sailing in May and June 1815, and the remainder leaving port from July through December. Of the total, nineteen vessels sailed for the Pacific, three rounded Good Hope, and the remainder went to the Brazil Banks and Walvis Bay. New Bedford entered nineteen vessels over the two-year period, with seven sailing to the Pacific, six of these under flags of William Rotch and Samuel Rodman, with Hathaway's ship *Swift*, the other. In the Rotch fleet were the veterans from Dunkirk, the ships *Barclay*, *Diana*, *Maria*, and *Phoebe Ann*, and the brigs *Sally* and *Mary*.[1]

In the Nantucket fleet, newly acquired ships were the *Ganges*, built at Haverhill in 1808; the *Globe*, launched at the North River in 1815;

1. Starbuck, *A History of The American Whale Fishery*, pp. 216–220.

the *Martha*, built the same year in the same place; and the *William Penn*, also completed that year. One of the fleet, the *Tarquin*, fell in with a disabled Portuguese frigate off the coast of Brazil and towed her into port. Captain James Bunker was rewarded by 900 bbls. of whale oil and permission to conduct whaling voyages in Portuguese waters for three years.

By 1818 eight new whaleships had been added to Nantucket's fleet, now totalling fifty-six, while New Bedford's total fleet numbered twenty-five. By 1829 Nantucket's fleet had grown to sixty-nine vessels, while New Bedford exceeded forty. During this time England's fleets were steadily dwindling, with London showing the greatest decline.

By December 1820 Nantucket's whaling fleet totalled seventy-two whaleships, and once again she was the leading port in the Southern Whale Fishery. The phenomenon of her recovery was noted by a contemporary journal:

When we consider the numerous other vessels engaged in the coasting trade of the Island, the small number of inhabitants it contains, and the Island itself but a speck on the bordering waters of our Republic; and moreover, that almost the whole of the shipping was captured or destroyed so recently as the last war, we are struck with admiration at the invincible hardihood and enterprise of this little active, industrious and friendly community, whose harpoons have penetrated with success every nook and corner of every ocean.[2]

The rising prosperity of the United States brought an increasing demand for whale oil and sperm candles. The growing cities of New York, Boston, Philadelphia, Baltimore, and New Orleans were lucrative markets, while the lifting of British restrictions on American trade to the West Indies added to the market, and the European trade increased. Another product of the industry, whalebone, was in use for umbrella ribs, buggy whips, canes, corset stays, and hoop skirts. Lighting the lamps of lighthouses and beacons, as well as lightships, along the coast of this nation brought additional demands for both whale and sperm oils. Industrial use included oil for paints, tanning hides, and lubrication. Improved designs in whale oil lamps, created by Argand and Miles,

2. *Niles Register*, Baltimore, Maryland, December 2, 1820.

provided a market for whale oil in rural communities and small towns as well as in the cities.

While many of her wandering whalemen returned home to Nantucket there were still a number who found their way to England to accept berths as shipmasters and mates in British South Seasmen. Among these, Captain Charles B. Worth had a colorful career, son of Captain Benjamin Worth, one of the most famous from this island, Captain Charles B. Worth made several voyages out of London. In 1823, while master of the ship *Rochester*, he was one of the first whalemen to provision his ship at Rotumah Island, in the South Pacific.

Captain Charles Worth died in mid-career, under circumstances that demonstrated the dangers always present in whaling but seldom reported in detail. While in his boat, approaching a sperm whale, his knee came in contact with the sharp edge of a harpoon's barb, inflicting a deep wound. Soon after his return to the ship, an infection developed, which spread rapidly. The ship put into the nearest port for medical aid. But blood poisoning had advanced beyond the physician's treatment and the young shipmaster died at the age of thirty-seven. When the letter bearing the melancholy news reached Nantucket in December 1826, the widow of Captain Worth (Ann Young, an English girl, whom he had married in London) was visiting the father, Captain Benjamin Worth, "and, with her child was anxiously awaiting the news of his safe arrival."[3]

Another Nantucketer, serving in British whaleships, who met an untimely end, was Albert Folger, mate of the famous *Seringapatam*. He jumped into sea off Cape Horn to attempt to rescue a boy who had fallen overboard but in the attempt lost his life "as it was intensely cold."[4]

It would appear that Captain William Swain had the longest continuous service of an American whaling master in the British Southern Whale Fishery. Born in Nantucket in 1777, he went to sea at the age of twelve, and on a voyage to Holland made his way to England and soon after entered the employ of Samuel Enderby and Sons. For the next thirty years he sailed for the Enderby firm. He was on an armed London

3. Marine Column, *The Inquirer*, Nantucket, Mass., December 16, 1826.
4. *The Inquirer*, January 26, 1824.

whaler and during her wars with France once engaged a French privateer and sank her.

One of the whaleships commanded by Captain Swain made literary history—the Enderby ship *Sarah & Elizabeth*, of London. It was on board this vessel that on June 1, 1832, Thomas Beale transferred at the Bonin Islands from the London whaler *Kent*, as the ship's surgeon. As a result of these voyages, Beale wrote *The Natural History of the Sperm Whale*, a book which Herman Melville used extensively in his research and warmly praised. During the voyage, Beale recounted an experience that well demonstrated the prowess of Captain Swain, who was then in his fifty-fifth year. Beale's description follows:

On the morning of the 18th of June, 1832, while we were still fishing in the "off-shore ground" of Japan, we fell in with an immense sperm whale, which happened to be just the sort of one we required to complete our cargo. Three boats were immediately lowered to give him chase; but the whale, from some cause or other, appeared wild in his actions long before it had seen any of our boats . . . different in his actions to most other large whales, because it never went steadily upon one course . . . he dodged us about until near four P.M., at which time the men were dreadfully exhausted from their exertions in the chase, which had been conducted under a broiling sun, with the thermometer standing in the shade at 93°. About half-past four, however, Captain Swain contrived, by the most subtle management and great physical exertions, to get near the monster, when he immediately struck him with the harpoon with his own hands, and, before he had time to recover from the blow, he managed with his usual dexterity to give him two fatal wounds with the lance, which caused the blood to flow from the blow-hole in abundance.

The whale, after the last lance, immediately descended below the surface, and the captain felt certain he was going to "sound," but in this he was much mistaken . . . he arose to the surface with great velocity, and striking the boat with the fore part of his head threw it high into the air with the men and everything contained therein, fracturing it to atoms and scattering the crew widely about. While the men were endeavoring to save themselves from drowning by clinging to their oars and pieces of the wreck of the boat, the enormous animal was swimming round and round them, appearing as if meditating an attack with his flukes . . . a few strokes of his ponderous tail would soon have destroyed his enemies; but this was not attempted . . . After they had remained in the water about three-quarters of an hour, assisting themselves by clinging to pieces of the wreck, one of the other boats arrived and took them in . . . Captain Swain, with twelve men in one boat, therefore made another attack upon the whale with the lance . . . the whale went into its flurry and soon died, when, to the dismay of the boat's crews . . . it sunk, and never rose again—an occurrence which is not very infrequent, owing of course to the greater specific gravity of the individual, perhaps from the greater development of bony and musular structures. Such were the adventures of that day, in the evening of which the crews returned to the ship, worn out and dispirited, having lost a favorite boat, with the whole of her instruments, besides the last whale wanted to complete the cargo, and worth at least £500! [5]

A young British whaleman, under the tutelage of a Nantucket shipmaster, became a captain in his own right. Captain George Rule not only rose to command, but was among those who reversed the tide of whaling migration by eventually coming to Nantucket and making the Island his home.

Few mariners had a more adventuresome life than Captain Rule's. He was born in Berwick, Scotland, in 1781, and at the age of eleven he ran away to sea, shipping on board a vessel out of Leith, bound for the Baltic. This was followed by a career as a Royal Navy transport sailor which took Lord Keith's army to Egypt to fight Napoleon. Following the Peace of Amiens, young Rule made his way to London to ship out as a third mate on the whaleship *Earl St. Vincent*, spoke the whaleship *Fame* of Nantucket, and exchanged places with the second mate, John Upham, who was anxious to return to London (an English lass, no doubt, in the offing). Upham later became a noted commander of British ships. Rule came to Nantucket and was signed on the ship *Lydia*, Captain Paul Ray, as second mate. He made two voyages in this well-known whaler, owned by Zenas Coffin. Upon his return from his second voyage he married a Nantucket girl, Rebecca Gardner.[6]

5. Thomas Beale, *The Natural History of The Sperm Whale* (London, 1839), pp. 173–176.
6. *The Inquirer*, August 19, 1859. Interview with Mrs. George C. Rule, July 16, 1950.

Following two more voyages—on the *Hope* and *Gardner*, out of Nantucket—Rule accepted a berth as a mate of the Rotch ship *Barclay*, Captain Gideon Randall. It was on this voyage the ship was captured by a Spanish privateer off the coast of Peru, and then retaken by Captain Porter in the U.S. frigate *Essex*. As has been noted, the *Barclay* managed to slip through the British blockade and reach New Bedford. The dull times for American whaling in the 1812 War induced Rule to go to England, where he obtained command of the London whaleships *Spring Grove* and *Fanny* on successive voyages. Upon retirement, he sailed for America to again take up residence in Nantucket, where he first purchased a farm and then, upon the death of his wife, "came to town" to build an English-style residence on Gay street and spend his declining years, death coming in 1859 at the age of seventy-eight.

In this period of Nantucket's history (the 1850's), with its own whaling destiny coming to a close the town had a large number of retired shipmasters. As already mentioned, a goodly number of them were returning from England and France, such as Captain Coffin Whippey, Captain Archaelus Hammond, Captain Paul West, Captain Alexander Macy, and Captain George Rule. Here, as one obituary notice mentioned, "they spent the evenings of their lives within their quiet domestic circles, surrounded by their families and friends."

Captain Elisha Clark was another Nantucket whaling master successful in British whalers, his last command being the *Orion* of London. When he returned home he took up residence in the family homestead on Farmer Street where five of his brothers had also been born—all to become masters of whaleships.

Among the returnees were those who had commanded French whaleships—Seth Cathcart, who had taken the *Massachusetts* out of Le Havre; Reuben Swain, of the *Superior*, Dunkirk; Timothy Upham and James Bunker, both of whom, during their careers, had been masters of the ship *Archimedes*, of Le Havre, on successive voyages; Jethro Coffin, of the *Georges*; Latham Paddock, of the *Bourbon*; Shubael Russell, of the *George & Albert*. The stories these men could have recounted will never be known. They retired to useful lives in their Island homes, living quietly on the money they had saved while voyaging in ships from across the world.

Several of the descendants of French mariners who had married Nantucket girls in Dunkirk came to live in Nantucket. One of these, Lewis Imbert, after hard work, became a merchant. There was also a mysterious figure named Paul Francois Thomas, who was said to be the illegitimate son of a French noblewoman and a Nantucket whaling master, but whose brief appearance on Nantucket has not been documented and remains a legend.

Captain Francis Baxter played an important part in the second revival period in French whaling (1830), and became a leader in the industry to such an extent that he was knighted by the King of France and became Sir Francis Baxter. His brother, David Baxter, had a similar success, and returned to Nantucket when a comparatively young man. He built a handsome home on India Street, which still is one of the best examples of the 1845 neoclassicist style of construction. There was some question as to his first wife and children in France—whether there was ever a divorce. In any event, the Baxter family were convinced the separation was a legal one. Captain David (later a merchant captain), remarried upon his return to Nantucket and in his will left his estate to his brother's children.[7]

In the summer of 1839 a list of Nantucket shipmasters who were still living in retirement on Nantucket was printed in the Island's newspaper. This was no ordinary list—those recorded were mariners who had commanded whaleships before 1800 and numbered twenty-one, representing ships from six different ports in the United States, Nova Scotia, London and Dunkirk. The list reads:

ON NANTUCKET

NAME	AGE	SHIP	DATE	PORT
Paul Worth	87	*Sally*	1787	Halifax
Micajah Gardner	81	*Active*	1790	Dunkirk
George Pollard	80	*Minerva*	1791	Boston
Coffin Whippey	79	*Canton*	1790	Dunkirk
Peter Myrick	78	*Minerva*	1793	Nantucket
Simeon Starbuck	72	*Tryal*	1797	Nantucket
Paul Ray	76	*Judith*	1792	Dunkirk

7. Captain David Baxter, Obituary, *The Inquirer and Mirror*, March 12, 1898.

ON NANTUCKET

NAME	AGE	SHIP	DATE	PORT
Obed Fitch	75	*William Penn*	1792	Dunkirk
Jonathan Briggs	75	*Harmony*	1792	Dunkirk
Benjamin Worth	76	*Hector*	1797	Nantucket
Hezekiah Barnard	73	*Alliance*	1797	Nantucket
William Joy	73	*Fox*	1794	Nantucket
Reuben Joy	75	*Ranger*	1794	Nantucket
Levi Starbuck	73	*Harlequin*	1799	Nantucket
Charles Gardner	73	*London*	1797	London
Griffin Barney	72	*Barclay*	1797	New Bedford
Joseph Allen	70	*Lee*	1798	Nantucket

LIVING OFF NANTUCKET

NAME	AGE	SHIP	DATE	PORT
Valentine Pease Martha's Vineyard	82	*Alliance*	1794	Nantucket
Levi Joy (Hudson)	76	*American Hero*	1797	Hudson
Andrew Barnard Fairfield, Me.	77	*Lion*	1793	Nantucket
Robert Gardner Cincinnati, Ohio	72	*Warren*	1795	Nantucket
Alpheus Coffin Fairfield, Me.	75	*Diana*	1792	Dunkirk[8]

The list of returning Nantucket shipmasters would hardly be complete without mention of Captain James Josiah Coffin who was an active mariner for nearly sixty years. Entering the employ of the British Whale Fishery in 1786, he continued sailing out of English ports for the next forty years—a remarkable record of sustained activity. His finest achievement came in September 1824 when at the age of sixty-nine, and in command of the Bristol, England, whaler *Transit*, he discovered a group of uncharted islands between Japan and the Marianas to which he gave

8. "A List of Whaling Masters," *The Inquirer*, July 22, 1839.

the name Bonin Islands. Three of these he named Fisher, Kidd and South Islands, the first two for the owners of the ship, the Bristol merchants, Fisher, Kidd and Fisher.[9]

Between the two larger islands which he named Fisher and Kidd, Captain Coffin came upon a wide bay. Sailing up this bay for some four miles he came to an excellent harbor at Fisher's Island, which he called Coffin's Harbor. He placed it in latitude 20°30′ north and 141° east longitude, being "sheltered from all winds and excepting W.S.W., and has no current or swell."

Captain Coffin reported excellent fish and lobster were available here, also a kind of green cabbage, as well as quantities of wild pigeons and numerous terrapin or land turtle. He remarked on large trees growing at the islands, and observed that "nor did it appear the footsteps of man had ever before been impressed on any of these islands."

Being quickly aware of the advantages to whalemen offered by these islands—so ideally situated on the sea route to the newly explored "Japan Grounds"—Captain Coffin noted:

For ships now employed in the whale fishery, or bound from Canton to Port Jackson, or the northwest coast of America, they will furnish a valuable place of refreshment.

Some four years later Captain W. F. Beechey, with H.M.S. *Blossom* came upon the Bonins, and he duly recorded in his journal: "The southern cluster is evidently that in which a whaleship commanded by Captain Coffin anchored in 1823, who was first to communicate it to this country." Beechey renamed the group Bailey Islands, after the late President of the Royal Astronomical Society. The northern cluster he called Parry's Islands. He claimed the Bonins for England. Thus, Captain Coffin's discovery was recognized but his name not honored.

On the occasion of his seventieth birthday, Captain James J. Coffin struck a sperm whale on the "Japan Grounds" and took it to the ship. The voyage was most successful and he returned to his home port Bristol in 1825 with a full ship. Soon after his arrival in England he decided to retire from the sea. Coming back home to Nantucket, the last years of his life were spent in pleasant retirement with his family and friends.

9. Captain James J. Coffin's Obituary, *The Inquirer*, November 10, 1839.

When he died at the age of eighty-three in 1838, a friend of long acquaintance wrote of him as a "gentleman of rare knowledge . . . a descendant of John Gardner." The allusion by the Nantucketers to their ancestors was not without humor. A half century after the first group of settlers established themselves on the island, there had been so many intermarriages that most Nantucketers shared same great-grandparents and great-great-grandparents.

That powerful and successful figure in British merchantilism, Charles Jenkinson, through his successive roles as representative of British commerce by his office as President of the Board of Trade, as Lord Hawkesbury, and later as the Earl of Liverpool, had become a man of destiny. By his astute handling of affairs the British South Seas Whale Fishery had scored a resounding triumph in the last decades of the eighteenth century. Jenkinson believed he had laid the foundation for its continued success.

But even this most astute of political figures was not at first able to fully comprehend these Nantucketers, although their special genius intrigued him. Lord Hawkesbury saw them not as a colony but as individuals, just as their ability made them necessary adjuncts to Enderby and his associates. However, it is evident that he did not at first fully appreciate that important part of their nature—their devotion to their Island home. These whalemen were the descendants of a people who had created a maritime tradition. Sooner or later, the powerful summons of that tradition would turn them homeward again. This was why Rotch wanted them to come as a colony, rather than as individuals; this was why they were successful at Dartmouth; and at Dunkirk; and also why Greville wanted them as a group at Milford Haven—to utilize their cohesiveness as workers both at sea and ashore.

It is important to the better appreciation of the development of events to note that Hawkesbury finally recognized this characteristic of a unique people, and, having lost the first opportunity offered by William Rotch, was, at first, to support Charles Greville's proposal to settle the Nantucketers at Milford Haven. Through political circumstances this plan eventually lost Hawkesbury's continued support, and without it was ultimately doomed in its purpose.

In 1828, across the Atlantic, in his New Bedford home, the aging

Quaker merchant William Rotch, Sr., could look back over the years and feel the true fulfillment of his success. It was the Nantucketer who had won the eventual victory, and this knowledge lessened his early distaste for the cool, calculating, resourceful opponent—Lord Hawkesbury—now long since passed on.

New Bedford hummed with its whaling prosperity in 1828, and its progress now found it closely equalling the success of Nantucket—that Athens of American whaling which had provided the younger town with some of its most enterprising merchants and ship masters. As the patriarch of the whaling industry William Rotch was now nearing his ninetieth birthday, he was described by a contemporary:

A courtly, venerable looking gentleman . . . tall with long silvery locks, his dress of the true William Penn order—a drab beaver, drab suit, long coat and waist coat, knee breeches with silver buckles. His step is a little faltering but still graceful, as becoming one who had stood before ministers and kings in the Old World— . . . and in the old Friends' Meeting House on Spring Street, seated in the high seats at the "head of the meeting," his very presence is calculated to insure a respect for the principles of peace he so truly inculcated both by precept and example.[10]

Thus, friend Rotch had lived to see his faith in the whaling industry the proven factor of his ultimate triumph. He had never wavered in his belief that this was an industry which knew no national bounds—that it lighted the lamps of the world and oiled the machinery of the great industrial revolution, and therefore belonged to all nations as a common industry. In his courageous stand and continued advocacy of these principles he deserves to be known as America's first internationalist.

With the coming of the mid-1830's, whaling once again became an American-dominated industry. In 1825 there had been ninety British South Seas whaleships, but in ten years the number had dwindled to sixty-one, while American whaling had increased tenfold. As an example of the great surge of the New England ports in particular, a total of ninety-two whaleships arrived in American ports in the year 1835 alone. Of these thirty-eight returned to New Bedford, thirty-two to Nantucket,

10. Daniel Ricketson, *New Bedford of The Past*, (Boston, 1903), p. 6–7.

and the remainder to New London, Bristol, Newport, Warren, Falmouth, and several other ports.[11]

An editorial in Nantucket's newspaper, *The Inquirer*, October 7, 1823, commented on the advantages enjoyed by French and British whalers through government subsidies in bounties and premiums. The writer then observed:

But the facts ought to be known that, at this moment, there are employed in the Southern Whale Fishery from the port of Nantucket alone nearly twice as many ships as are engaged in that fishery from all the ports of France and England, and that most of the whaling ships belonging to both these countries are commanded by persons from Nantucket and New Bedford, who have been drawn from their homes by foreign allurements.

British South Seas whaling came to a virtual end with the advent of the 1840's. By this time over seven hundred whaleships were sailing from American ports. A London shipowner, addressing a committee from the House of Commons, commenting on the demise of Britain's Southern Whale Fishery, stated the case fairly:

What advantage do American whaleships have over British? . . . because they take a greater interest on it; they have better information for their people in it, and they are nearer the locality and it is more their business . . . Then, the great increase of American ships in the trade arises from the fact that their situation is better; they are more skilled in it, and they follow it with more perseverence than the British . . . Yes, they make more of a business of it, and they can do it at less expense than we can.[12]

In 1843, Britain's protective duty on sperm was reduced to £15 15 shillings per ton and on whale oil to £6–6 per ton. Six years later the duties were repealed entirely. The idea of repealing the duties was to permit the importation of American oil and thus diminish the cost to British customers. Between 1849 and 1860, Britain imported oil and bone from the United States to the amount of £3,744,460.[13]

11. Marine Columns of *The Inquirer*, January 9 and January 16, 1836.
12. Mr. Blythe, "Address to a Select Committee of The House of Commons," excerpts reprinted by *The Inquirer*, November 20, 1860.
13. Ibid.

The pursuit of whales and destiny now came full cycle. Not only had America regained her supremacy but she had completely outdistanced her chief competitor, Britain. And the Nantucket Quaker whalemen, in playing their vitally important roles on the great stage of the oceans of the world, had completed their part in the maritime drama. In pursuing their own destiny as well as the whale, both as individuals and as a people, they gave to America and Europe a saga as strange and as exciting as will be found in the pages of maritime history.

Appendix

STATE OF THE AMERICAN WHALEFISHERY: 1787–1790

	No. of Vessels Annually Northern Fishery	Their tonnage	No. of Vessels in Southern Fishery	Their tonnage	No. of Seamen	Barrels of Spermaceti	Barrels of Whale Oil
Nantucket	18	1,350	18	2,700	487	3,800	8,260
Wellfleet and other Ports on Cape Cod	12	720	4	400	212	—	1,920
Dartmouth (Bedford)	45	2,700	5	750	650	2,700	1,750
Cape Ann	—	—	2	350	28	—	1,200
Plymouth	1	60	—	—	13	100	—
Martha's Vineyard	2	120	1	100	39	220	—
Boston	6	450	—	—	78	360	—
Dorchester and Wareham	7	420	1	90	104	800	—
Total	91	5,820	31	4,390	1,611	7,980	13,130

An account of the industry at this time by a contemporary writer would have added much to the knowledge of the period. Pitkin, in his *View of the Commerce of the United States*, and Jefferson, in the above table, have given us statistical facts of considerable importance. Beale, in his *The Natural History of the Sperm Whale* (London, 1839), and Scoresby's *Arctic Regions and the Northern Whale Fishery* offer much about the British whalemen and the whale, with little of the American whaler and his enduring enterprise.

Following are the vessels out of Dartmouth, Nova Scotia, during the years 1785–1791, when such a remarkable whaling record was established:

VESSEL	MASTERS*
Argo	Ebenezer Bailey
Nancy	Jonathan Barnard, John Sprague
Prince William Henry	Thomas Brock, Matthew Pinkham
Africa	Zachary Bunker, Ransom Jones, Nathaniel Macy
Falkland	Peleg Bunker
Colony	William Chadwick

VESSEL	MASTERS*
Neptune	John Chase
Joseph	Joseph Clasby
Lucretia	Jonathan Coffin, 2nd, Philip Fosdick, Thomas Hiller
Harriett	Brown Coffin
Aurora	Eben Coffin, Peleg Hussey, Paul Worth
Ark	David Coleman, Elisha Pinkham
Dartmouth	Solomon Coleman, Jr., Albert Hussey, Barnabas Swain
Nancy	Andrew Coleman
Romulus	Obed Bunker, John Chadwick, Zimri Chase, Matthew Pinkham
Ann	Stephen Gardner
Neptune	William Paddock
Charlotte & Statira	Daniel Kelley, Abner Briggs
Charles	Laban Russell, David Baxter
Lively	John Chadwick
Somerset	Daniel Kelley
Hero	Valentine Pease
Rachel	Obed Barnard
Ores	Timothy Folger
Resource	Joseph Clasby
Brothers	Ebenezer Bailey
Susannah	Edmund Macy
Watson	Daniel Ray

* A whaleship would often change masters on successive voyages.

WHALESHIPS IN THE PACIFIC OCEAN: 1790–1793

Rounding Cape Horn, and in the Pacific Ocean off the west coast of South America, during the season 1790–1793, were the following whaleships from London, Nantucket, Dunkirk, and Bedford:

Redbridge, Capt. Samuel Kelley, London, England.
Minerva, Capt. Seth Coffin, Nantucket, U.S.A.
Greenwich, Capt. John Lock, London, England.
Dolphin, Capt. Andrew Swain, London, England.
Judith, Capt. Paul Ray, Dunkirk, France.
Liberty, Capt. Paul Bunker, Hudson, N.Y.

Warren, Capt. Robert Meader, Nantucket, U.S.A.
Harlequin, Capt. Benjamin Whippey, Nantucket, U.S.A.
Rebecca, Capt. Seth Folger, Nantucket, U.S.A.
Hector, Capt. Thomas Brock, Nantucket, U.S.A.
Friendship, Capt. Obed Barnard, London, England.
Washington, Capt .George Bunker, Nantucket, U.S.A.
Britannia, Capt. Thomas Melville, London, England.
Salamander, Capt. John Nichols, London, England.
Countess, Capt. Henry Delano, London, England.
Lydia, Capt. Benjamin Clark, Dunkirk, France.
Morse, Capt. Abijah Lock, London, England.
Fonthill, Capt. Elisha Pinkham, London, England.
Atlantic, Capt. John Bassett, London, England.
Speedy, Capt. John Lock, London, England.
Lucy, Capt. Robert Inott, Dunkirk, France.
York, Capt. Jonathan Daggett, London, England.
Kitty, Capt. Matthew Swain, London, England.
Spencer, Capt. Eber Bunker, London, England.
London, Capt. Joshua Coffin, London.
Hope, Capt. Mark Clark, Boston U.S.A.
New Hope, Capt. Jethro Daggett, London.
Kent, Captain Paul Pease, London.
Emilia, Capt. James Shields, London.
Liberty, Capt. Tristram Clark, London.
Ann, Capt. Prince Coleman, Dunkirk.
Bedford, Capt. Laban Coffin, Dunkirk.
Beaver, Capt. Paul Worth, Nantucket, U.S.A.
Rebecca, Capt. Joseph Kersey, Bedford, U.S.A.
Aurora, Capt. Parker Butler, London.
Stormont, Capt. Reuben Ellis, London.
Ospray, Capt. Latham Paddock, Dunkirk.
Penelope, Capt. Isaiah Worth, Dunkirk.
Mary, Capt. George Whippey, Dunkirk.
Venus, Capt. Daniel Coffin, London.
Chase, Capt. Charles Clark, London.

Of the total: 23 British whalers, eight from Dunkirk, seven from Nantucket, and one each from Bedford, Hudson, and Boston—totalling 41 whaleships—32 were under the command of Nantucket masters.

ARTICLES OF AGREEMENT BETWEEN WHALESHIPS IN THE GALAPAGOS ISLANDS—1801

1st—The ship which Discovers a School of Whales will immediately hoist his Ensign to the Miz. Topmast head . . .

2nd—If in a School of Whales our Lines get foul of each other we will use every endeavour to clear them and Not cut each Other's lines without Mutual Consent.

3rd—If a boat is in Danger of losing her line the Nearest Boat is to Give his and without any Exception or Claim whatever.

4th—If a boat Strikes a Whale & parts & then Strikes another, the first whale so Struck becomes the property of Any Boats who shall kill her; but must Return the Craft (harpoon). Yet, if the first boat is following his whale & hath not been fast to another he Claims the whale, even though she is killed by Another Ship's boat.

5th—If a ship falls in with Dead whales not having any Craft in them, a Union Jack will be hoisted for the benefit of those to whom they may belong.

6th—If only a Single whale is in sight no Signal will be Made & if more than one Ship's boats are after her tis agreed to render the Fast boat every assistance, & should any Damage be Sustained in Consequence, it will be Made Good by the Ship who takes the Whale.

7th—If Ships be in want of Assistance in the Night, a Tryangular Light will denote it where best to be seen. Every other Ship who sees it will Render her Every Assistance they Conveniently can.

8th—If any Dispute should occur respecting the Claim of a whale or whales, such Disputants shall Chuse two Disinterested Masters, the two so Choseth to Appoint a Third. The award of Eather Two Shall be conclusive.

9th—In case of falling in with a Stoven Boat tis agreed to Render her all Convenient assistance it being the intent of our present Resolves to promote our Mutual Interest and preserve Amity.

10th—It is further Agreed that each Commander will Malse Known the Intent of our Present Regulations in order that no Collusion may hereafter ensue.

Seringapatum of London, Captain Wm. A. Day
Favourite of London, Captain James Keith
Aurora of Milford Haven, Captain Andrew Myrick
Leo of Nantucket, Captain Joseph Allen

Euphrates of London, Captain Ian Frazier
Neutrality of Nantucket, Captain Thaddeus Folger
Hannah & Eliza of Milford Haven, Captain Micajah Gardner

The original manuscript of this "Agreement" is in the author's collection of whaling material.

THE DUNKIRK FLEET

The whaling fleet out of Dunkirk maintained by the Rotch firm over the period from 1786 through 1805 included vessels taking part also in the Nantucket–Dunkirk–Bedford triangle operation. They comprise craft engaged in whaling in the North and South Atlantic, Indian and Pacific Oceans, and were as follows:

SHIP	CAPTAINS
Penelope	Clothier Allen, Elisha Folger, Jethro Gardner, Isaiah Worth, Archaelus Hammond
Ann	Jonathan Barnard
Ardent	Peleg Bunker, Prince Bunker
Pioneer	Benjamin Clark
Juno	George Clark
Resolution	Prince Coleman
Maria	Bartlett Coffin, William Mooers
Bedford	Laban Coffin
United States (second)	Thaddeus Coffin, William Fitch, Uriah Swain, Joseph Wyer
William Penn	Obed Fitch
States	Stephen Gardner
Edward	Micajah Gardner, Peter Brock
Boston Packet	Nathaniel Gardner
Industry	Francis Baxter
Harmony	Jonathan Briggs, David Starbuck
Brutus	Francis Bunker, Reuben May
Lydia	Benjamin Clark, 2nd, William Mooers, Timothy Wyer, Micajah Gardner
Diana	Alphes Coffin, Timothy Long, Valentine Swain, Thaddeus Swain
Phebe	Edward Coffin
Dauphin	Paul Coffin
Brothers	Shubael Coffin, David Swain
Janus	Benjamin Coleman
Friends	Jedediah Fitch, Abel Rawson

SHIP	CAPTAINS
Favorite	David Folger
Thomas	Seth Folger
Hebe	Amaziah Gardner, Francis Macy
America	Tristram Gardner
Fame	Benjamin Glover, Peleg Hussey
Cyrus	Archaelus Hammond, Paul West
Maria & Eliza	Abishai Hayden
Mercury	Benjamin Glover
Ganges	Charles Harris
Susan	Barzillai Hussey
United States (first)	Benjamin Hussey
Fox	Ebenezer Hussey
Mary	Isaiah Worth, Coffin Whippey
Swan	Silas Jones, William Long
Ospray	Benjamin Paddock
Falkland	Obed Paddock
Judith	Paul Ray
Hope	Edward Starbuck
Hudson	Matthew Starbuck
Canton	Coffin Whippey
Necker	George Whippey
South Carolina	James Whippey
Seine	Jonathan Worth

PETITION TO THE FRENCH GOVERNMENT

On November 13, 1793, the firm of Lubbert Freres and Fils of Bordeaux, which had sold wine to William Rotch, submitted a petition to the French government as follows:

> To the Representatives of the People:
> Citizens:—If among the foreign vessels detained by the embargo there are any which deserve particular considerations and a speedy release, no doubt the three following American ships have a right to that claim:
>
> The *Penelope*, Captain Hammond,
> The *Ann*, Captain Coleman, and
> The *Lydia*, Captain Gardner.
>
> These ships belong to William Rotch and Samuel Rodman, merchants at Nantucket, a province of the United States.

The law requires that neutral vessels should export but a quantity of goods equal in value to that which they have imported.

These three ships have brought to Dunkerque three considerable cargoes of whale oil and other products of that fish to the amount of 250 or 300 thousand livres, which we have the most authentic proofs of.

One of these ships, the *Ann*, Captain Coleman, not meeting with a freight for the colonies in August last, Citizen Rotch, a partner of the said house of Nantucket, who is settled at Dunkerque, then present at Bordeaux, gave us an order to load her here for their account with 198½ tuns of red wine fit for St. Domingo, amounting according to invoice to £162,780.17.

It is evident that the ship *Ann* is entitled not to be unloaded and to a speedy expedition.

1st. Because she brought to France a cargo of articles of the most indispensable necessity, for our manufactories and consumption, and which by far exceed the amount of that she has in return.

2nd. Because this cargo is for the supplying of an important colony, which can subsist but by the assistance of neutral vessels.

We beg leave to observe to the representatives of the people, that Citizen Rotch is one of those industrious Quakers of Nantucket, who came to settle at Dunkerque in order to establish in that place the most valuable whale fishery and to whom government always granted a particular protection.

Citizen Rotch has even lately obtained from the Republick a premium of about 38,000 livres for the three valuable cargoes he has imported.

Being invested with his power, we hope the representatives of the people will be so kind as to suffer Captain Coleman to proceed on his voyage without further hindrance.

(Signed) LUBBERT FRERES & FILS.

Bordeaux the 13th day of November, 1793, in the second year of the French republick, one and indivisible.

The Rotch firm recovered only $5,326.67 in damages, but it was at least some reimbursement.

WHALESHIPS IN THE FALKLAND ISLANDS

In the year 1752 there arrived at Newport, Rhode Island, a young man named Aaron Lopez. He had been born in Portugal and had come to this country to join his father and older brother, and father-in-law, all a closely-knit family of merchants. Due to business reasons, Aaron had used the Christian given name of Edward in Portugal but in Newport he returned to his original name of Aaron. Naturalized at Taunton, Massachusetts in 1754, he soon became a well known shipper and during the next decade built up a lucrative trade with the West Indies. Most important, however, was the sperm candle works which his father-in-law Jacob Rodriques Rivera had launched at Newport, the first in this country. As has already been recorded, the attempt to establish a monopoly through the United Company of Spermaceti Candlers, found Lopez joining Obadiah Brown of Providence, Henry Lloyd of Boston, and the Rotches of Nantucket and Bedford. For several years this was a profitable venture. Lopez naturally invested in whaling and eventually became an associate with Francis Rotch in Bedford, Leonard Jarvis in Fairhaven, and Henry Lloyd of Boston.

Just what vessels were first to use the Falkland Islands as bases for whaling is not as yet proven, but there is evidence that the New England whalers were first there. In his famous speech "On Concilliation With America," delivered in 1775, in the House of Commons, Edmund Burke, in reference to the American whalemen, made this important statement: "The Falkland Islands, which seemed too remote and romantic an object for the grasp of national ambition, is but a stage and resting place in the progress of their victorious industry."

From existing documents, it is evident that Aaron Lopez was in close touch with the Rotches on Nantucket (the association extending from 1754, or soon after his arrival at Newport) and that he was aware of the latest information regarding whaling voyages. The Falkland Islands were probably reached by the whaleships as early as 1772, and the Nantucketers, with their propensity as pioneers, were soon aware of the potential whaling grounds.

In a letter received in May 1774, Aaron Lopez learned of the loss of one of his ships, bound for the Falklands—the *Leviathan*, under Captain Lathrop. As was the usual case, the unwelcome facts came through news from another whaleship. Seated in his counting house on Thames Street, Newport, Lopez read the following from one of his agents, Samuel Hart, Jr., who was writing from Barbados:

I am sorry to acquaint you that Captain Wm. Mooers, in the brig Bedford, belonging to Mr. Wm. Rotch, of Nantucket, who arrived here 29th last month (February) from the Coast of Brazil, gives the following particulars:

That about February 7, in Latitude 25° south, Longitude 40° west, he fell in with a Portuguese Snow . . . . . from Rio de Janeiro, aboard of whom was Captain Lathrop, who informed that his brig was taken from him by the Portuguese, and desired him not to go nigh the coast or he would also be taken. Capt. Mooers seems to think the Snow was bound for Lisbon . . . . when he came up to her . . . . the Capt. of the Snow forbid Capt. Mooers coming on board . . . . therefore was obliged to pass on, although each of the vessels had whaleboats in tow.

Just when Aaron Lopez received the full news of the capture of his whaleship is not clear, but there does exist an official report from Captain Lathrop, dated

November 15, 1773, with a postscript dated December 4, in which a complete account of the *Leviathan*'s misfortune is given. Upon reaching the coast of Brazil, Captain Lathrop had found his vessel "very leaky in the upper works and . . . knowing Rio de Janeiro to be a port where many English vessels put in for supplies . . . with the advice of my Headmen thought it proper to come in." To his surprise and consternation, the *Leviathan* was promptly seized by the Governor on grounds that he had sent a boat on shore before arriving at the official port of entry—Rio itself. The boat and crew, incidentally, were seized as soon as it reached the shore and were never permitted to return to the ship, which then sailed on to Rio de Janeiro, where she was promptly taken over by the authorities.

Captain Lathrop, however, soon learned the real reason for his detention and he wrote: "Notwithstanding (His Excellency, the Governor, informed) I might be of great service to this nation in showing them how I caught the Spermacetie Whale."

In the interim, the *Leviathan* was virtually interned and the Governor declared she would remain so until he heard from Lisbon, where he had written for advice. A whaleship out of Rio, owned by Joseph Fonseca, was offered to sail in company with the *Leviathan* when she was allowed to sail, so that the Portuguese would be instructed in the art of capturing the sperm whale. What was reported by Captain Mooers in the *Bedford* was that Captain Lathrop and his crew's carrying out the stipulation as to this cruise, in order to recover their vessel. As a strong and revealing appraisal of his situation, Captain Lathrop stated:

There is no doubt our voyage must have been prosperous if (it) had not been detained here, for I have the greatest confirmation that man would desire, but I shall not say any further concerning this—I am almost Crazy to meet with such a disappointment. . . . I wish I could write you what they mean to do but it is past Man's Art. . . . We are all in Health on board & well guarded as much as Englishmen would guard Pirates. God knows I don't fear nor love them . . . Please to give my compliments to Capt'n Allen & Love to my Family, as I have not wrote to them.

It was another year before Captain Lathrop and the *Leviathan* was able to reach home. He never recovered the boat's crew which had first gone ashore just below the harbor of Rio. The coming of the American Revolution so checked any colonial activity out of Newport that the ship returned to Bedford, and its future history has been lost.

Insofar as existing documents are concerned, whaleships from Nantucket, and Bedford were at the Falklands before the outbreak of the Revolution, and a history of Truro, on Cape Cod, states that Captains Smith and Collins of Wellfleet took whaling vessels into those islands after being advised by the British Admiral Montague that whales were seen in great numbers off the Falklands.

But it is with that remarkable effort of the whaling merchants of Nantucket, Bedford, and Newport to use the Falklands as a base for their fleet of whaleships that the first evidence of the importance of these islands to whaling is revealed.

Both Lopez and Francis Rotch, at the outbreak of the Revolution, in their ill-starred effort to send a fleet of a dozen whaleships to the Falklands as a headquarters, the oil to be taken directly to England, were basing their ambitious plan on the experience of their own whaling masters at the islands.

In a letter of instruction to Captain Lock of the brigantine *Minerva*, in September, 1775, Rotch and Lopez carefully advised a course from Newport to the Cape de Verde islands, thence south and west, across the Atlantic to the Coast of Brazil. From this point, his letter reads:

We would recommend your cruizing Southward after Spermaceti Whales.... until you get in the latitudes of the Falkland Islands... We have certain & very late information that between the latitudes of 45° & 48°, South, near Soundings, there is a great plenty of Spermaceti Whales than at any other spot we have heard of.... We think it expedient to strictly forbid your going into any Port or Harbor within... the Brazil Coast... as it may be extremely dangerous to fall into their hands.... We mean for you to cruise as before mentioned until such time as you have filled your casks, or until by other necessities, from bad weather or accidents you may have occasion to make some port, you are for many reasons (particularly that of Insurance) at all events to proceed for Port Egmont in the Falkland Islands, where are with all the rest of the vessels upon these voyages to rendezvous & recruit... & where you will find our Francis Rotch.

Besides the presence of Francis Rotch in the Falklands the commanders of the whaleships were also to find agents of the firm of Sampson Mears, at the Dutch island of St. Eustatius in the West Indies; Mayne and Co., at Lisbon, Portugal; Blackburn and Sanchez at Madeira; Livingston and Turnbull at Gibraltar; Mahoney and Woulfe at Teneriffe; Duff and Welsh at Cadiz, and George Haley, the London agent.

The fleet of seventeen vessels was successful in getting to sea, but on the voyage across the Atlantic five were taken off the Azores by the British frigates *Renown* and *Experiment*, and the frigate *Niger* was dispatched to capture as many of the others she could find at the Falklands. Only a few ships managed to complete their voyage.

From scattered pieces of evidence it is known that several of the Rotch–Lopez–Jarvis fleet did reach safe haven following their voyages to the Falklands. The *Fox* under Captain Silas Butler, arrived in Barbados in June, 1776. "The vessel was... in bad order which, added to the present unhappy dispute which exist in your place, made him deem it advisable to land his cargo here." (The letter, written to Francis Rotch in Nantucket, also stated that the oil would be sold there "to the best advantage." The prices were high—£35 per ton for head oil,

£25 per ton of body oil, and £18 per ton for right whale oil. The Nantucket whaleship *Amazon*, Captain Mooers, on her record voyage to the South Atlantic, was captured by a British man-of-war *Atlantic*, but released by the good offices of Thomson and Seed, the same firm which had taken care of the *Fox* and its cargo.)

Another whaler to make her way safely back home after voyaging to the Falklands was the brig *Lydia*, which Captain Thomas Folger got into Providence early in 1777. In one of the few documents existing of that period which gives details of the sale of cargo, lays, etc., the total proceeds amounted to £1228: 12: 10¾, of which amount £329: 3 was the total deduction, including the amount £268: 5 for the shares or lays to Captain Folger and crew, the latter ranging from 1/16th to 1/34th. It was carefully noted that shares of "Negroes Ebenezer and Caesar" were deducted because being slaves their lays were "incorporated with the owner's oil."

As the astute Rotch addressed Lord Germaine on his petition, he gave as argument a strong point by declaring the oil obtained at the Falklands would be shipped" . . . immediately for London . . . no other market calling for the quantity or affording so good a price for the proceeds of the Sperm Fishery." He further stated that he was now loading supplies aboard his vessels "in the River Thames . . . which are near ready to sail for the Falkland Islands." With justifiable pride, Rotch declared that the American whalemen "were and are the only people to kill the Sperm Whale."

From the obvious necessity of the moment, Rotch, in order to save the ships, stipulated the oil they would obtain would be brought back to England. The fleet for which he sought protection, as well as similar privileges, consisted of the following vessels:

		TUNS	MASTER
Ship	*Jacob*	160	Matthew Cornell
Ship	*Africa*	150	Joseph Ripley
Ship	*Cleopatra*	150	James Fitch
Brig	*Lydia*	120	Thomas Folger
Brig	*Charlotte*	120	John Woodman
Brig	*Ann*	110	John Darling
Brig	*Fox*	120	Silas Butler
Brig	*George*	100	George Whippey
Brig	*Mermaid*	100	Lokbury Blackman
Brig	*Minerva*	140	Ephraim Pease
Sloop	*Dartmouth*	80	Peter Pease
Sloop	*Delight*	80	Benj. Norton
Sloop	*Nelly*	80	Wm. Norton
Brig	*Royal Charlotte*	100	Nathaniel Hathaway

THOMAS MELVILLE'S LETTER TO ENDERBY AND SONS

Ship *Britannia*
Sydney, Port Jackson
November 22, 1791

Messrs. Samuel Enderby & Sons.

Gentlemen:

I have the pleasure to inform you of our safe arrival in Port Jackson in New South Wales, October 13, after a passage of fifty-five days from the Cape of Good Hope. We was only six weeks from the Cape (Good Hope) to Van Diemen's Land, but met with contrary winds after we doubled Van Diemen's Land, which made our Passage longer than I expected.

We parted Company with our Agent the next day after we left the Cape of Good Hope and never saw him again till we arrived at Port Jackson, both in one day. The *Allemarle* and us sailed much alike. The *Admiral Barrington* arrived three days after us; I am very well myself, thank God, and all the Crew are in high Spirits. We lost in all on our Passage from England twenty-one convicts, and one Soldier. We had one Birth on our Passage from the Cape.

I try'd to make, and did make the Island of Amsterdam, and made it in Long. of 76° 4^{m} 14^{s} East of Greenwich by a good Lunar Observation. My intention was to run close to it to discover whether the Seal Business might not have been carried on there, but the weather was so bad, and thick weather coming on, I did not think it prudent to attempt it, likewise lose a night's run, and a fair wind blowing.

The day before we made it, we saw two Schoals of Sperm Whales. After we doubled the South Cape of Van Diemen's Land, we saw a Large Sperm whale off Maria's Island, but we did not see any more, being thick weather and blowing hard till within Fifteen Leagues of the Latitude of Port Jackson. Within three leagues of the shore we saw Sperm Whales a great Plenty. We sailed through different Shoals of them, from twelve o'clock in the day till after sunset; all around the Horizon, as far as I could see from the Masthead.

In fact, I saw a very great Prospect in making our Fishery upon this Coast, and establishing a Fishery here. Our people was in the highest Spirits at so great a sight, and I was determined, as soon as I got in, and got clear of my live Lumber, to make all Possible dispatch on the Fishery on this Coast.

On our arrival here I waited upon His Excellency Governor Phillips, and delivered my letters to him. I had the Mortification to find he wanted to dispatch me, with my convicts to Norfolk Island, and likewise wanted to purchase our Vessel to stay in the Country, which I refused to do. I immediately told him the Secret of seeing the Whales, thinking that he would get me off going to Norfolk Island that there was a prospect of establishing a fishery here, and might be of Service to the Colony and left him.

I waited upon him two hours afterwards with a Box directed to him. He took me into a private Room. He told me that he would render me every Service that lay in his Power; that next morning he would dispatch every long-boat in the Fleet to take our Convicts out and take our Stores out immediately, which he did accordingly. And he everything to dispatch us on the Fishery.

Captain King used all his Interest in the business, also. He gave his kind respects to you.

The secret of seeing the Whales, our Sailors could not keep from the rest of the Whalers here. The news put them all to the Stir, but we have the pleasure to say we was the first ship ready for sea; notwithstanding they had been some of them a month arrived before us.

We went out in Company with the *William & Anne*, the 11th day after our Arrival. The next day after we went out we had very bad weather, and fell in with a very great number of Sperm Whales. At Sun rising in the morning we could see them all round the Horizon. We run through them in different Bodies till two o'clock in the afternoon, when the weather abated a little, but a very high Sea running.

We lowered away two boats and Bunker followed the Example. In less than two hours we had seven whales killed, but unfortunately a heavy gale came on from the S.W. and took the Ship aback with a squall that the ship could only fetch two of them, the rest we was obliged to cut from and make the best of our way on board to save the Boats and Crews. The *William and Anne* saved one, and we took the other, and rode by them all night with a heavy gale of wind.

The next morning it moderated. We took her in. She made 12 barrels. We saw large Whales next day, but was not able to lower away our Boats. We saw Whales everyday for a week after, but the weather being so bad we could not attempt to lower a boat down.

We cruized fifteen days in all, having left our sixty Shakes of Butts on Shore with the *Gorgon*'s cooper to selt up in our Absence, which Captain Parker was so kind as to let us have, and wanting to purchase more casks of Mr. Calvert's Ships, and having no prospect of getting any good Weather. I thought it most Proudent to come in and refit the Ship, and compleat my Casks, and fill my water, and by that time the weather would be more moderate.

The day after we came in, the *Mary Anne* came in off a Cruize, having met with very bad weather, shipped a Sea, and washed her Try-works overboard. He informed me he left the *Matilda* in a Harbour to the Northward; and the *Salamander* had killed a 40-barrel whale and lost her by bad weather.

There is nothing against make a voyage on this Coast but the weather, which I expect will be better next Month. I think to make another Month's trial of it. If a Voyage can be got upon this Coast, it will make it shorter than going to Peru, and the Governor has been very attentive to sending Greens for refreshment to our Crew at different times. Capt. Parker has been kind, and has given me every assistance that lay in his Power. He carries our Longboat home, as we cannot sell her here. He will dispose of her for you, or leave her at Portsmouth. He will wait upon you at his arrival in London.

Captain Ball of the *Supply*, who is the bearer of this letter, has likewise been very kind and rendered us every Service, that lay in his Power. He will wait upon you likewise.

The Colony is all alive expecting there will be a Rendezvous for the Fishermen. We shall be ready to sail on Tuesday the 22nd on a Cruze. The *Matilda* has since arrived here; she saw the *Salamander* four days ago. She had seen more whales, but durst not lower their boats down. She has been into Harbour twice. We have the pleasure to say we killed the first 4 whales on this Coast.

I have enclosed you the certificate for the Convicts, and receipts for the Stores. Captain Nepean has paid every attention to me, and has been so kind as to left us have a Cooker. He dines with me tomorrow. I am collecting you some beautiful Birds, and Land Animals, and other Curiosities for you.

The Ship remains light and Strong, and in good condition. I will write you by the *Gorgon*, man of war. She sails about a month of Six weeks time.

I am, Sirs,
Your Humble Servant
Thos. Melville.

BRITISH SOUTHERN WHALE FISHERY

VESSELS SAILING IN 1793

Vessel's Name	Master	Tonnage	Owners	Home Port	Whaling Grounds
Leviathan	Stavers	303	P. Mellish	London	Africa
Mercury	Day	240	Day & Co.	London	East Coast Africa
Mercury	Anderson	246	J. Mather & Co.	London	Africa
Eliza	Ellis	233	A. & B. Champion	London	Pacific Ocean
Sally	Ellis	171	Allen & Co.	London	East Coast Africa
Friendship	Barton	217	J. Mather & Co.	London	Africa
London	Coffin	247	A. & B. Champion	London	Pacific Ocean
Adventure	Keen	237	A. & B. Champion	London	Africa and Brazil
Good Intent	Starbuck	269	Symes & Co.	London	Pacific Ocean
Edward	Birnie	226	Terry & Co.	London	Pacific Ocean
Swift	Ray	174	Guillam & Co.	London	East Coast Africa
Herald	Wheatley	152	Curtis & Co.	London	Sealing
William	Folger	302	S. Enderby & Sons	London	New South Wales
Atlantic	Law	222	S. Enderby & Sons	London	Pacific Ocean
Kingston	Aiken	120	Baker & Co.	London	Guinea
Brothers	Flemming	260	Aspinal	London	Africa
Pilgrim	Gardner	290	Birch & Co.	London	Africa
Lively	Blythe	144	Cox & Co.	London	Brazil & South Georgia

Vessel's Name	Master	Tonnage	Owners	Home Port	Whaling Grounds
Kitty	Swain	251	Swain & Co.	London	South Georgia
Fox	Hopper	173	Lucas & Co.	London	South Georgia
Bellisarius	Anderson	336	J. Mather & Co.	London	Pacific Ocean
Alderney	Halcrow	270	Curtis & Co.	London	Patagonia
Southampton	Hart	226	Poor & Co.	S'hampton	Brazil
Fanny	Turnbull	242	Wardell & Co.	London	Pacific Ocean
Anne	Pitman	217	Wardell & Co.	London	South Georgia
Mary	Watson	241	Dunn & Co.	London	Pacific Ocean
Yorke	Gardner	333	Yorke & Co.	London	Pacific Ocean
Active	McKinly	178	Wardell & Co.	London	Brazil
Minerva	Jones	156	Jones & Co.	London	Brazil
Spy	Fitch	175	Calvert & Co.	London	Brazil
Fonthill	Daggett	266	Shodbread & Co.	London	Pacific Ocean
Friends	Gardner	320	A. &. B Champion	London	Pacific Ocean
Speedy	Melville	313	S. Enderby & Sons	London	New South Wales
Resolution	Lock	317	P. Mellish & Co.	London	New South Wales
Nelly	Luce	199	Jones & Co.	Bristol	New South Wales
Lucas	Folger	190	Lucas & Co.	London	New South Wales
Trelawney	Hillman	350	Jones & Co.	Bristol	Pacific Ocean
Minerva	Wallace	167	Bell & Co.	Hull	Brazil
Lightning	Cooke	254	Price & Co.	Bristol	Pacific Ocean
Barbara	Skiff	241	Lucas & Co.	London	Pacific Ocean
VESSELS PREPARING TO SAIL AFTER JAN. 1, 1794					
Salamander	Nicholl	303	P. Mellish & Co.	London	New South Wales
Triumph	Barton	246	J. Mather & Co.	London	Africa
New Hope	Lock	222	Yorke & Co.	London	Brazil
Mentor	Lucas	219	Lucas & Co.	London	Walwich Bay
Sparrow	Clark	147	Dowson & Co.	London	Africa
Lord Hawkesbury	Mackay	219	Bennett & Co.	London	South Georgia
Harbour Grace	Dixon	154	Joine & Co.	Bristol	Falkland Islands
Aurora	Butler	213	Teast & Co.	Bristol	South Atlantic
Joseph	Scott	189	Guillam & Co.	London	Brazil
Matty	———	——	Yorke & Co.	London	———

Recapitulation
40 Vessels Sailed in 1793
18 Vessels Sailed in 1792 not yet Returned
10 Vessels Preparing to Sail After Jan. 1, 1794
—
68 Total number in British Southern Whale Fishery

S. Enderby & Sons
London, Jan. 10, 1794
[Melville letter and Table from Chalmers Papers, Public Library of New South Wales, Sydney, Australia.]

SOUTHERN WHALE FISHE[illegible]

VESSELS RETURNING IN THE YE[illegible]

Vessel's Name	Tonnage	Master	Sperm Oil in Tuns	Wh[illegible] Oil Tu[illegible]
Emilia	268	Shields	84	—
Prince Wm. Henry	230	Swift	120	10[illegible]
Greenwich	286	Quested	143	—
Venus	297	Coffin	167	9[illegible]
Liberty	276	Buxton	200	12[illegible]
Prince of Wales	318	Bolton	125	3[illegible]
Stormont	233	Wright	109	2[illegible]
British Tar	347	Fitch	219	—
Rasper	175	Gage	88	[illegible]
Planter	304	Hales	22	12[illegible]
Lively	244	Brown	65	14[illegible]
Chaser	374	Clark	116	11[illegible]
Hercules	230	Coleman	48	[illegible]
Sydenham	152	Eckstein	—	2[illegible]
St. Andrew	119	Ingrace	—	—
Lily	171	Ellis	10	1[illegible]
Adventure	237	Keen	7	20[illegible]
Swift	172	Ray	—	10[illegible]
Arnold	58	Wheatley	—	—
Kingston	120	Aiken	—	8[illegible]
Lively	144	Blythe	—	8[illegible]
Kitty	251	Swain	—	14[illegible]
Mary	241	Watson	—	18[illegible]
Active	178	McKinstry	—	11[illegible]
Minerva	156	Jones	—	14[illegible]
Hetty	199	Luce	—	12[illegible]
Lucas	190	Folger	—	10[illegible]
Chesterfield	180	Alt	—	5[illegible]

M GREAT BRITAIN

4, WITH ACCOUNT OF CARGOES

s	Cwt. Bone	Owners	Whaling Grounds
60	—	Enderby	Coast of Peru, Pacific Ocean
—	60	Watson	Coast of Peru, Pacific Ocean
00	—	Enderby	Coast of Peru, Pacific Ocean
—	20	Champion	Coast of Peru, Pacific Ocean
—	10	Lucas	Coast of Peru, Pacific Ocean
00	20	Mather	Coast of Peru, Pacific Ocean
00	20	Wardell	Coast of Peru, Pacific Ocean
—	—	Mangles	Coast of Peru, Pacific Ocean
00	5	Enderby	Coast of Peru, Pacific Ocean
—	110	Hall	Delagoa Bay, Indian Ocean
—	60	Wilton	Coast of Peru, Pacific Ocean
—	70	Fiott	Coast of Peru, Pacific Ocean
00	—	Wilton	Coast of Peru, Pacific Ocean
00	—	Teast	Patagonia, South America
00	—	Teast	Patagonia, South America
—	—	Guillam	Cape of Good Hope, Indian Ocean
—	110	Champion	Walwich Bay, Africa, West Coast
—	60	Guillam	Brazil, South Atlantic
00	—	Curtis	Patagonia, South America
—	60	Baker	Brazil Banks, S. Atlantic
—	50	Coxe	Brazil Banks, S. Atlantic
—	100	Swain	Brazil Banks, S. Atlantic
00	—	Dunn	South Georgia Island
60	—	Wardell	South Georgia Island
—	—	Jones	South Georgia Island
—	100	Cave	Brazil Banks, S. Atlantic
—	75	Lucas	Brazil Banks, S. Atlantic
00	40	Duncan	Walwich Bay, Africa, West Coast

SOUTHERN WHALE FISH

VESSELS RETURNING IN THE Y

Vessel's Name	Ton-nage	Master	Sperm Oil in Tuns	W Oi T
Eliza	233	Ellis	145	
Rattler	317	Colnett	48	-
Friendship	218	Bunker	—	11
Triumph	246	Barton	—	14
Lord Hawkesbury	219	Mackay	8	11
Leviathan	303	Stavers	—	20
Brothers	260	Flemming	—	21
Minerva	167	Wallis	—	1

Thirty-Six Vessels Brought Home in 1794 Cargoes Value
Follows:

Sperm Oil—1,724 Tuns at £50 per tun	£86,200
Whale Oil—2,875 Tuns at £21 per tun	60,375
Whale Bone—1,640 cwt. at £5 per cwt.	9,020
Seal Skins—65,460 at 2/– each	6,546
Ambergris—125 lbs. at 5/– per oz.	312
	£162,453

[From Chalmers Papers, Public Library of New S
Wales, Sydney, Australia, and B.T. 6/95 and B.T. 6/
P.R.O., London.]

M GREAT BRITAIN (*Continued*)

., WITH ACCOUNT OF CARGOES

	Cwt. Bone	Owners	Whaling Grounds
–	—	Champion	Coast of Peru, Pacific Ocean
–	—	Enderby	Coast of Peru, Pacific Ocean
–	80	Mather	Walwich Bay, Africa, West Coast
–	100	Mather	Walwich Bay, Africa, West Coast
–	75	Bennett	Walwich Bay, Africa, West Coast
–	160	Mellish	Walwich Bay, Africa, West Coast
–	180	Aspinal	Walwich Bay, Africa, West Coast
–	75	Bell	Walwich Bay, Africa, West Coast

THE AMERICAN WHALING FLEET ENGAGED IN THE SOUTHERN WHALE FISHERY DURING THE YEARS 1817–1819

The following listings reveal how the American fleet, led by Nantucket and New Bedford, had surpassed the British fleet, thus restoring the supremacy in the Southern Whale Fishery to America.

Nantucket Fleet	78 ships, barks, brigs, schooners.
New Bedford Fleet	42 ships, barks, brigs.
Other Ports	25 ships, barks, brigs, schooners.
Total Fleet	145

A LIST OF WHALING VESSELS BELONGING TO NANTUCKET—1819

Vessel	Master	Owners	Whaling Ground
Ark (ship)	Reuben Clasby	Jethro Mitchell	Pacific Ocean
Alert (brig)	George Meader	George Myrick	Pacific Ocean
Atlas (ship)	Robert Joy	Francis Joy & Co.	Pacific Ocean
Aurora (ship)	Daniel Russell	G. Folger & P. Macy	Pacific Ocean
Atlantic (ship)	Barzillai Coffin	Paul Macy & Co.	Pacific Ocean
Boston (ship)	Frederick Barnard	Jethro Mitchell	Brazil
Barclay (ship)	Peter Coffin	Griffin Barney & Co.	Pacific Ocean
Brothers (ship)	Alexander Bunker	Obed Mitchell & Co.	Pacific Ocean
Betsy (brig)	Libni Gardner	Thomas McCleave	Brazil
Charles (ship)	Abraham Swain	J. Cartwright & Co.	Pacific Ocean
Columbus (ship)	Daniel Folger	Gardner & Folger	Pacific Ocean
Criterion (ship)	Shubael Brown	Tristram Hussey & Co.	Pacific Ocean
Diana (brig)	Calvin Bunker	Coffin & Hussey	Atlantic Ocean
Dauphin (ship)	Seth Pinkham	Gilbert Coffin & Co.	Pacific Ocean
Emeline (brig)	Joseph Plaskett	Paul Gardner & Plaskett	Brazil
Essex (ship)	George Pollard	Paul Macy & Co.	Pacific Ocean
Eagle (ship)	William H. Coffin	Thomas Starbuck & Sons	Pacific Ocean
Equator (ship)	Elisha Folger, Jr.	Barnard & Macy	Pacific Ocean
Foster (ship)	Shubael Chase	Paul Mitchell	Pacific Ocean
Factor (ship)	Wm. Fitzgerald	Reuben Baxter & Co.	Pacific Ocean
Eagle (brig)	Joseph McCleave	Reuben Baxter & Co.	Patagonia
Edward (brig)	William Paddack	George Clark	Iceland
Francis (ship)	Timothy Fitzgerald	Gideon Golger & Co.	Pacific Ocean
Franklin (ship)	Elihu Coffin	Cartwright & Folger	Pacific Ocean
Gen. Jackson (brig)	Henry Cottle	Macy & Coffin	S. Atlantic

Gideon (bark)	John R. Caswell	Reuben Baxter	Brazil
Ganges (ship)	Isaiah Ray	Gideon Gardner	Pacific Ocean
George (ship)	John Fitch	Obed Mitchell	Pacific Ocean
Globe (ship)	George W. Gardner	Christopher Mitchell	Pacific Ocean
Golden Farmer (ship)	Peter Coffin	Nath'l Rand & Co.	Pacific Ocean
George Porter (ship)	David Cottle	David Pease & Co.	Brazil
Governor Strong (ship)	Obed Fitch	Valentine Swain & Co.	Pacific Ocean
Hero (ship)	James Russell	Thomas Starbuck & Sons	Pacific Ocean
Huntress (schooner)	Chris. Burdick	Burdick & Co.	S. Atlantic
Hysco (ship)	Amiel Coffin	Zenas Coffin	Pacific Ocean
Industry (ship)	Amaziah Gardner	Valentine Swain	Brazil
Independence (ship)	George Swain	Zenas Coffin	Pacific Ocean
Independence 2d (ship)	George Barrett	Aaron Mitchell	Pacific Ocean
Improvement (ship)	Obadiah Coffin	Gilbert Coffin & Co.	Pacific Ocean
Juno (schooner)	Abraham Pollard	Mitchell & Co.	Atlantic
John Adams (ship)	Peter Paddack	Barnard & Macy	Pacific Ocean
Hesper (ship)	Reuben Joy, Jr.	Griffin Barney	Pacific Ocean
John Adams 2d (ship)	David Easton	Easton & Austin	Brazil
John Jay (ship)	John Bunker	Zenas Coffin	Pacific Ocean
Kingston (ship)	Alexander Perry	Uriah Folger & Co.	Pacific Ocean
Leander (ship)	Ariel Coffin	Gardner, Macy & Co.	Pacific Ocean
Lady Adams (ship)	Shubael Hussey	Obed Mitchell	Pacific Ocean
Leo (ship)	William Joy	Francis Joy & Son	Brazil
Lydia (ship)	Elias Ceely	Zenas Coffin & Co.	Pacific Ocean
Lima (ship)	Albert Clark	Christopher Mitchell	Pacific Ocean
Martha (ship)	Reuben Weeks	Mitchell & Gardner	Pacific Ocean
Maro (ship)	Joseph Allen	Mitchell & Gardner	Pacific Ocean
Minerva (ship)	Silvanus Coffin	Barker & Chase	Pacific Ocean
North America (ship)	Obed Wyer	Tristram Hussey & Co.	Pacific Ocean
Pacific (ship)	Benjamin Whippy	Paul Mitchell	Pacific Ocean
Paragon (ship)	William Perkins	Jenkins & Co.	Pacific Ocean
Peru (ship)	David Harris	Coffin & Macy	Pacific Ocean
Peruvian (ship)	Christopher Wyer	Christopher Mitchell	Pacific Ocean
Planter (ship)	George B. Chase	Jared & Gilbert Coffin	Pacific Ocean
President (ship)	Jonathan Swain	Paul Gardner & Sons	Pacific Ocean
Prince George (brig)	George Luce	Mitchell & Cary	Pacific Ocean
Rambler (ship)	Benjamin Worth	Aaron Mitchell	Pacific Ocean
Reaper (ship)	Jedidiah Fitch	P. Gardner & Sons	Pacific Ocean
Roxana (ship)	Francis Coffin, 3d	Peter Myrick & Co.	Brazil
Ruby (ship)	Obed Ray	Jethro Mitchell	Brazil
Sally (ship)	Thomas Paddack	Barker & Barney	Brazil
Sea Lion (ship)	Benjamin Folger	J. Jenkins & Co.	Cape of Good Hope
Samuel (ship)	Hezikiah Pinkham	Jethro Mitchell	Brazil
South America (ship)	Joseph Earl	Barney & Macy	Pacific Ocean
States (ship)	David Swain	Zenas Coffin	Pacific Ocean
Tarquin (ship)	Micajah Gardner	Aaron Mitchell	Brazil
Thomas (ship)	John Brown	S. & O. Macy	Pacific Ocean

Two Brothers (ship)	George Worth	Obed Mitchell	Pacific Ocean
Vulture (ship)	Jesse Coffin	M. Barney & Co.	Pacific Ocean
Weymouth (ship)	William Chadwick	Jonathan Mooers & Co.	Pacific Ocean
William (ship)	Obed Luce	Barker & Macy	Pacific Ocean
William & Nancy (brig)	Tristram Folger	S. & O. Macy	Patagonia
Washington (ship)	Reuben Swain 2d	Zenas Coffin	Pacific Ocean

A LIST OF WHALING VESSELS
BELONGING TO NEW BEDFORD—1819

Vessel	Master	Owners	Whaling Ground
George & Susan (ship)	G. Randall	G. & J. J. Howland	Patagonia
Ann Alexander (ship)	L. Snow	George Howland	
Mary (brig)	B. Howland	Wm. Rotch & Sons	S. Atlantic Ocean
Martha (ship)	S. West	Seth Russell & Sons	Brazil
Milwood (ship)	D. Wilcox	Seth Russell & Sons	S. Atlantic Ocean
Maria (ship)	W. Swain	Samuel Rodman	Pacific Ocean
Orion (brig)	G. Tobey	Wm. Rotch & Sons	Delagoa Bay
President (brig)	S. Clark	Wm. Rotch & Sons	Cape de Verdes
Phebe Ann (ship)	T. Covill	George Howland	Pacific Ocean
Richmond (ship)	C. Earle	Isaac Howland, Jr.	Patagonia
William & Eliza (ship)	W. Randall	Wm. Rotch & Sons	Pacific Ocean
Winslow (ship)	B. Chase	Wm. Rotch & Sons	Pacific Ocean
Wm. Thatcher (ship)	H. Tucker	Wm. Rotch & Sons	S. Atlantic Ocean
Augustus (ship)	S. Butler	S. Russell & Sons	Patagonia
Barclay (ship)	H. Coffin	Wm. Rotch & Sons	Pacific Ocean
Balaena (ship)	Edmund Gardner	Wm. Rotch & Sons	Pacific Ocean
Commodore Decatur (brig)	W. Tucker	G. & J. Howland	Patagonia
Charles (ship)	B. Coffin	Samuel Rodman	Pacific Ocean
Gleaner (brig)	D. Leslie	John A. Parker	Patagonia
Golconda (ship)	F. Bennett	George Howland	Patagonia
Independence (ship)	J. Perry	Wm. Rotch & Sons	Pacific Ocean
Juno (brig)	—. Spooner	Seth Russell	Brazil
Mercator (ship)	J. Swain	Seth Russell	Pacific Ocean
Ospray (ship)	J. Drew	Wm. Rotch & Sons	Pacific Ocean
Persia (ship)	T. Cross	G. Grinnell, Jr.	Pacific Ocean
Pindus (ship)	G. Barrett	———	Pacific Ocean
Russell (ship)	—. Arthur	Seth Russell & Sons	Pacific Ocean
Triton (ship)	Z. Wood	G. & J. Howland	Pacific Ocean
Victory (ship)	R. Bunker	Wm. Rotch & Sons	Patagonia
Alliance (brig)	W. Butler	———	Brazil
Cornelia (brig)	A. Gardner	G. & J. Howland	Patagonia
Dragon (brig)	F. Chadwick	J. A. Parker	Patagonia

Francis (ship)	E. Howland	Wm. Rotch & Sons	Brazil
Iris (ship)	G. Hathaway	T. S. & N. Hathaway	Pacific Ocean
Midas (ship)	—. Williams	John Coggeshall	Patagonia
Pacific (ship)	—. West	T. Hazard	Patagonia
Swift (ship)	—. Price	Wm. Rotch & Sons	Pacific Ocean
Timoleon (ship)	G. Randall	I. Howland, Jr.	Brazil
Lorenzo (ship)	—. Coffin	———	Brazil
Minerva Smyth (ship)	D. McKenzie	I. Howland, Jr.	Pacific Ocean
Parnassa (ship)	—. Covell	Samuel Rodman	Pacific Ocean
Russell (ship)	J. Arthur	Seth Russell	Pacific Ocean

Index

Active (ship), 286
Acts for Encouragement of The Southern Whale Fishery: 26 Geo. III, Cap. 50, 82, 87, 115–116; 28 Geo. III, Cap. 20, 116; 42 Geo. III, Cap. 77, 194
Acushnet River, 171–172, 249, 251
Adams, John, 9, 10, 13, 17, 19, 22, 23, 24, 25, 29, 41, 136, 138, 270
Adams, Samuel, 41
Adventure (ship), 89, 93, 95, 119, 154, 276
Africa (East Coast), 51, 113, 130, 162
Africa (West Coast), 113, 130, 162
Africa (ship), 94
Aicken, James, 308
Aiken, Capt. John, 95, 118
Aiken, Capt. Thomas, 398
Aiken, Capt. William, 96
Alaska (Russian), 147
Albemarle Island, 296, 344, 349
Albion (ship), 189, 190
Alexander (ship), 316
Alfred (ship), 193
Algerian Corsairs, 101
Allen, Capt. Joseph, 279, 362, 367
Alliance (ship), 359
Alligator (ship), 320
Alsop, John, 29
Alt, Capt. Matthew B., 186
Ambergris, 114–115
America (ship), 145
American Whale Fisheries, 30, 113, 137, 317, 371
American Hero (ship), 281
Amherst, Lord, 77
Amsterdam, Holland, 1, 29, 97
Amsterdam Island, 152, 153
Anglo-French War, 263–266, 285
Anglo-Spanish Convention, 150, 152, 154
Ann (ship), 108, 142, 162, 174, 237, 238, 242, 255–256, 316
Ann, 2nd (ship), 256
Ann Alexander (ship), 322–323
Ann Delicia (ship), 93
Ann & Sally (ship), 265
Anna Josepha (ship), 271, 272
Antigua, West Indies, 286
Antipodes Islands, 305, 306
Apollo (ship), 371
Apprentices, Whaling, 131
Arbuthnot, Admiral, 50
Archimedes (ship), 376
Ardent (ship), 170, 291
Argo (ship), 90, 206, 293–297
Argonaut (ship), 155
Ark (ship), 355
Arnold, James, 260
Arret, French, 133–134
Articles of Agreement, 279–280
"Artificers," 219
Asia (ship), 230
Assembly, French (National), 159
Assembly, Nova Scotia, 35, 88
Astrea (ship), 118
Atlantic (ship), 129, 130, 179, 266, 267, 326, 339, 344, 345, 347
Atlas (ship), 291–292, 300, 326, 328, 331, 339, 345

Attempt (ship), 96
Auburn, New York, 355
Auckland Islands, 312
Audacious (ship), 96
Aurora (ship), London, 96
Aurora (ship), Milford Haven, 218, 242, 279
Australia, 118, 155, 174, 175, 176–194, 238, 242, 299–313

Bailey Islands, 379
Bainbridge, Commodore, 335, 336
Baleine (ship), 171
Baltic, 375
Baltimore, Maryland, 4
Baltimore Whig, 323
Barbara (ship), 154
Barclay, Robert, 6, 40, 59, 233
Barclay (ship), 162, 233, 254, 340, 342, 343–348, 371, 376
Barker, Eliza (ship), 371
Barker, Elizabeth, 107, 143
Barker, Josiah, 27, 143
Barnard, Capt. Andrew, 378
Barnard, Capt. Frederick, 404
Barnard, Capt. Hezikiah, 378
Barnard, Capt. James, 327
Barnard, Capt. Jonathan, 204
Barnard, Capt. Libni, 204
Barnard, Capt. Obed, 90, 387
Barnard, Peter, 30
Barnard, Capt. William, 316
Barnard, Capt. Zaccheus, 341
Barney, Capt. Griffin, 254, 378
Barney, Capt. James, 255
Barney, Capt. Jonathan, 316
Barney, Capt. William, 129, 316
Barrallier, Jean-Louis, 198, 225, 236, 246
Barrett, Capt. Francis, 93
Barrington, Nova Scotia, 28, 188
Barton, Capt. H., 95
Bass, Capt. George, 312
Bass Straits, 303, 310
Bassett, Capt. John, 129, 130
Baudin, Capt. Nicholas, 303
Baxter, Capt. David, 129, 344–345, 377
Baxter, Captain (Sir) Francis, 320, 377
Beale, Dr. Thomas, 374
Beaufoy, Harry, 59, 61
Beaver (brig), 38, 130
Beaver (ship), 266, 271
Bedford, Massachusetts, 92, 104, 130, 145, 172, 174, 213, 229, 231, 232, 233, 237–238, 256, 282
Bedford (ship), 10, 12, 80, 101, 110, 145, 169
Beechy, Capt. W. F., 379
Belvedere (ship), 277
Bengal, India, 179, 186
Benjamin (ship), 230
Bennett, Daniel, 94, 95, 186, 303
Bennett Island, 94
Betsy (ship), 266, 271
Black Princess (privateer), 188
Bligh, Lieut. William, 192
Blonde, H.M.S., 364
Blossom, H.M.S., 379
Bly, Isaac, 340, 342–343
Board of Trade, British, 18, 19, 21, 25, 54, 61, 62, 69, 76–77, 79, 81, 82, 114, 120, 123, 130, 141, 146, 174, 197, 198, 201, 208, 211
Bolton, Capt. F., 95
Bon Homme Richard, U.S.N., 146
Bonin Islands, 379
Bordeaux, France, 101, 145
Borrowdale (ship), 119
Boston, Massachusetts, 4, 6, 31, 42, 92, 115, 130, 149
Botany Bay, 118, 119, 176, 303
Bounties, Whaling (British), 18, 25, 51–57, 71–73, 77–78
Bounties, Whaling (French), 140, 174
Bounties, Whaling (United States), 31, 42, 58

Bounty, H.M.S., 177, 192
Bourbon (ship), 376
Bowdoin, James, 25
Bowen, Lieut., 190, 191, 311
Boyd, Julian, 134, 137, 140
Boylston, Thomas, 45, 115, 138, 145
Brazil Banks, 9, 10, 17, 41, 88, 90, 92, 101, 113, 116, 118, 162
Breck, Samuel, 32
Bridgeport, England, 44
Bristol, England, 6, 44, 45, 73, 92, 125, 186, 378, 379
Bristow, Captain Abraham, 312–313
Britain, Great, 13, 18, 21, 24, 31, 32, 59, 61, 75, 78, 108, 109, 120, 174
Britannia (Enderby ship), 177, 180, 181–183, 187
Britannia (Nantucket ship), 8
Britannia (St. Barbe ship), 183–185
Britannia (Salem ship), 371
British Ministry, 41
Brock, Capt. Andrew, 242
Brock, Capt. Thomas, 130
Broken Bay, Australia, 189, 242
Brook, Watson and Co., 145
Brothers (ship), 57, 188, 274, 302, 316
Brothers, 2nd (ship), 319
Broughton, Lieut. W. R., 179
Brown, Capt. R., 95
Brown, Capt. Thomas, 41
Brutus (ship), 170
Buenos Aires, 329, 346
Bunker, Alfred, 359
Bunker, Capt. Christopher, 290
Bunker, Cromwell, 256
Bunker, David, 30
Bunker, Capt. Eber, 118, 176, 181–183, 187–195, 242, 305, 321
Bunker, Capt. George, 130, 271
Bunker, Isaac, 188
Bunker, Capt. Isaiah, 230
Bunker, Capt. James, 188, 376
Bunker, Col. Laurence E., 194
Bunker, Capt. Obed, 230, 291
Bunker, Capt. Owen, 57, 93, 188, 301, 306
Bunker, Capt. Peleg, 170, 291
Bunker, Capt. Simeon, 188
Bunker, Thaddeus, 306
Bunker, Capt. Tristram, 129, 291, 306
Bunker, Capt. Uriah, 275
Bunker, Capt. Uriel, 243, 276, 320
Bunker's Group (Islands), 190
Burke, Edmund, 6
Buxton and Enderby, 28

Caesar (ship), 281
Cahoon, Capt. Isaiah, 229
Calcutta, India, 179
California, 156, 276
Callao, Peru, 263, 272, 292, 339
Calonne, M. de, 98, 138–139
Calvert and Co., 176
Calvery and Co., 186
Campbell and Co., 191
Canton, China, 147, 148, 186, 300–301
Canton (ship), 57, 100, 121, 145, 154, 170, 172, 176, 300, 306, 316, 355
Cape Corrientes, 267
Cape of Good Hope, 76, 78, 79, 89, 116, 162, 180, 212
Cape Horn, 76, 78, 79, 89, 114, 116, 123, 143, 147, 149, 150, 155, 156, 265, 273, 275, 328, 336
Cape St. Vincent, 257, 264
Cape Town, 113, 176, 183, 265
Carmarthen, Francis, Marquis, 62, 77, 121
Carnot, M. de, 104
Carrier (ship), 371
Cary, Capt. Nathaniel, 146, 361
Carysford, H.M.S., 253
Castle Hall, Milford, 241, 244–245
Castle Pill, Milford, 198, 203

Castre, Marshall de, 98
Cathcart, Capt. Seth, 376
Catherine (ship), 341, 346, 347
Cereno, Don Benito, 312
Chardon, M., 137, 140
Charles (ship) (Rotch), 171, 243, 276, 331, 332, 339
Charles (ship) (British), 366
Charles Island, 295, 343–344
Charles Street, Milford, 225
Charles & Henry (ship), 223
Charleston, South Carolina, 4
Charlton (ship), 279, 297, 341, 344, 348, 349
Chadwick, Capt. John, 12, 205
Chadwick, Capt. Jonathan, 90
Chadwick, Capt. William, 92
Chalmers, George, 81, 112, 123, 207, 211
Champion, Alexander, 16, 17, 52, 56, 89, 93, 95, 103, 114, 118
Champion, Benjamin, 16, 89, 93, 96, 103, 118, 153, 156, 186, 188
Champion, Richard, 6
Champion and Dickason, 6, 8, 40, 75, 105, 255
Champlin, Christopher, 102
Chance (ship), 110
Chase, Capt, Joseph, 146
Chase, Capt. Latham, 90
Chase, Capt. Peter, 307–310
Chase, Lieut. Reuben, 146
Chase, Capt. Samuel R., 312
Chase, Capt. Shubael, 292
Chaser (ship), 154
Chatham, H.M.S., 179
Chatham Island (Galapagos), 157, 266
Chatham, Lady, 82
Chathams Island, 185
Cherub, H.M.S., 346, 350
Chesterfield (ship), 185
Chile, 263, 337–338
Chili (ship), 327, 331
China Trade (Furs), 148, 154, 183
Claghorn, George, 233
Clark, Capt. Albert, 327
Clark, Capt. Benjamin, 130
Clark, Capt. Edward, 243, 297
Clark, Capt. Elisha, 376
Clark, Capt. Obed, 325
Clark, Capt. Thomas, 154
Clark, Capt. Tristram, 95, 96
Clark, Capt. William, 327, 330
Clasby, Capt. Joseph, 243
Clasby, Capt. R., 327
Cloudy Bay, New Zealand, 278
Clousier, Jacques and Co., 136
Coast of Peru, 170, 181, 269, 271, 277, 325
Cochran, Thomas, 33
Coffin, Capt. Alexander, 15, 29, 30, 41, 268–271
Coffin, Capt. Amiel, 405
Coffin, Capt. Bartlett, 15, 111, 112
Coffin, Capt. Benjamin, 406
Coffin, Capt. Brown, 386
Coffin, Capt. Daniel, 42, 57, 89, 93, 118
Coffin, Capt. David, 292, 295
Coffin, Capt. Edward, 229
Coffin, Capt. Francis, 92
Coffin, Capt. Frederick, 202, 222, 292, 362
Coffin, Capt. Hezikiah, 38
Coffin, Admiral Sir Isaac, 320
Coffin, Capt. James Josiah, 378, 379
Coffin, Jared, 30
Coffin, Capt. Jonathan, 90, 205
Coffin, Capt. Joshua, 93, 114, 115, 129, 387
Coffin, Capt. Josiah, 30
Coffin, Capt. Laban, 145, 169
Coffin, Capt. Micajah, 145, 146, 315, 316–317
Coffin, Capt. Paul, 93, 145, 317
Coffin, Capt. Richard, 15
Coffin, Capt. Shubael, 145

Coffin, Capt. Shubael, Jr., 315
Coffin, Capt. Solomon, 274, 326
Coffin, Capt. Thaddeus, 104, 108, 145
Coffin, Capt. Thomas, 145
Coffin, Capt. Uriah, 25
Coffin, Capt. Zebdial, 286
Coffin, Capt. Zenas, 146, 316–317, 319, 375
Coffyn, François, 97, 98, 104, 108
Coleman, Capt. Andrew, 386
Coleman, Capt. Benjamin, 153, 230, 316
Coleman, Capt. David, 205
Coleman, Capt. John B., 223
Coleman, Capt. Prince, 108, 129, 142, 237, 255, 297
Coleman, Capt. Seth, 205
Coleman, Seth, 223
Coleman, Capt. Stephen, 154
Coles, Willet, 307
Collingwood, Lord, 188, 323
Colnett, Capt. James, 155, 156–157
Colonial Whaling, 1, 5, 315
Columbia (ship), 149, 356–357
Comet (ship), 327–330
Commerce (ship), 265, 272
Commercial Treaty, Anglo-French, 133
Commissioners of His Majesty's Customs, 12, 55
Committee For The Consideration of all Matters Relating to Trade and Foreign Plantations (*see* Board of Trade)
Committee of South Sea Whalers From London, 131, 201
Commonwealth of Massachusetts, 1, 4, 5, 23
Congress of the United States, 23, 24
Constitution of the United States, 171, 261, 322
Continental Congress, 4, 10, 29
Cook, Capt. James, 81, 113, 121, 147, 155
Cooper, Dr. William, 39
Coquimbo, Chile, 276, 340
Corn Bread (American), 245
Cornwall (ship), 271
Cornwallis, Lord Charles, 179
Cotterel, Stephen, 61, 62, 63
Cottle, Capt. Obed, 277, 316
Cowes, Isle of Wight, 73
Criterion (ship), 302, 308–310, 327, 330
Crozet Islands, 305
Cumberland, Richard, 33, 35, 88, 219
Cumberland (ship), 355
Curling, Robert and Co., 96
Curtis, Timothy, 94, 118
Curtis, William, 94, 118, 151, 152
Curtis and Co., 93, 186
Custom House, London, 58, 125
Customs Commissioners, London, 45, 51, 71, 75, 78, 82
Cutting, John Brown, 138
Cyrus (ship), 170, 273–279, 297, 344, 355

Daedalus (store ship), 186
Daggett, Capt. Jethro, 96, 153
Daggett, Capt. Thomas, 129
Dartmouth, England, 44
Dartmouth, Massachusetts, 250, 252
Dartmouth, Nova Scotia, 32, 34, 42, 43, 53, 87, 88, 89, 90, 92, 136, 188, 197, 201–208, 221–223, 380, 385–386
Dartmouth (brig), 252
Dauphin (ship), 103, 104, 145, 230, 231, 255
Dauphin, 2nd (ship), 320
Dayton, Hezidiah, 29
De Bauque Freres, 101, 102, 103, 110, 145, 171, 228, 255
De Castre, Marshall, 100
Deal, England, 128
Delagoa Bay, 113, 130, 162, 174, 230, 269, 274, 291

Delano, Capt. Abishai, 89, 92, 288, 356–357
Delano, Capt. Amasa, 311–312
Delano, Capt. Henry, 41, 57, 92, 130, 356–357
Delano, Capt. Thomas, Sr., 356
Delano, Capt. Thomas, Jr., 93, 288, 356–357
Delaware (ship), 255
Denby, Lord, 77
Derwent River, 190, 192, 309
Desolation Island (Kerguelen), 186, 230
Diana (ship), 45, 145, 162, 255, 281, 301, 344, 371
Dickason, Thomas, 16, 17, 171
Dieppe, France, 171
Digby, Admiral, 50
Directory, French, 282, 285, 290
Discovery (British ship), 179
Discovery (Spanish ship), 186
Dixon, Capt. George, 148, 149
"Doctors' Commons," 94
Dolphin (ship), 93, 96
Dominique, W.I., 323
Dorr, Capt. Ebenezer, 301
Dover, England, 105
Downes, Lieut. John, 338, 340, 345, 346–349
Downing Street, 20
Duke of Kent (ship), 243
Duke of Portland (ship), 305
Duke of York (ship), 120
Dundas, Henry, 2, 6, 76, 77, 81, 116, 154
Dunkirk, France, 83, 97, 98–104, 106–111, 130, 133–146, 159, 162–174, 188, 203, 207, 213, 226–235, 237, 243, 254, 255, 256, 282, 283, 290, 316–317, 371, 377, 380
Durfy, Elijah, 130
Dusky Bay, New Zealand, 182, 183, 185
Duty on Whale Oil, British, 4, 13, 34, 382
Duty on Whale Oil, Colonial, 34, 87, 92
Duty on Whale Oil, French, 108, 146
Duty on Whale Oil, United States, 105

Eagle (brig), 288
Earl St. Vincent (ship), 375
East India Company, 76, 78, 79–85, 89, 116, 119, 148, 152, 154, 175, 176, 179, 183, 187
East Indies, 368
Edward (ship), 107, 142, 230, 231, 348
Edwards, Capt. Edward, 177
El Plumier (ship), 271
Eleanor (brig), 149
Eliza (ship), 230, 255
Elizabeth (ship), 154, 191–192, 242
Elizabeth & Margaret (ship), 122
Elkins, John, 315
Elligood (ship), 303
Elliott, Grey, 67–69
Ellis, Capt. Reuben, 96
Emilia (ship), 93, 123–129, 143–144, 281, 355
Emily (ship), 366–370
Endeavor (ship), 312
Enderby, Charles, 15, 16
Enderby, George, 16
Enderby, Samuel, Sr., 9, 52, 56, 115, 146
Enderby, Samuel, Jr., 16, 54, 112, 123, 147, 153, 154
Enderby and Sons, 6, 8, 9, 15, 16, 40, 45, 48, 49, 57, 73, 75, 89, 92, 93, 95, 105, 122, 123–125, 129, 146, 152, 155, 156, 180, 186, 200, 312–313, 380
Endymion, H.M.S., 351
England, 1, 25, 28
English Channel, 45, 105

"Enquiries," to Gov. Parr, 33
Enterprise (ship), 193, 281
Equator, 116, 154
Essex (U.S. frigate), 193, 330, 335–350, 355
Estates General, French, 142
Etches, John, 148
Etches, Richard C., 120, 148
Etches, William, 148
Euphemia (ship), 271
Euphrates (ship), 279
Europe, Whale Oil Market, 22, 39, 100
Experiment, H.M.S., 7
Experiment (ship), 95, 96
Eyre, Admiral George, 363

Fairy (snow), 301, 303
Falkland Islands, 8, 16, 25, 51, 62, 74, 95, 100, 101, 102, 113, 122, 129, 273, 275
Falkland (ship), 5, 7, 142, 144, 162, 290, 291
Falmouth, England, 44, 121, 153, 183
Fame (ship), 162, 276, 319, 375
Fanning, Capt. Edmund, 306–307, 311–312
Fanny (ship), 129, 376
Fanshaw, Captain, 253
Farragut, David Glasgow, 347–348
Farrar, Eliza (Rotch), 107, 172, 173, 228, 235, 237
"Father of Australian Whaling," 194
Favorite (ship), 142, 145, 169, 229, 255, 355
Favourite (ship), London, 279, 280
Favourite (ship), Nantucket, 305–307
Federal George (ship), 287
Federalists, 285
Fenn, Cecilia, 306
Fenn, Mary, 306
Ferguson, Capt. D., 95
Fiji Islands, 304, 305, 308
Fisher, Kidd and South Islands, 379
Fisher's Island, 379
Fishery Bill (British), 66
Fitch, Capt. Jedidiah, 273, 356
Fitch, Capt. Obed, 142, 286
Fitch, Capt. Timothy, 93
Fitzgerald, Nathaniel, 326, 327, 329
Flinders, Capt. M., 190, 303
Fly (sealer), 312
Folger, Abiel Coleman, 198, 226, 246
Folger, Capt. Albert, 373
Folger, Capt. Christopher, 288–290
Folger, Capt. David, 142, 169, 229
Folger, Capt. Elisha, 108, 297
Folger, Capt. George, 45, 145
Folger, Capt. Henry, 96, 293
Folger, Capt. Mayhew, 185
Folger, Capt. Noah P., 361–362
Folger, Capt. Obed, 290
Folger, Capt. Seth, 130, 170
Folger, Capt. Solomon, 326
Folger, Capt. Thaddeus, 279
Folger, Capt. Thomas, 242, 272
Folger, Timothy, 27, 32, 34, 42, 88, 198, 205, 208, 214
Folger, Timothy, Jr., 205, 208, 218, 220, 223, 224, 226
Folger, Capt. Tristram, 90
Folger, Walter, Jr., 325
Folger, Capt. William, 92, 154, 186
Fonthill (ship), 154
Foster, John, 204
Foveaux Straits, 305
Fowey, England, 44
Fox, Charles James, 151, 152
Fox (ship), 93, 95, 170, 287–288
France, 20, 31, 66, 83, 97, 108, 142, 150, 162, 204
France, Ministry of, 98, 100
Francis (sloop), 185
Franklin, Benjamin, 98, 108
Frazier, Capt. Ian, 279

Frederick (ship), 193, 323
French Government, 109, 133–146, 159, 286
French Republic, 226
French Spoilation Claims, 287
Friends Meeting House, Milford Haven, 232, 246
Friends Meeting House, Nantucket, 107
Friendship (ship), 89, 92

Gage, Capt. Matthew, 92, 95
Gage, Capt. Thomas, 57, 95, 96, 152
Galapagos Islands, 156, 267, 277, 291, 295, 296, 298, 342, 343–346
Gamble, Lieut. John, 348
Ganges (ship), 170, 290, 371
Gardner, Capt. Amaziah, 229, 272
Gardner, Capt. Andrew, 288
Gardner, Capt. Calvin, 360
Gardner, Capt. Charles, 293
Gardner, Capt. Francis, 96, 154
Gardner, Capt. George W., 277
Gardner, Capt. Gideon, 275
Gardner, Capt. Grafton, 332, 339
Gardner, Capt. Henry, 360
Gardner, Capt. Jared, 255, 280, 301
Gardner, Capt. Micajah, 142, 145, 230–231, 238, 242, 279, 280, 286, 293, 361
Gardner, Capt. Paul, 308–309
Gardner, Paul, and Co., 145, 305
Gardner, Reuben, 168, 276
Gardner, Capt. Shubael, 97, 98, 142, 168
Gardner, Capt. Silas, 288
Gardner, Capt. Stephen, 90, 230, 255
Gardner, Capt. Tristram, 110
Gardner (ship), 327, 331, 376
Gelston, Cotton, 30
General Boyd (brig), 301
General Court of Massachusetts, 9, 21, 27, 30
Geographe, Le (ship), 303
George III, King, 150, 234, 235
George (ship), 327, 331, 338, 339, 361
George & Albert (ship), 376
Georgiana (ship), 280, 345–346, 348
Gibraltar, 257, 259, 323
Gilbert, T., 95, 96
Globe (ship), 371
Glover, Capt. Benjamin, 226
Goldsmith, Capt. William, 57
Goliath, H.M.S., 266
Good Intent (ship), 95, 154, 169
Gracechurch Street, London, 40, 61
Grand Sachem (ship), 242, 281, 355
Grant, Lieut. James, 272, 273
Great Britain, 87, 179, 204, 211
Greenland (Northern) Whaling, 6, 18, 51, 52, 54, 78, 89, 90, 94, 113, 115
Greenwich, England, 10, 37
Greenwich (ship), 92, 129, 130, 267, 341, 346, 348
Grenville, Lord, 62, 116, 154, 174, 201
Greville, Charles F., 84, 197, 199–214, 217–218, 235–236, 240, 245, 246, 380
Greville, Robert Fulke, 247
Greyhound (ship), 146, 230, 291
Grieve, David, 205
Grose, Lieut. Gov. Francis, 183, 185
Grosvenor Square, 20
Guadaloupe Island, 286, 287, 323
Guayaquil, Ecuador, 127, 343, 348
Guernsey Lily (ship), 96
Guillaume, T., 95, 96
Gwinn, Capt. James, 41, 237, 238, 242, 320, 356

Haden, Capt. Zophar, 365
Haiti, 230
Halcyon (ship), 301
Hales, Capt. George, 301
Halifax, Nova Scotia, 32, 33, 42, 92, 136

Hall, John, 96
Hall and Downing, 95, 120
Hamilton, Lady, 197, 198, 235, 244, 245
Hamilton, Sir William, 197–199, 203, 235, 245
Hammond, Capt. Archaelus, 125, 127, 143, 144, 170, 273, 281, 353, 354, 376
Hannah & Eliza (ship), 238, 242, 272, 279, 280, 292, 302
Hannah & Sally (ship), 310
Harlequin (ship), 229, 265, 276
Harmony (ship), 144, 168, 170, 229
Harpooner (ship), 154, 171
Harpooners (head-men), 40, 53, 73
Harrax, Capt. Charles, 170
Harriet (ship), 308
Harrington (brig), 192
Hasselburgh, Capt. Frederick, 312
Hathaway, T. S. and N., 260
Haverfordwest, Wales, 244
Hawaiian Feathered Cape, 364
Hawaiian Islands, 361, 362, 363
Hawes, Capt. John, 117, 169, 228, 230
Hawk (brigantine), 146
Hawkesbury, Lord (Charles Jenkinson), 21, 62, 82, 83, 89, 98, 112, 116, 120, 122, 130, 143, 146, 147, 149, 152, 154, 186, 198, 200, 204, 207, 211, 213, 225, 260–261, 380, 381
Hayden, Capt. Abishai, 110, 142
Hayden, Capt. Zophar, 103
Hayley, George, 6, 7, 8, 12, 17, 40, 94
Hayley, Madame, 8, 12, 24, 101, 102, 145
Hazard, Thomas, Jr., 260, 288
Head Matter (Spermacetti), 101, 113, 115
Hebe (ship), 170, 229
Hector (ship), 130, 255, 344, 346, 347
Hercules (ship), 154, 355
Hero (ship), Le Havre, 171
Hero (ship), London, 92, 93, 154, 229
Hero (ship), Nantucket, 265, 367
Hero (ship), Peruvian, 297
Heywood, Capt. Peter, 346
Hibernia (brig), 92
Hillman, Capt. R., 95, 96
Hillyar, Capt. James, 346
Holland, 6, 17, 150
Holmes, Benjamin F., 205–206
Homberg and Homberg, Freres, 146, 162
Honduras Packet (ship), 301, 302
"Honorable John Company," 81
Hood Island, 343
Hope (ship), merchantman, 186, 301
Hope (ship), whaler, 130, 142, 143–144, 154, 168, 291, 376
Hopper, Capt. J., 95, 122
House of Commons, 5, 76, 151, 382
House of Lords, 152
Howland, Cornelius, 287
Howland, Isaac and Co., 260
Hudson, New York, 6, 28, 29–30, 42, 92, 130, 268
Hudson River, 6, 29
Hudson (ship), 145, 275
Hull, England, 186
Humphreys, David, 257
Hunter, Capt. John, 189, 302, 303
Hussey, Rev. Alfred Rodman, 235
Hussey, Capt. Ammiel, 243
Hussey, Capt. Barzillai, 255
Hussey, Capt. Benjamin, 25, 102, 103, 169, 171, 228, 233, 279, 282, 283
Hussey, Capt. Ebenezer, 170
Hussey, Capt. Isaiah, 168, 230
Hussey, Joseph, 255
Hussey, Phebe, 279
Hydra (ship), 366, 367

Imbert, Lewis, 377
Impregnable, H.M.S., 327

Independence (ship), 304–306, 320
India, 149, 185
India Board, 116, 154
Indian Ocean, 113, 152, 176
Indispensable (ship), 327
Industrial Revolution (British), 18
Industry (brig), 92, 206
Industry (ship), 10, 12, 13
Integrity (cutter), 311
Intrepid (British sealer), 120
Intrepid (Spanish ship), 186
Irish, Capt. W., 186
Isle of Wight, 157

Jacmel, Haiti, 323
Jaffrey, Patrick, 102, 145
Jane Maria (brig), 312
Janus (ship), 255
Japan Grounds, 362, 367
Jarvis, Leonard, 7
Jasper (schooner), 92
Jay, John, 24, 32, 108, 135, 139, 225, 285
Jay Treaty, 285
Jefferson, Thomas, 9, 13, 32, 108, 109, 133–141, 225, 320
Jefferson (ship), 242
Jenkins, Capt. David, 323
Jenkins, Marshall, 29
Jenkins, Capt. Seth, 5, 6, 29
Jenkins, Capt. Thomas, 29
Jenkinson, Charles (Lord Hawkesbury), 21, 22, 50, 54, 62, 63, 64–70, 75, 76, 77–85, 116, 141, 236, 380
Johanna (ship), 286
John Jay (ship), 277, 316, 317–319
John & James (ship), 327
John & Susannah (ship), 153
Johnson, Capt. Robert, 312
Jones, Capt. John Paul, 8, 146
Jones, Capt. Ransom, 93, 94, 95
Jones, Capt. Silas, 145, 177, 256, 355
Johnny-Cake Hill, 260
Joseph (brig), 206
Joy, Capt. David, 279
Joy, Capt. Francis and Co., 315
Joy, Capt. Obed, 292, 326, 328, 331
Joy, Capt. Reuben, 289, 291–292
Joy, Capt. Samuel, 292
Juan Fernandez Island, 123, 126
Jubilee, H.M.S., 266
Judith (ship), 168, 229
Juno (ship), 281

Kamchatka, 147
Kamehameha, King, 362, 363–364
Kangaroo Island, 303, 304
Keen, Capt. Shadrach, 154
Keith, Capt. James, 279, 280
Kelley, Capt. Daniel, 90
Kempton, Capt. Josiah, 255
Kent (ship), 57, 89, 93, 119, 290, 374
King, Gov. Philip G., 182, 190, 300, 302, 310, 311
King George (cutter), 312
King George Sound Co., 148
Kingston (ship), 129
Kitty (ship), 134, 271

Lady Adams (ship), 297
Lady Gage (ship), 93
Lady Penrhyn (ship), 118, 119
L'Aigle (ship), 362–363
Lady Nelson (ship), 192, 272
Lafayette, Marquis de, 134, 135, 138
Lambert, M., 133, 134
La Perouse, Comte de, 148, 180
Lastre, Don Francisco, 338
Lay System, 17, 77, 156
Leander, H.M.S., 242
Leard, Capt. John, 120
Le Havre, France, 146, 162, 170, 256, 282, 290, 303, 361, 376
Leith, William, 184, 185
LeMesurier and Secretan, 96
Leo (ship), 276, 277, 279, 316, 327
Liberty (ship), 95, 96, 129
Licenses, East India Co., 79

Licenses, South Seas Co., 115, 116
Lighting, Whale Oil, 18
Limehouse Hole, 274
Lima, Peru, 127, 263, 280, 292, 312, 328, 343
Lima (ship), 277, 293, 327, 331
L'Importune (frigate), 287
Liverpool, England, 73, 186
Lively (ship), 90
Livingston, Robert R., 271, 286
Lloyds of London, 17, 18
Lobos Islands, 127, 129, 277
Lock, Capt. Abijah, 154
Lock, Capt. John, 92, 129–130, 186, 316
London Assurance Co., 103
"London Club," 281, 332
London, England, 1, 5, 7, 10, 13, 19, 24, 27, 31, 35, 41, 42, 48, 54, 58, 73, 90, 92, 105, 113, 130, 141, 146, 171, 176, 188, 377
London Whale Oil Merchants, 6–9, 15, 18, 40, 51, 52, 58, 87, 89, 92, 93, 105, 204, 209, 210, 213
London (ship), 93
Long, Capt. Peleg, 96
Long, Capt. Simeon, 330
Looe, England, 44
Lopez, Aaron, 7
Lord, Simeon, 272, 305, 306–310
Lord Hawkesbury (ship), 93, 144, 168, 357
Lord Nelson Hotel, Milford, 236
L'Orient, France, 171
L'Orient (flagship), 244
Louis XVI, King, 142, 159
Lovejoy, Capt. John, 96
Loyalists, American, 4, 5, 22, 23, 29, 32, 60, 96, 223
Luzerne, M. de, 133–134, 137
Lucas, Isaac and Co., 94, 95
Lucas and Spencer, 93, 118, 186
Lucas (ship), 93, 95, 114
Lucretia (ship), 33, 90
Lydia (ship), Coffin, 317–318
Lydia (ship), London, 266
Lydia (ship), Rotch, 130, 146, 255, 281, 375
Lyme, England, 44
Lynx (sealer), 312

Macquarie, Gov., 193–194
Macquarie (ship), 193
Macy, Capt. Alexander, 361, 376
Macy, Capt. Edmund, 204
Macy, Capt. Francis, 170
Macy, Capt. John, 327, 344
Macy, Nathaniel, 304
Macy, Obed, 10, 306, 333, 354
Macy, Capt. Reuben, 170
Madagascar, 93, 119, 162
Madison, President James, 324, 350
Madrid, Spain, 122, 257
Maitland (ship), 243, 276
Malaspina, Don Alexandro, 186
Manila, Philippines, 193
Maria (ship), 35, 39, 40, 74, 100, 108, 110–112, 121, 145, 162, 277, 292, 295, 371
Maria & Eliza (ship), 142
Maro (ship), 362, 367
Marquesas Islands, 154, 193
Marsden, Rev. Samuel, 186
Marshall, Capt. Samuel, 154
Marsillac, John, 163–165
Martha (ship), 372
Martha's Vineyard, Massachusetts, 154
Martinez, Don Estaban José, 149
Mary (ship), Rotch, 104, 145, 174, 227, 371
Mary (ship), British, 325
Mary (brigantine), 254
Mary Ann (ship), 176, 181, 182, 327
Massachusetts, 1, 4, 5, 28, 31, 42, 105
Massachusetts (ship), 376
Massafuero Island, 123, 126, 341

Mather, James, 156, 186
Matilda (ship), 176, 181
Mayflower (sloop), 255
Mayhew, Capt. Constant, 154
Meader, Capt. Robert, 387
Mears, Capt. John, 150
Medley, The, of New Bedford, 228
Mediterranean Sea, 9, 323
Melville, Herman, 179, 223, 312, 374
Melville, Capt. Thomas, 180, 181–183, 186
Mellish, James, 94, 186, 361
Memorandum, William Rotch, 38, 61, 84, 163–164, 227
Memorandum on the Southern Whale Fishery, 63
Memorial, Nantucket to Massachusetts, 27, 28
"Memorials," London Whaling Merchants, 52, 53, 58, 61, 62, 73–75, 76, 88, 115–116, 122, 149, 152, 155
Mentor (ship), 57
Mercury (ship), 42, 57, 162, 226
Mexico, 149, 267
Middleton, Capt. Thomas, 122
Milford, H.M.S., 246
Milford Haven, Wales, 45, 197–215, 217–226, 235–247, 280, 320, 380
Mill Cove, Dartmouth, Nova Scotia, 34, 91
Mill Prison, England, 188
Minerva (ship), 265, 277, 281, 286–287
Mirabeau, Comte de, 163, 166–167
Mitchell, Aaron, 287
Mitchell and Gardner, 286, 287
Mitchell, Capt. Christopher, 108
Mitchell, Richard, 27, 315
Mocha Island, 275
Molucca Islands, 368
Monroe, James, 285
Montezuma (ship), 344–345, 346
Montgomery, James, 96
Monticello (ship), 327, 331
Montmorin, M., 133
Mooers, Capt. Shubael, 145
Mooers, Capt. William, 10, 12, 30, 40, 80, 100, 105, 110, 111, 145, 169, 171, 233, 256, 282, 290
Morgan, Charles W., 260
Moss (ship), 154, 266
Mott, Capt. W., 265
Mozambique Channel, 93
Munroe, Capt. Mark, 176
Murrell, Capt. Joseph, 312
Myrick, Capt. Andrew, 218, 242, 279

Nancy (ship), 15, 288
Nantucket Colony (Whaling), 37, 41, 62, 83, 88, 197, 200, 225
Nantucket Monthly Meeting, 221
"Nantucket Navigators," 30
Nantucket Proposal, 59–70
Napoleon Bonaparte, 290
Narborough Island, 344, 349
Naru Island (Pleasant), 301
National Assembly (France), 142, 159, 163–168, 207
Naturalist, Le (ship), 303
Navy, Royal, 1, 4, 16, 17, 18
Necker, M., 98, 134, 135, 138, 140
Necker (ship), 170
Nelson, Lord, 188, 198, 244, 245
Nereus, H.M.S., 346
Nereus (ship), 129
Nereyda (privateer), 340–341
Neutrality for Nantucket, 27–28, 43, 49, 350–351
Neutrality (ship), 276, 277, 279, 293
New Bedford, Massachusetts, 169, 249–261, 287, 320–322, 376, 381
New England, 381
New Holland, Australia, 119, 147, 175, 177, 182, 183
New Hope (ship), 95, 96, 129, 153
New London, Connecticut, 382
New Nantucket, 89

New South Wales, 175, 176, 179, 181, 183, 187, 272, 301, 302, 304, 308
New York City, 4
New Zealand, 182, 183, 185, 186, 189–190, 238, 242, 302
New Zealander (ship), 348
Newport, Rhode Island, 6, 7, 252, 371
Nicol, John, 129
Nicoll, Capt. John, 176
Niles Register, Baltimore, 330
Nimble (ship), 96, 154
Nimrod, H.M.S., 327
Nimrod (ship), 340–341, 344
Nimrod (armed brig), 331
No Duty On Tea (brig), 253
Nootka Sound, Vancouver Island, 147, 148–150, 155, 186
Norfolk, Virginia, 104
Norfolk Island, 176, 182, 186, 305
North Pacific, 147, 148–150
North River, Massachusetts, 112
Northern Whale Fishery (Greenland), 6
Northwest Coast, 147, 148, 154, 179, 186, 301
Nostra Senora de Bethlehem (ship), 271
Nova Scotia, 28, 29, 30–32, 34, 39, 42, 49, 53, 74, 75, 87, 88, 93, 136, 137, 188, 204, 207, 210, 215, 219, 377
Nova Scotian, newspaper, 33
Nymph (ship), 154

Observations on Rotch Proposals, 67–69
Observations on The Whale Fishery, Thomas Jefferson, 133, 135, 136, 138
Ocean (ship), 273, 312
Ogle and Co., 95
Olive Branch (ship), 265
Orion (ship), 376
Osnaburg Island, 181
Ospray (ship), 108, 168, 172, 174, 227, 268
Oswego (ship), 281
Otter (brig), 301
Oxford Row, 18

Pacific Ocean, 116, 126–127, 129, 144, 157, 175, 263
Paddack, Capt. Benjamin, 108, 168, 355, 358
Paddack, Capt. David, 144, 344
Paddack, Capt. Jonathan, 305–307
Paddack, Capt. Paul, 357
Paddack, Capt. Silas, 92
Paddack, Capt. Stephen, 30
Paddack, Capt. Thomas, 327, 331
Paddock, Capt. Latham, 376
Page, Capt. Benjamin, 129, 263
Paita, Peru, 129, 263
Palladium (ship), 361
Palliser, Admiral Sir Hugh, 76
Pandora (ship), 177
"Paper Received From Mr. Roach," 49, 54, 62
Paris, France, 18, 98, 133, 134, 170, 174
Parker, Daniel, 129
Parker, John Avery, 260
Parliament, Act for Milford, 199
Parliament, British, 5, 57, 75, 151
Parr, Gov. John, 32–34, 42, 87, 88, 201, 220
Parr (ship), 90
Parry's Islands, 379
Partridge (ship), 361
Patagonia, South America, 113, 120
Patrickson, Capt. Thomas, 186, 300
Paul's Wharf, London, 95
Pease, Capt. Noah, 362
Pease, Capt. Paul, 41, 57, 89, 120
Pease, Capt. Valentine, 92
Pegasus (ship), 192
Peggy (sloop), 92

Pembrokeshire, Wales, 197
Pendleton, Capt. Isaac, 304
Penelope (ship), 105, 108, 110, 142, 162, 168
Penguin Island, South America, 122
"Pere Winslow" (Jeremiah), 228, 282, 290
Perseverance (ship), 288, 312, 341
Perseverenda (ship), 327, 331, 341, 344
Peruvian Coast, 129
Petition (Quakers to French), 164–165
Phebe (ship), 229
Phoebe, H.M.S., 346, 350
Phoebe Ann (ship), 276, 292, 371
Philadelphia, Pennsylvania, 4, 129
Philadelphia (ship), 186
Phillip, Gov. Arthur, 119, 176, 179, 180, 183, 187
Phillips, James, 43, 83
Pinckney, Charles, 285
Pinkham, Capt. Andrew, 257–260, 325
Pinkham, Capt. Elisha, 89, 93, 95, 119, 154, 276, 320, 355
Pinkham, Capt. Matthew, 243, 355–356
Pinkham, Capt. William, 92
Pitcairn Island, 185
Pitt, William (the Younger), 19, 20, 21, 40, 55, 59–60, 76, 77, 78, 79, 82, 83, 100, 121, 150, 152, 154, 155, 209, 217
Pitt (ship), 15
Plymouth, England, 44, 188
Plymouth, Massachusetts, 188
Poinsett, Joel R., 329–330, 338
Policy (ship), 345, 347
Polly (ship), 276, 316
Pomona (ship), 188, 189
Poole, England, 44, 73
Port Desire, South America, 122
Port Egmont, Falklands, 7
Port Jackson (Sydney), 175, 176, 180, 183, 302
Port Roseway, Nova Scotia, 29, 75
Porter, Capt. David, 330, 335–350
Portlock, Capt. Nathaniel, 148–149
Portsmouth, England, 7, 156, 176, 233
Portugal, 77
Portuguese, 77, 149
Post Office Bay (Charles Island), 296, 344
Premiums, British, 78, 92, 113, 115
Premiums, French, 147
President (ship), Bedford, 257–260
President (ship), Nantucket, 326–330
Prince de Neufchatel (privateer), 351
Prince of Wales (ship), 118, 119
Prince William Henry (ship), 154
Priory Pill, Milford, 198
Privy Council, British, 21, 34, 49, 51, 61, 77, 78, 87
"Proposals," of William Rotch, 50, 59, 99
Providence, Rhode Island, 15, 186, 301
Providence (sloop), 8
Provincial Assembly, Nova Scotia, 34

Quakers, Nantucket, 4, 26, 27, 44, 57, 104, 108, 163–168, 171, 203, 243
Quaker (ship), 95
Quasi-War with France, 285–290
Queen (ship), 95

Rachel (brig), 90
Racoon, H.M.S., 350
Ramsdell, Capt. Benjamin, 154
Randall, George, 347
Randall, Capt. Gideon, 342–343, 376
Ranger (ship), 93, 95, 118, 315, 355
Rasper (ship), 95, 152–153, 316

Ratcliff Cross, 274
Ratcliff Meeting, 26
Rattler (ship), 156–157
Raven, Capt. William, 96, 183–185
Rawson, Capt. Abel, 229
Ray, Capt. Abel, 229
Ray, Capt. Daniel, 90
Ray, Capt. Isaiah, 327, 331
Ray, Capt. Paul, 229, 375
Ray, Capt. William, 205
Rebecca (ship), Nantucket, 130, 288, 327
Rebecca (ship), Bedford, 130, 255
Recruit, H.M.S., 326
Redbridge (ship), 272
Reliance (privateer), 288
Renown (ship), 7, 341
Resolution (ship), 154, 186
Resource (schooner), 206
Resource (ship), 243
Revolution, American, 1, 4, 7, 10, 25, 28, 60, 93, 101, 162
Revolution French, 142, 156, 159, 162–168, 170, 173, 209, 215
Ricketson, Daniel and Sons, 287
Rio de Janeiro, Brazil, 128, 156, 176, 274, 310
Risdon Cove, Tasmania, 190
Rising States (ship), 25
River Plate Whaling Grounds, 113
Robert Street, Milford, 225
Robeson, Andrew, 260
Robespierre, 173, 228
Robinson, Benjamin, 204
Rochester (ship), 373
Rodman, Elizabeth (Rotch), 172
Rodman, Samuel, 40, 43, 73, 99, 105, 141, 143, 171, 172, 218, 233, 260
Romulus (ship), 90, 206, 243, 355
Rose, George, 76, 84
Rose (ship), 302, 341
Rotch, Benjamin, 35, 37, 39, 43, 70, 97, 105, 107, 108, 134, 135, 143, 163, 169, 171, 173, 198, 209, 228, 234–235, 243–247
Rotch, Eliza, 107, 233, 234, 237
Rotch, Elizabeth (Barker), 143, 172, 173, 198
Rotch, Francis, 7, 8, 9, 24, 40, 73, 95, 97, 101–103, 108, 121, 133, 138, 140, 141, 145, 168, 230, 231, 233, 251
Rotch, Joseph, Sr., 7, 172, 249–252
Rotch, Joseph, Jr., 45, 251
Rotch, Thomas, 174, 226
Rotch, William, Sr., 7, 8, 10, 24, 27, 35, 37, 38–50, 57, 59–70, 73, 75, 82, 83–85, 89, 97–99, 102, 104, 105–112, 135, 141, 142, 143, 146, 159, 163–168, 169, 171, 200, 218, 226–228, 249, 260–261, 380, 381
Rotch, William, Jr., 43, 45, 49, 99, 105, 146, 231, 260
Rotch, William and Sons, 15, 101, 255, 326, 371
Rotherhithe, London, 105
Royal Bounty (ship), 355, 357
Royal Navy, 1, 4, 16–18, 94, 236, 252, 263–265, 272, 375
Royal Society of London, 115
Ruby (ship), 291
Rule, Capt. George, 375
Russell, Charles, 367–368
Russell, George, 154, 376
Russell, Capt. James, 367
Russell, Joseph, 7, 172, 249–251, 252
Russell, Capt. Joseph, 154
Russell, Capt. Laban, 243, 366–370
Russell, Mary Hayden, 366–370
Russell, Seth, 146, 287
Russell, Seth and Co., 260
Russell, Capt. Shubael, 376
Russell, Capt. Sylvanus, 292
Russell, Capt. William, 369, 370
Russian Hemp, 48

Sag Harbor, New York, 92, 371
Salamander (ship), 176, 181, 186
Salem, Massachusetts, 371
Sally (brig), 371
Sally (ship), New Bedford, 154, 297, 325–326
Sally (ship), New Haven, 307
Sally (brigantine), 90
Samuel (ship), 275
Sandalwood Trade, 304, 307–308
Sandwich Islands (Hawaiian), 364–366
Sandwich (ship), 93
Sanford, Frederick, 127
San Lorenzo Island, 342
Sappho (ship), 95, 122
Sarah (ship), 171, 313
Sarah & Elizabeth (ship), 374
"Sarah's Bosom," Aucklands, 313
Saucy Ben (ship), 96
Scorpion, H.M.S., 274, 359
Scorpion (ship), 291
Scott, Captain, 16, 25
Scott, Job, 226
Scurr, Capt. Abel, 328, 329, 330
Scurvy, 128, 155–156
Sea Elephant Oil, 114, 121
Seal Fishery, 120–124, 129, 152, 183–185
Seal Skins, 90, 95, 114, 120–124, 153, 184, 302–313
Seine (ship), 170
Serena, General, 329
Seringapatam (ship), 193, 279, 341, 348, 350, 373
Sever, Capt. William, 118
Shakespeare (ship), 367
Sherborn, Nantucket, 1, 10, 29, 105
Shields, Capt. James, 93, 125–129, 144
Shooters Hill, London, 37, 38, 105
Siberian Coast, 148
Siddons, Capt. Richard, 312
Sierra Leone (ship), 218
Sims, W., 95
Sirius (ship), 341
Sir Andrew Hammond (ship), 349
Slade, Capt. William, 243
Smith, Owen Folger, 305–306, 310
Smith, Richard, 7
Snow, Capt. Loum, 322–323
Society of Friends (Quakers), 4, 5, 9, 27, 33, 34, 37, 63, 98, 100, 163–168, 204
South America, 122, 291, 292
South Atlantic, 9, 16, 51, 113
South Carolina (ship), 170
South Georgia Island, 114, 121, 357
South Island, New Zealand, 182, 305
South Sea Company, 76, 78, 81, 89, 115, 148, 152, 154, 195
South Seasmen, 9
Southampton, England, 44, 186
Southampton (ship), 96
Spain, 122, 149
Spanish Coast (Patagonia), 122, 123
Spanish Claims, Northwest Coast, 149–150
Spanish Convention, 150–151
Sparrow (ship), 154
Speedwell (ship), 358
Speedy (ship), 186, 316
Spencer (ship), 93, 118, 188
Spencer and Lucas, 122, 188
Sperm Whales, 101
Spermacetti Candles, 4, 5, 19, 101, 254
Spermacetti Oil, 18, 19, 31, 57, 89, 111, 113, 115, 146, 147, 174, 207, 237, 245
Springrove (ship), 376
St. Barbe, John, 17, 52, 54, 96, 176, 200, 210
St. Helena Island, 231, 264, 277
St. Jago, Cape de Verdes, 129
St. Mary's Island, 272, 275, 312, 337
St. Paul's Island, 301
St. Paul's Wharf, 125

Stackpole, E. A., 7, 29
Stanhope, Lord, 120, 217
Starbuck, Abigail, 234
Starbuck, Alexander, 4, 10, 13, 27, 350
Starbuck, Daniel, 205
Starbuck, Capt. David, 104, 144
Starbuck, Capt. Edward, 154
Starbuck, Lucretia, 218
Starbuck, Capt. Obed, 363
Starbuck, Capt. Reuben, 317, 318
Starbuck, Samuel, Sr., 32, 33, 42, 88, 198, 205, 208, 220, 221, 234
Starbuck, Samuel, Jr., 33, 34, 205, 208, 213, 244
Starbuck, Capt. Valentine, 362–366
Starbuck Island, 363
Staten Land (Island), 126, 129
States Harbor (Falklands), 121
Stavers, Capt. William, 348
Sterling (ship), 277
Stewart Island, New Zealand, 305
Stokes, Charles, 198, 201, 203
Stormont (ship), 96
Straits of Le Maire, 126, 336
Straits of Magellan, 89, 116, 149
Street Lighting (Whale Oil), 18, 20
Sukey (ship), 142, 276, 277, 293, 327, 341, 344
"Sunday Islands" (the Snares), New Zealand, 184
Superior (ship), 376
Susan (ship), 255
Swan (ship), 145, 355
Swain, Capt. Andrew, 96, 118, 357
Swain, Capt. Barnabas, 205
Swain, Capt. Calvin, 233
Swain, Capt. David, 233
Swain, Capt. George, 355
Swain, Capt. Matthew, 93, 95
Swain, Capt. Reuben, 376
Swain, Capt. Solomon, 293
Swain, Capt. Thaddeus, 143, 144
Swain, Capt. Uriah, 104, 121, 145, 355
Swain, Capt. Valentine, 145
Swain, Capt. William, 93, 315, 355, 373, 374
Swain, Capt. Zaccheus, 288
Swift, Gen. Joseph Gardner, 357
Swift (ship), 154, 371
Sydney, Lord, 116, 148, 187, 220
Sydney (Port Jackson), Australia, 119, 175, 182, 183, 185, 186, 189, 191, 193, 201, 299–309
Syren (ship), 192, 362

Table Bay, South Africa, 113, 212
Table Mountain, South Africa, 113
Talcahuano, Chile, 112, 327, 330
Tallyrand, M., 290
Tarquin (ship), 372
Tarrett and Clark, 186
Tasman Sea, 182, 185, 308
Tasmania (Van Dieman's Land), 119, 180, 189, 190, 300
Teast, Sydenham, 154
Thames (ship), 341
Thames River, London, 10, 13, 37, 94, 105, 125, 189, 274
Thomas (gunboat), 329–330
Thomas (ship), 170
Thompson, John, 96
Thompson, Miss Margaret, 188
Thomson, Surgeon, James, 310
Thornton, Capt. G., 95–96
Three Friends (ship), 229
Thuellesen, Peter and Co., 146
Thurston, John T., 29
Tierra del Fuego, 121
Tiger (ship), 95
Tingmouth, England, 44
Tombez (Tumbez), 277, 346
Tongatabu, Tonga Islands, 304
Topaz (ship), 185
Tower of London, 10
Townsend, Capt. Isaiah, 305
Trafalgar, Battle of, 188, 323
Transit (ship), 378

Treaty of Paris, 1, 10, 13, 16, 22, 23, 39, 40, 93, 286, 290
Trelawney (ship), 265
Tres Marias Islands, 267
Triumph (ship), 89, 118
Tryal (ship), 312
Tumbez, Peru (Tombez), 277, 346
Tupper, Dr. Benjamin, 282

Uniacke, Richard, 33, 215, 219
Union (brig), 304–306
Union (ship), 255, 256, 265, 316
United States (frigate), 323
United States (ship), 25, 74, 95, 121, 168
United States of America, 24, 27, 39, 75, 101, 102, 103, 106, 290, 353, 372
Upham, John, 375
Upham, Capt. Timothy, 376

Valparaiso, Chile, 126, 130, 271, 280, 339–340, 349
Van Dieman's Land, 180, 300
Vancouver, Capt. George, 179, 183, 186
Vancouver Island, 149
Venus (ship), 93, 312
Vergennes, Comte de, 98, 104
Versailles, France, 98, 104, 163
Victorie (ship), 171
Victory, H.M.S., 323
Ville de Paris (ship), 142, 168, 256
Vulture (privateer), 292
Vulture (ship), 242

Walvis Bay (Woolwich) (Walfish), 113, 130, 162, 169, 229, 315, 316
Wagstaff, Thomas, 40
Walker (ship), 340, 358
Wapping (London), 95, 96
Wareham (ship), 237, 238, 272
Warren (ship), 153, 255
Warwick (ship), 45, 48, 100, 105, 108
Washington, President George, 135, 141
Washington (ship), 130, 271
Washington (sloop), 149
Waterford Packet (ship), 93
Waterman, Stephen, 205
Watson (sloop), 90
Watson and Co., 301
Weatherhead, Capt. Matthew, 176
Wellington (ship), 193
West, Benjamin, 234
West, Capt. Paul, 273–279, 297–299, 358–360, 376
West, Capt. Silas, 358
West, Capt. Stephen, 340, 341, 358
West India Trade, 23, 26
West Indies, 22, 26, 31, 265
Whalebone, 90, 111, 170
Whalefishery Bill, 76, 82
Whalefishery Protection:
Act For The Encouragement of The Southern Whale Fishery: 26 Geo. III, Cap. 50, 82, 87, 115–116; 28 Geo. III, Cap. 20, 116; 42 Geo. III, Cap. 77, 194
Whale Oil, 18, 19, 31, 111, 153
Whale Oil, prices, 133, 170
Whale Oil, use, 140
Wheatley, Nathaniel, 15
Whippey, Capt. Coffin, 108, 170, 242, 281, 287, 316, 355, 376
Whippey, Capt. George, 145, 168, 170
Whippey, Capt. James, 57, 100, 104, 170
Whitby, England, 71
Whitney, Daniel, 305, 307
Wilkes, John, 102
William Street, New Bedford, 260
William (ship), 186, 265–268
William & Ann (ship), 176, 181–183
William Penn (ship), 230, 372
Winslow, Jeremiah ("Pere"), 228, 233, 282, 290

Winslow (ship), 276, 297
Wolfe, Capt. John, 120
Woolwich Bay (Walvis bay), 130, 315–316
Woolwich Dockyard, 156
Worth Capt. Benjamin, 302, 338, 339, 373
Worth, Capt. Charles, 373
Worth, Capt. John, 142, 168
Worth, Shubael, 30
Wyer, Capt. Obed, 326, 339, 345–346

Young, Ann, 373
Young States (ship), 104, 108, 355
Young, Sir William, 152
Yorke, Thomas, 95, 129, 153
Yorke (ship), 129